BIOCHEMISTRY RESEARCH TRENDS

BIOCHEMISTRY RESEARCH UPDATES

BIOCHEMISTRY RESEARCH TRENDS

Additional books in this series can be found on Nova's website under the Series tab.

Additional E-books in this series can be found on Nova's website under the E-books tab.

BIOCHEMISTRY RESEARCH TRENDS

BIOCHEMISTRY RESEARCH UPDATES

SIMON J. BAGINSKI
EDITOR

Nova Science Publishers, Inc.
New York

For permission to use material from this book please contact us:
Telephone 631-231-7269; Fax 631-231-8175
Web Site: http://www.novapublishers.com

NOTICE TO THE READER

Additional color graphics may be available in the e-book version of this book.

Library of Congress Cataloging-in-Publication Data

Biochemistry research updates / editor, Simon J. Baginski.
p. ; cm.
Includes bibliographical references and index.
ISBN 978-1-61209-700-8 (hardcover)
1. Biochemistry--Research. I. Baginski, Simon J.
[DNLM: 1. Biochemical Phenomena. QU 34]
QD415.B47 2011
572.072--dc23

2011026362

Published by Nova Science Publishers, Inc. †New York

CONTENTS

PREFACE

This book provides a study of biochemistry research trends across a broad spectrum of applications. Topics discussed include the biochemical, physiological and therapeutic aspects of some components of the large class of lipids; streptavidin-biotin binding and immunomodulatory constructs; cardiac opioid peptides and classical opioid receptor systems; angiotensin converting enzyme inhibitors; phylogenomic analyses of the superprotein families of different metabolic systems; aptamers as the new biorecognition element for proteomic biosensing; the interactions of microorganisms with heavy oils and urinary adrenaline measurement in a forest environment setting.

Chapter 1 - This chapter approaches biochemical, physiological and therapeutic aspects of some components of the large class of lipids. The authors begin with a brief history of essential FA and finish presenting the main techniques used in research to analyse the cell content of fat. Lipids are not only the best form of energy storage, cell membrane constituent, or hormones precursors. Lipids and their metabolites participate in intra and intercellular signalling pathways and regulate gene expression. Recent studies on cytotoxicity of FA are discussed. A significant part of the content herein presented is product of the scientific investigation of the own authors and new results, not yet published, are also shown.

Chapter 2 - Several immunomodulatory constructs based on streptavidin-biotin binding have been described that exploit the modularity and flexibility of this system. In general, a streptavidin core serves as the carrier of targeting and effector molecules. Targeting agents, such as whole antibodies or their fragments and natural ligands, deliver antigen to specific receptors on leukocytes. This targeted delivery can both enhance and fine-tune immune responses, which are directed against streptavidin itself or its cargo. Due to its tetrameric structure, streptavidin-based targeting complexes can crosslink cell surface receptors. This property enables such constructs to initiate signaling events in the target cells. Thus, in addition to improving antigen uptake, activation of APCs can be achieved. By using liposome- or nanobead-bound streptavidin, one can both boost the crosslinking effect and target different cell populations, specialized in the uptake of particulate antigens. To generate streptavidin-based constructs, various molecular biological and chemical methods have been applied. In a preferred setup, monobiotinylated components are used, in order to ensure the formation of complexes with controllable stoichiometry. This strategy also allows the combination of distinct targeting and antigenic moieties in the same complex. Here the authors review various approaches of streptavidin-based immunotargeting and their effects on immune responses.

Chapter 3 - Streptavidin is a ~53kDa tetrameric protein purified from the bacterium *Streptomyces avidinii.* It is well known to contain four identical subunits, each containing a single binding site that binds very tightly to the small molecule biotin, also known as vitamin H. This high affinity (K_{assoc}= $10^{14}M^{-1}$) has been applied widely in biotechnology and biosensors. This review will cover the latest trends in the application of streptavidin-biotin system in DNA sensors and immunosensors using electrochemical and optical techniques. Streptavidin is sometimes used in connection with magnetic beads in DNA sensing. By immobilizing the biotinylated DNA probes on streptavidin-conjugated magnetic beads, the target DNA can be easily isolated from a complex matrix using an external magnet. Electrochemical and fluorescence-based immunosensors utilize the biotinylated antibodies with streptavidin-conjugated enzymes (horseradish peroxidase, alkaline phosphatase, etc.) or streptavidin-conjugated fluorescent dyes as the source of analytical signals. Surface plasmon resonance (SPR) and Localized SPR, which provide label-free analysis of dynamic surface events, are valuable optical tools for studying interactions between biomolecules. In immunosensors, SPR is typically applied to detect the formation of a sandwich-type immune complex comprised of a capturing antibody immobilized on a sensor surface, the target biomolecule, and a detecting antibody. The well-studied streptavidin-biotin system serves as an excellent model for the development of LSPR biosensors using metal nanoparticles. This review will also describe the recent fundamental spectroscopic studies that reveal key relationships governing the LSPR spectral location and its sensitivity to the local environment, including changes on the substrate due to binding events. The immobilization of streptavidin on gold nanoparticles is detected with the changes in the peak wavelength and intensity of the LSPR absorption band. Further attachment of the biotinylated ligands on streptavidin-conjugated gold nanoparticles can be detected at high sensitivity and selectivity. The results from these biosensor studies guide the design of new sensing experiments, illustrated through applications of nanomaterials, in which researchers use both the streptavidin-biotin affinity and the advanced transduction techniques to detect molecules of chemical and biological relevance.

Chapter 4 - The endogenous opioidergic system is known to express both receptors and peptides within the cardiovascular system, including within the heart itself. Precursors for all three families of endogenous opioid peptides (EOPs), prodynorphin (PDYN), proenkephalin (PPE) and proopiomelanocortin (POMC) are generated in the heart and alternative processing results in a myriad of mature peptides. Biosynthesis of the opioid peptide precursors, prodynorphin and proenkephalin, occurs predominantly in ventricular cardiomyocytes, not in the atria; whereas, processing of proopiomelanocortin to the cardiac β-endorphin peptide occurs in atrial myocytes. PPE has been shown to be more highly expressed in the adult heart and higher in the ventricles than in the atria, with left-sided preference. Similarly, POMC has been reported to only be detectable in the adult heart and to be preferentially localized to the ventricles compared to the atria. However, PDYN is reported to only be detected in the atria. Similar to the biosynthesis of cardiac peptides from their prohormone precursors, cardiac opioid peptides are spatially distributed; levels of enkephalins are higher in ventricular tissue and β-endorphin peptide is predominantly found in atrial tissue. In atrial tissue, enkephalins are packaged in vesicles, but are found unpackaged in ventricles.

Chapter 5 - Hypertension, a major cause of heart-diseases is caused by malfunction of one or more of the complex set of mechanisms controlling blood pressure. In last four decades specific agents interacting with these regulatory mechanisms have become available.

Introduction in early eighties of Captopril, a rationally designed potent inhibitor of Angiotensin converting enzymes (ACE), heralded a new era in antihypertensive therapy. Since then rapid strides have been made in understanding of the active site and the spatial requirements for ACE inhibitors. This chapter has been aimed to report the existing knowledge and to enable researchers all over the world to exploit further with an objective to develop better therapeutic agents. Bio receptors being chiral, interactions with organic molecules are strongly influenced by the chirality of the substrate. The two enantiomers of a drug are some times known to elicit differential and in some cases even opposite responses, amply heightened by work on ACE inhibitors. The present trend, in drug research, is to design and synthesize molecules with designated chirality. ACE inhibitors represent a class of rationally designed drugs based on the comprehensive knowledge of the enzyme and its binding sites and had led to great revolution in antihypertensive therapy. ACE is a Zn^{++} containing metalloenzyme similar to carboxypeptidases, whose binding sites have been studied properly and a hypothetical model incorporating all the binding sites have been developed. Based on the hypothetical model for ACE binding sites and the role which it plays in living organism, attempts have been made to develop potent inhibitors of ACE with an objective to control blood pressure. Early researches on ACE inhibitors started with snake venom peptides, showing antihypertensive activity. Though these could not be used as drugs due to high toxicity but certainly opened an avenue to look for development of antihypertensive agents on these lines. Further researches coupled with structure activity relationship (SAR) studies led to development of various ACE inhibitors and ultimately in early eighty's first rationally designed drug "Captopril" was introduced in market. Since then various other drugs acting through same mechanism and controlling blood pressure effectively have come out. The chapter entitled "Angiotensin converting enzymes inhibitors: A class of potent antihypertensive agents" is based on literature reports and intended to summarized the available knowledge and to give an update picture of major advances towards the design and synthesis of angiotensin converting enzymes inhibitors and their role in controlling blood pressure and shall undoubtedly be valuable to readers working in the area. This chapter gives information about the designing and developing of ACE inhibitors reported till date with future prospectives to achieve further progress as for as designing and development of even better ACE inhibitors is concerned.

Chapter 6 - Since a metabolic pathway with similar function has similar evolutionary relationship, the operation of metabolic pathways is essential to the survival of organisms, and metabolism related genes usually constitute only 10–20% of the total number of genes, metabolic pathway-based analysis can minimize the effect of the gene number, and thus allow more precise investigation of evolutionary relationships. Herein, the authors have shown that protein superfamilies of archaea were diverged and metabolically converged together within the genomes of this domain as a consequence of long evolutionary and diverged metabolic events. However, a close phylogenetic proximity between archaea and bacteria was observed for sharing metabolic functions by protein family divergence. A close phylogenetic boundary between methanogens and methylotrophs was revealed the reverse methanogenesis hypothesis. Energetic metabolism of methanogens are shared their ancestral behaviors with primitive proteobacteria. A mosaic nature of some functional domains of protein superfamilies was observed between mesophilic methanogens and thermophilic archaea/halophilic archaea, which may be resulted due to selection pressures. Overall results revealed that the phylogenetic analysis of protein superfamilies obtained from archaeal domain provided a standpoint for the evolutionary history of some key metabolic modules among methanogens. Therefore, molecular evolutionary hypothesis from this finding would provide a new horizon for growth of research in advancement of metabolomics and metabolic engineering in methanogens.

Chapter 7 - Aptamers are single stranded artificial nucleic acid ligands that can be generated against almost any kind of target, such as ions, metabolites aminoacids, drugs, toxins, proteins or whole cells. They are isolated from combinatorial libraries of synthetic nucleic acids by an iterative process of adsorption, recovery and amplification, know as SELEX (Systematic Evolution of Ligands by EXponential enrichment) process. Aptamers, the nucleic acid equivalent to antibodies, are easy to synthesise, is not required the use of animals for its synthesis, for this reason it can be developed again toxins and small molecules that do not produce immune response in animals and can be tuned for affinity in closer to assay conditions permitting recognition out of the physiological state. So, aptamers posses numerous advantages that make them preferred candidates as biorecognition elements. In view of the advantages and simple structure of aptamers, they have been used in a wide range of applications such as therapeutics, diagnosis, chromatography, environmental detection, among other.

Chapter 8 - The objective of this chapter is to demonstrate the basic characteristics of halophilic strain TM-1 and *Bacillus* SP-5 which were isolated from a reservoir of the Shengli oil field in East China and evaluate the effects of strains TM-1 and SP-5 on different heavy oils. Strain TM-1 is a halophilic gram-positive coccus and strain SP-5 is gram-positive rod with spores. The cells of strains TM-1 and SP-5 were grown at temperatures up to 58℃ in the neutral to alkaline pH range. They could grow steadily, and produce various metabolites, have various surface features and use different organic substrates (acetate, D-glucose, fructose, glycerol, maltose, pyruvate, starch, sucrose, and xylose). Laboratory studies have demonstrated that the strains TM-1 and SP-5 affected different heavy oils; converted and degraded various components and changed the physical and chemical properties of heavy oils. The bioconversion of heavy oils leads to an enrichment in lighter hydrocarbons and an overall redistribution of these hydrocarbons. The interactions of microorganisms with heavy oils are variable and depend on the microbial species and the chemical compositions of heavy oils.

Chapter 9 - Humans have enjoyed forest environments for ages because of the quiet atmosphere, beautiful scenery, mild climate, and fresh, clean air. Empirically, forest environments may reduce stress and have relaxing effect on human. Adrenaline is released from the adrenal medulla, and adrenaline levels increase under circumstances of novelty, anticipation, unpredictability, and general emotional arousal. Measurement of free adrenaline in urine provides a reliable measure of the circulating concentration of adrenaline in the bloodstream and thus is a measure of sympathoadrenal medulla activity. To explore the relaxing effect of forest environments on humans, the authors investigated the effect of trips to forest parks on urinary adrenaline and noradrenaline in both male and female subjects. The authors found that three-day/two-night trips to forest parks significantly reduce the concentration of urinary adrenaline and/or noradrenaline. Moreover, a day trip to a forest park also has a similar effect on urinary adrenaline. The Profile of Mood States (POMS) test showed that forest environments significantly increased the score for vigor and decreased the scores for anxiety, depression, anger, confusion, and fatigue suggesting that the subjects were under conditions of lower stress during the forest trips, which support the findings on urinary adrenaline. Other studies have reported that forest environments reduce the concentration of cortisol in saliva, reduce prefrontal cerebral activity, reduce blood pressure, and stabilize autonomic nervous activity in humans, which also support our findings on urinary adrenaline. These findings indicate that forest environments may reduce stress and have relaxing effect on human by reducing adrenaline level.

Chapter 10 - Here, a microarray-based spectroscopic assay with three readout principles, fluorescence, surface resonance Raman (SERS) and resonance light scattering (RLS), for studying kinase functionality and screening kinase inhibitors has been reported. In this assay, the phosphorylation and inhibition events are marked by biotinylated anti-phosphoserinen antibodies, and gold nanoparticles are attached to the antibodies by standard streptavidin-biotin chemistry, followed by silver deposition for RLS and SERS signal enhancement. The streptavidin conjugated fluorescein is used as fluorescent probe. The utility of this assay to high-throughput screening was demonstrated with the interactions of the α-catalytic subunit of cyclic adenosine 5′-monophosphate (cAMP) dependent protein kinase (PKA) with a commercial inhibitor library, a collection of 80 potential kinase inhibitors, and satisfactory results were obtained. In addition, quantitative determination of binding strength and the inhibiting type (type I) of these inhibitors are also demonstrated by the adenosine 5′-triphosphate (ATP) competing assays.

In: Biochemistry Research Updates
Editor: Simon J. Baginski

ISBN 978-1-61209-700-8

Chapter 1

LIPIDS AND CELL FUNCTION

Anna Karenina Azevedo-Martins[1], Thais Martins de Lima Salgado[2], Renata Gorjão[3], Érica Portioli Silva[2], Jarlei Fiamoncini[2], Maria Fernanda Cury-Boaventura[3], Elaine Hatanaka,[3] and Rui Curi[2]

[1]School of Arts, Science and Humanities, University of São Paulo, Brazil
[2]Institute of Biomedical Sciences, University of São Paulo, Brazil
[3]University Cruzeiro do Sul, Brazil

ABSTRACT

This chapter approaches biochemical, physiological and therapeutic aspects of some components of the large class of lipids. We begin with a brief history of essential FA and finish presenting the main techniques used in research to analyse the cell content of fat. Lipids are not only the best form of energy storage, cell membrane constituent, or hormones precursors. Lipids and their metabolites participate in intra and intercellular signalling pathways and regulate gene expression. Recent studies on cytotoxicity of FA are discussed. A significant part of the content herein presented is product of the scientific investigation of the own authors and new results, not yet published, are also shown.

1929: THE ESSENTIALITY OF THE FATTY ACIDS

In 1854, Grösmann first described a 20-carbon long saturated fatty acid (FA) derived from peanut (also known as groundnut) oil, which was called arachidic acid, due to the species *Arachis hypogaea*, that in turn was probably named because of the cobweb aspect of its roots. In 1909, Hartley characterized a 20-carbon FA containing four double bonds, isolated from liver lecithin, which was later named arachidonic acid [1, 2]. Up to 1929, FA were regarded exclusively as an efficient form of storing energy. The study of food components at the end of the 1800s and beginning of 1900s revealed various essential substances, particularly vitamins. It was during the study of vitamin E that the first clues for the discovery of the essential FA appeared. As vitamin E is lipossoluble, diets prepared lacking this vitamin had to be also devoid of lipids. It was then found out that animals that eat lipid-poor diets tend to develop a specific syndrome, different from that known for any vitamin. The results of these studies were published in 1927, by Evans and Burr, in 1929 and 1930 by George and Mildred Burr [3-5]. These researchers not only showed that certain lipids are essential, describing the specific deficiency symptoms, but also pointed out linolenic acid as the main essential FA, which is capable of reverting the syndrome. Among linolenic acid deficiency symptoms, Burr and Burr described *"abnormal scaly condition of skin between the 70th and 90th day"* after specific diet was initiated with 21-day-old rats. Later *"the tail may become inflamed and swollen and the whole tail soon is heavily scaled and ridged. Haemorrhagic spots may arise in the skin throughout the entire length of the tail. The swelling of the tip may gradually be replaced by a true necrosis, resulting in the loss of 1-3 cm of the tail. The hind feet become red and somewhat swollen at times, in some cases with large scales over the dorsal surfaces. The hair on the back of the body become fills with dandruff. There is a tendency to lose the hair, especially about the face, back and throat. Sores often appear on the skin. The skin of the face especially seems to become sore at times and the irritation causes the animal to rub the face continually with his fore feet"* [4, 5]. Other symptoms described include: abnormal kidneys, appearance of blood in urine, cessation of growth, alterations or cessation of ovulation, lowered copulation capacity and infertility in males, high consumption of water, high skin permeability to water (in both directions), elevated respiratory quotient and high metabolic rate [4-6]. When diet lacking fat was given to weanling rats growth ceased in few weeks, a deadline weight was reached in 5 months and animals then died in another 3-4 months. The symptoms could be reverted by the addition of lard, corn oil, linseed oil, butter and coconut oil, although the last three fats were less efficient. The analysis of FA profile led the authors to point out linolenic acid as being essential. Later it was found that the same syndrome appeared on infant fed diluted skim milk plus sugar, which proved that FA can also be essential for humans [7].

Kurzrok e Lieb, who were working with artificial insemination, found that the human uterus could contract or relax upon instillation of semen, made the next breakthrough in the study of lipids. Euf von Euler and Goldblatt independently described during the thirties that prostate extracts caused contractions in uterus and intestine and lowered the blood pressure. Von Euler named the substance responsible for the effect *prostaglandin* (from prostate) having associated it with FA fraction from the prostate. It was in 1958 that the first prostaglandins were crystallized: PGE1 and PGF1, by Sjövall and Bergström. Soon afterwards other six prostaglandins were also purified and characterized. It was also Bergström, in collaboration with van Dorph, who discovered the connection between the essential polyunsaturated FA (PUFA) and prostaglandins, after managing to produce prostaglandin E and F from labelled FA incubated in homogenized sheep glands, in 1964 [8, 9].

Other important discoveries made in the sixties include the inhibitory effect of aspirin and indomethacin on the biosynthesis of prostaglandins, the antisecretory and the protective effect of prostaglandin in the stomach, and the abortion and labor-induction potential of prostaglandins. In a study carried out in 1968, once again with FA deficient mice, it became clear that other FA, not only linoleic acid (9,12-18:2), could abolish symptoms, such as polyenoic 17-carbon FA, gamma-linolenic (6,9,12-18:3) and *bis*-homo gamma-linolenic acid (8,11,14-20:3), thereby expanding the class of the so-called essential FA [6, 8, 9]. Indeed, it was found that columbinoic acid, which can be converted to leukotrienes, but not to prostaglandins and thromboxanes, can revert the deficiency syndrome, therefore excluding the importance of cyclooxygenase-derived eicosanoids in this pathological condition [10]. The decades of 1970 and 1980 were rich in publications describing other eicosanoids and lipoxins, such as some leukotrienes and prostacyclin. The various metabolic routes related to eicosanoid synthesis and breakdown were elucidated, including both enzymatic and non-enzymatic reactions [8]. Although the discovery of many prostaglandin, thromboxane and leukotriene receptors has happened during the passed decades, these had been found to be exclusively extracellular until recent work characterizing nuclear receptors – the PPAR (peroxissome proliferator activated receptor). With this class of receptors it has become clear that eicosanoids can act both in their original or metabolised forms, upon various different external or internal receptors, signalling both through classic cAMP, GTP, phosphorylation and calcium pathways and directly on gene transcription control. The recent studies on biological effects of FA and eicosanoids derived from them are now enabling researchers to understand the essential nature and the physiological importance of these compounds.

STRUCTURE AND BIOCHEMISTRY OF LIPIDS

Lipids are broadly defined as any water-insoluble (lipophilic) or nonpolar compounds of biological origin, such as fats, oils, waxes, cholesterol, sterols, fat-soluble vitamins (such as vitamins A, D, E and K), monoglycerides, diglycerides, phospholipids, sphingolipids, glycolipids and terpenoids (eg. retinoids and steroids). Some lipids are linear aliphatic molecules, while others have ring structures. Some are aromatic, while others are not. Some are flexible, while others are rigid. These structural differences are extremely important to define the metabolism and function of the lipids.

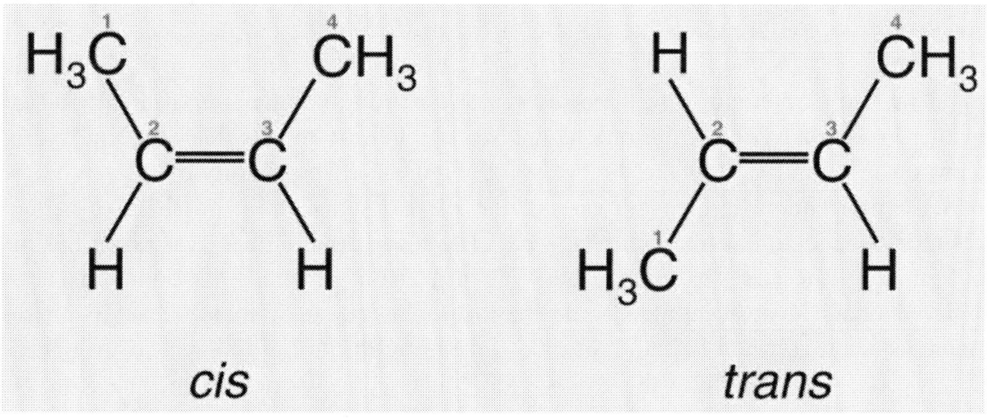

Figure 1. *Cis* and *trans* configuration of double bonds. A *cis* configuration means that adjacent hydrogen atoms are on the same side of the double bond. In *trans* configuration, adjacent hydrogen atoms are on the opposite side of the double bond.

Lipids have many key biological functions, such as structural components of cell membranes, energy storage sources and intermediates in signaling pathways. Lipids originate entirely or in part from two distinct types of biochemical subunits or "building blocks":

ketoacyl and isoprene groups. Using this approach, lipids may be divided into eight categories: fatty acyls, glycerolipids, glycerophospholipids, sphingolipids, saccharolipids and polyketides (derived from condensation of ketoacyl subunits); and sterol lipids and prenol lipids (derived from condensation of isoprene subunits).

Fatty Acids

Fatty acids are long-chain hydrocarbon molecules containing a carboxylic acid moiety at one end that can be represented by the formula RCO_2H. Usually the R group comprises a non-ramified hydrocarbon chain with even number of carbon atoms, that can contain carbon-carbon double bonds or not. The carboxylic acid moiety constitutes the polar portion of the molecule and the R chain is the non-polar portion.

The length of the hydrocarbon chain determines their classification in:

- Short chain FA: formed by 2 or 4 carbon atoms
- Medium chain FA: formed by 6 to 14 carbon atoms
- Long chain FA: formed by more than 14 carbon atoms
- Very long chain FA: formed by 20 and more carbon atoms

The presence of double bonds in the hydrocarbon chain determines their classification in:

- Saturated FA: contain no double bonds in their structure
- Unsaturated FA: contain one (monounsaturated) or more than one (polyunsaturated) double bonds in their structure

The hydrocarbon chain of saturated FA exists mainly as an extended form, once this linear and flexible conformation is the state of minor energy for the molecule. On the other hand, unsaturated FA contain rigid bends in their hydrocarbon chain because double bonds do not spin. The two carbon atoms in the chain that are bound to either side of the double bond can occur in a *cis* or *trans* configuration. A *cis* configuration means that adjacent hydrogen atoms are on the same side of the double bond (Figure 1). The rigidity of the double bond freezes its conformation and, in the case of the *cis* isomer, causes the chain to bend and restricts the conformational freedom of the FA. The more double bonds the chain has in the *cis* configuration, the less flexibility it has. When a chain has many *cis* bonds, it becomes quite curved in its most accessible conformations. The effect of this is that, in restricted environments, such as when FA are part of a phospholipid in a lipid bilayer, or triglycerides in lipid droplets, *cis* bonds limit the ability of FA to be closely packed, and therefore could affect the melting temperature of the membrane or the fat. A *trans* configuration, by contrast, means that the next two hydrogen atoms are bound to opposite sides of the double bond. As a result, they do not cause the chain to bend much, and their shape is similar to straight saturated FA.

Nomenclature

1. Fatty acids can be identified by their trivial names (or common names), which are the most frequent naming system used in the literature. These names do not follow any pattern, but are concise and generally unambiguous. They usually derive from the natural source of the specific FA. For example, palmitic acid was first identified in palm oil and oleic acid was found in olive oil.
2. Systematic names (or IUPAC names) derive from the standard IUPAC Rules for the Nomenclature of Organic Chemistry, published in 1979, along with a recommendation published specifically for lipids in 1977. Counting begins from the carboxylic acid end. The name will have a prefixe refering the number of carbon atoms in the hydrocarbon chain and will always end with the sufix oic, followed by the word acid. This notation is generally more verbose than common nomenclature, but has the advantage of being more technically clear and descriptive.
3. The numeric designations used for FA come from the number of carbon atoms, followed by the number of sites of unsaturation. For example, palmitic acid is a 16 carbon FA with no unsaturation and is designated by 16:0. This notation can be ambiguous, as some different FA can have the same numbers. Consequently, when ambiguity exists this notation is usually paired with either a Δx or n−x term.
4. In Δx (or delta-x) nomenclature, each double bond is indicated by Δx, where the double bond is located on the xth carbon–carbon bond, counting from the carboxylic acid end. Each double bond is preceded by a *cis*- or *trans*- prefix, indicating the conformation of the molecule around the bond. For example, linoleic acid is designated *cis*,*cis*-Δ9,Δ12 18:2.
5. n−x (also ω−x (omega-x)) nomenclature does not provide names for individual compounds, but is a shorthand way to categorize FA by their properties. A double bond is located on the xth carbon–carbon bond, counting from the terminal methyl carbon (designated as n or ω) towards the carbonyl carbon. For example, α-Linolenic acid is classified as a n−3 or ω-3 (omega-3) FA, and so it shares properties with other compounds of this type.

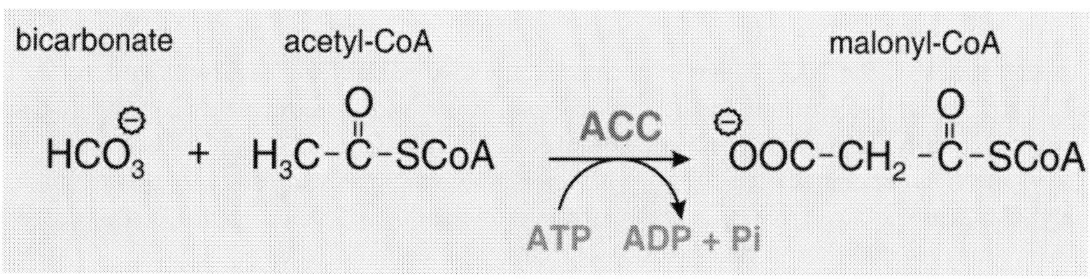

Figure 2. Synthesis of malonyl-CoA. Formation of malonyl-CoA is the first step required for fatty acid synthesis.

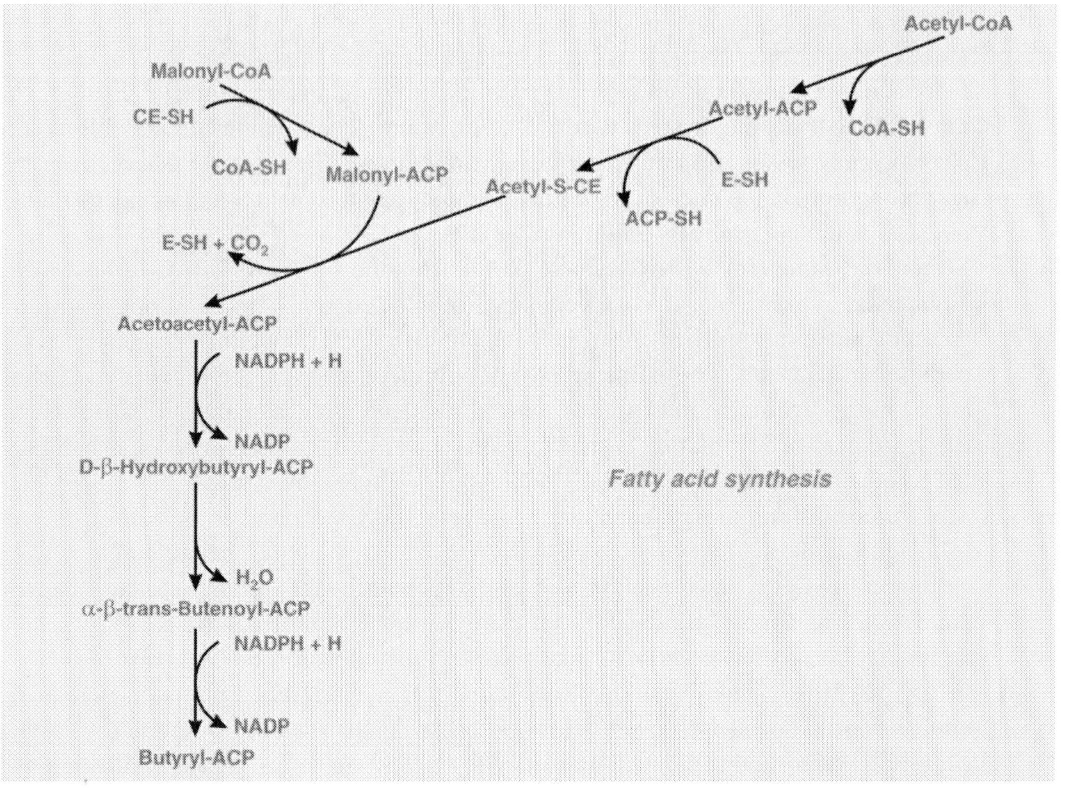

Figure 3. Fatty acid synthesis. Acetyl and malonyl groups are loaded onto acetyl transferase and malonyl transferase, respectively. The acetate unit that forms the base of the nascent chain is transferred first to the acyl carrier protein domain and then to the β-ketoacyl synthase. Attack by ACP on the carbonyl carbon of a malonyl unit on malonyl transferase forms malonyl-ACP. Decarboxylation leaves a reactive, transient carbanion that can attack the carbonyl carbon of the acetyl group on the β-ketoacyl synthase. Reduction of the keto group, dehydration, and saturation of the resulting double bond follow, leaving an acyl group on ACP.

Synthesis of Fatty Acids

The pathway for FA synthesis occurs in the cytoplasm and involves oxidation of NADPH. The synthesis of fats utilizes an activated two carbon intermediate acetyl-CoA that exists temporarily bound to the enzyme complex as malonyl-CoA. The synthesis of malonyl-CoA is the first committed step of FA synthesis and the enzyme that catalyzes this reaction, acetyl-CoA carboxylase (ACC), is the major site of regulation of FA synthesis. Like other enzymes that *trans*fer CO_2 to substrates, ACC requires a biotin co-factor.

The rate of FA synthesis is controlled by the equilibrium between monomeric ACC and polymeric ACC. The activity of ACC requires polymerization. This conformational change is enhanced by citrate and inhibited by long-chain FA. ACC is also controlled through hormone mediated phosphorylation. The acetyl groups that are the products of FA oxidation are linked to CoASH. The carrier of acetyl groups (and elongating acyl groups) during FA synthesis is a phosphopantetheine prosthetic group that is attached to a serine hydroxyl in the synthetic enzyme complex. The carrier portion of the synthetic complex is called acyl carrier protein,

ACP. The acetyl-CoA and malonyl-CoA are transferred to ACP by the action of acetyl-CoA transacylase and malonyl-CoA transacylase, respectively. The attachment of these carbon atoms to ACP allows them to enter the FA synthesis cycle.

The synthesis of FA from acetyl-CoA and malonyl-CoA is carried out by fatty acid synthase (FAS) (Figure 3). The active enzyme is a dimer of identical subunits. All of the reactions of FA synthesis are carried out by the multiple enzymatic activities of FAS. Fat synthesis involves 4 enzymatic activities. These are: β-keto-ACP synthase, β-keto-ACP reductase, 3-OH acyl-ACP dehydratase and enoyl-CoA reductase. The two reduction reactions require NADPH oxidation to NADP+.

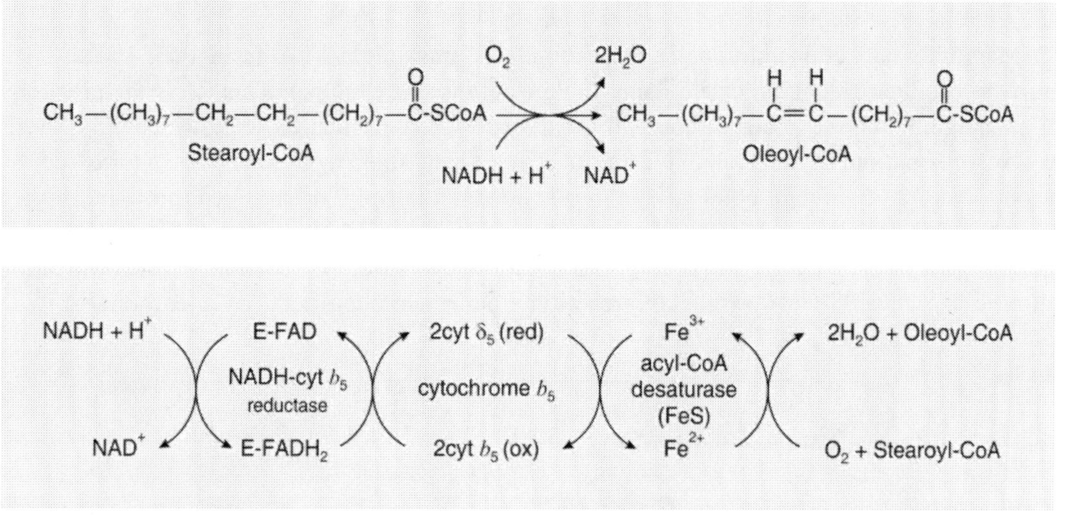

Figure 4. Fatty acid desaturation. The conversion of stearoyl-CoA to oleoyl-CoA in eukaryotes is catalyzed by stearoyl-CoA desaturase in a reaction sequence that also involves cytochrome b5 and cytochrome b5 reductase. Two electrons are passed from NADH through the chain of reactions as shown, and two electrons are also derived from the fatty acyl substrate. This is the only mean by which animals can synthesize FA with double bonds at positions beyond C-9.

The net result of this biosynthetic cycle is the synthesis of a four-carbon unit, a butyryl group, from two smaller building blocks. In the next cycle of the process, this butyryl-ACP condenses with another malonyl-ACP to make a six-carbon β-ketoacyl-ACP and CO_2. Subsequent reduction to a b-alcohol, dehydration, and another reduction yield a six-carbon saturated acyl-ACP. This cycle continues with the net addition of a two-carbon unit in each turn until the chain is 16 carbons long. The β-ketoacyl-ACP synthase cannot accommodate larger substrates, so the reaction cycle ends with a 16-carbon chain. Hydrolysis of the C16-acyl-ACP yields a palmitic acid and the free ACP.

Palmitate is then released from the enzyme and can undergo separate elongation and/or unsaturation to yield other FA molecules. Elongation and unsaturation of FA occur in both mitochondria and endoplasmic reticulum (microsomal membranes). The predominant site of these processes is the endoplasmic reticulum membranes. Elongation involves condensation of acyl-CoA groups with malonyl-CoA. The resultant product is two carbons longer (CO_2 is released from malonyl-CoA as in the FAS reaction) which undergoes reduction, dehydration and reduction yielding a saturated FA. The reduction reactions of elongation require NADPH as co-factor just as for the similar reactions catalyzed by FAS. Mitochondrial elongation involves acetyl-CoA units and is a reversal of oxidation except that the final reduction utilizes NADPH instead of $FADH_2$ as co-factor.

Desaturation occurs in the endoplasmic reticulum membranes in mammalian cells and involves 4 specific fatty acyl-CoA desaturases (non-heme iron containing enzymes) (Figure 4). These enzymes introduce unsaturation at C4, C5, C6 or C9. The electrons transferred from the oxidized FA during desaturation are transferred from the desaturases to cytochrome b5 and then NADH-cytochrome b5 reductase. These electrons are uncoupled from mitochondrial oxidative-phosphorylation and, therefore, do not yield ATP. Since these enzymes cannot introduce sites of unsaturation beyond C9 they cannot synthesize either linoleate (18:2Δ9,12) or linolenate (18:3Δ9,12,15). These FA must be acquired from the diet and are, therefore, referred to as essential FA.

Mammals can add additional double bonds to unsaturated FA from their diets. Their ability to produce arachidonic acid from linoleic acid is one example. This FA is the precursor for prostaglandins and other biologically active derivatives such as leukotrienes. Synthesis involves formation of a linoleoyl ester of CoA from dietary linoleic acid, followed by introduction of a double bond at the 6-position. The triply unsaturated product is then elongated (by malonyl-CoA with a decarboxylation step) to yield a 20-carbon FA with double bonds at the 8-, 11-, and 14-positions. A second desaturation reaction at the 5-position followed by an acyl-CoA synthetase reaction liberates the product, a 20-carbon FA with double bonds at the 5-, 8-, 11-, and 14-positions.

Triglycerides and Fatty Acids Oxidation

Fatty acids represent the main form of stored energy for many organisms. There are two important advantages to storing energy in the form of FA. (1) The carbons in FA (mostly -CH_2O groups) are almost completely reduced compared to the carbon in other simple biomolecules (sugars, amino acids). Therefore, oxidation of FA will yield more energy (in the form of ATP) then any other form of carbon. (2) Fatty acids are not generally hydrated as mono and polysaccharides are, and thus can pack more closely in storage tissues.

Although some of the fat in our diets is in the form of phospholipids, triacylglycerols are a major source of FA, as they are our principal stored energy reserve. Fatty acids are mobilized from adipocytes in response to hormones such as adrenaline, glucagon, and cortisol.

These signal molecules bind to receptors on the plasma membrane of adipose cells and lead to the activation of adenylyl cyclase, which forms cyclic AMP from ATP. In adipose cells, cAMP activates protein kinase A, which phosphorylates and activates a triacylglycerol lipase (also termed hormone-sensitive lipase) that hydrolyzes a FA from C-1 or C-3 of triacylglycerols. Subsequent actions of diacylglycerol lipase and monoacylglycerol lipase yield FA and glycerol. The cell then releases the FA into the blood, where they are carried (in complexes with serum albumin) to sites of utilization.

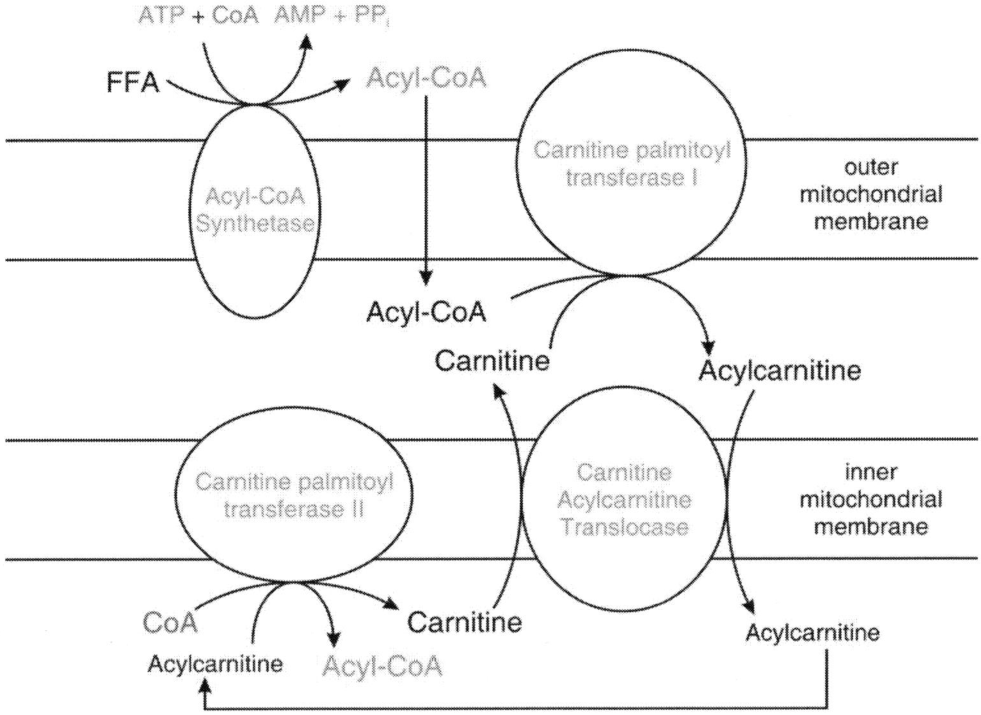

Figure 5. Transport of FA from the cytoplasm to the inner mitochondrial space for oxidation. Following activation to a fatty-CoA, the CoA is exchanged for carnitine by CPT I. The fatty-carnitine is then transported to the inside of the mitochondrion where a reversal exchange takes place through the action of CPT II. Once inside the mitochondrion the fatty-CoA is a substrate for the β-oxidation machinery.

Oxidation of FA occurs in the mitochondria, but they must be activated in the cytoplasm before being oxidized in the mitochondria. Activation is catalyzed by fatty acyl-CoA ligase (also called acyl-CoA synthetase or thiokinase). The net result of this activation process is the consumption of 2 molar equivalents of ATP:

Fatty acid + ATP + CoA ——> Acyl-CoA + PPi + AMP

Long-chain fatty acyl-CoA derivatives cannot be transported into the matrix directly. These long-chain derivatives must first be converted to acylcarnitine derivatives (Figure 5). Carnitine acyltransferase I, located on the outer side of the inner mitochondrial membrane, catalyzes the formation of the o-acylcarnitine, which is then transported across the inner membrane by a translocase. At this point, the acylcarnitine is passed to carnitine acyltransferase II on the matrix side of the inner membrane, which transfers the fatty acyl group back to CoA to re-form the fatty acyl-CoA, leaving free carnitine, which can return across the membrane via the translocase.

The process of FA oxidation is termed β-oxidation since it occurs through the sequential removal of 2-carbon units by oxidation at the β-carbon position of the fatty acyl-CoA molecule (Figure 6). Each round of β-oxidation produces one mole of NADH, one mole of $FADH_2$ and one mole of acetyl-CoA. The acetyl-CoA, the end product of each round of β-oxidation, then enters the Krebs cycle, where it is further oxidyzed to CO_2 with the concomitant generation of three moles of NADH, one mole of $FADH_2$ and one mole of ATP.

The NADH and $FADH_2$ generated during the fat oxidation and acetyl-CoA oxidation in the Krebs cycle can then enter the respiratory pathway for the production of ATP.

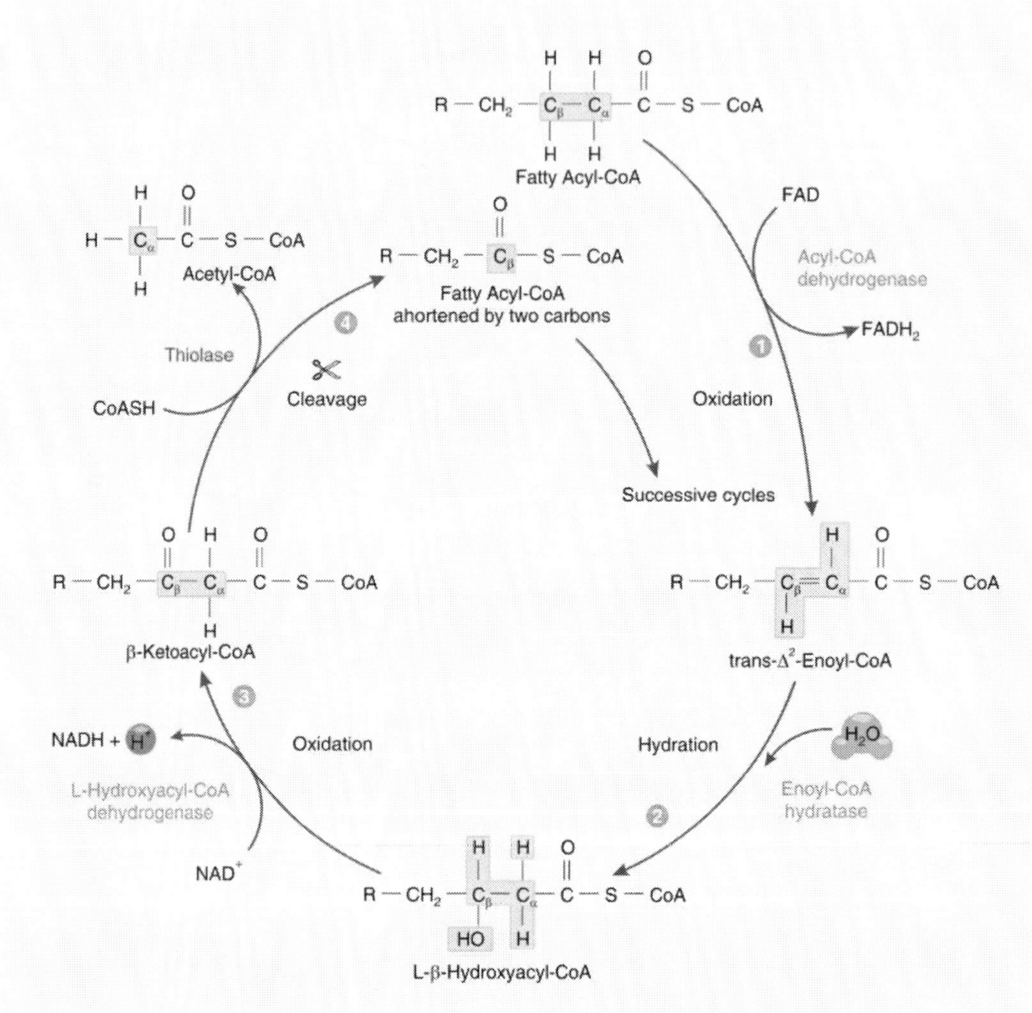

Figure 6. β-Oxidation of fatty acid. The β-oxidation of saturated FA involves a cycle of four enzyme-catalyzed reactions. Each cycle produces single molecules of $FADH_2$, NADH, and acetyl-CoA and yields a FA shortened by two carbons. (The delta [Δ] symbol connotes a double bond, and its superscript indicates the lower-numbered carbon involved.).

The oxidation of FA yields significantly more energy per carbon atom than does the oxidation of carbohydrates. The net result of the oxidation of one mole of palmitic acid (a 16-carbon FA) will be 106 moles of ATP (2 moles are used during the activation of the FA), as compared with 114 moles from an equivalent number of glucose carbon atoms.

Unsaturated FA are also catabolized by β-oxidation, but two additional mitochondrial enzymes - an isomerase and a reductase - are required to handle the *cis*-double bonds of naturally occurring FA.

Glycerolipids

Glycerolipids are mainly mono-, di- and tri-substituted glycerols. Glycerol (propane 1,2,3 triol) has three hydrophilic hydroxyl groups that can be easily replaced (Figure 7A and 7B). Fatty acids sterified to glycerol constitute the acylglycerol group that is formed by monoacylglycerols, diacylglycerols (DAG) and triacylglycerols. Triacylglycerides comprise the bulk of storage fat in animal tissues. The chemical formula is $R_1COO\text{-}CH_2CH(\text{-}OOCR_2)CH_2\text{-}OOCR_3$, where R_1, R_2, and R_3 are longer alkyl chains. The three FA R_1COOH, R_2COOH and R_3COOH can be all different, all the same, or only two the same. Chain lengths of the FA in naturally occurring triglycerides can be of varying lengths but 16, 18 and 20 carbons are the most common.

Triacylglycerols (TAG) are the main form of FA storage, but as adipocytes lack glycerol kinase, dihydroxyacetone phosphate (DHAP), produced during glycolysis, is the precursor for TAG synthesis in adipose tissue. This means that adipocytes must oxidize glucose in order to store FA in the form of TAG. Glycerol-3-phosphate acyltransferase then esterifies FA to glycerol-3-phosphate generating the monoacylglycerol phosphate structure called lysophosphatidic acid. The second reaction pathway utilizes the peroxisomal enzyme DHAP acyltransferase to fatty acylate DHAP to acyl-DHAP that is then reduced by the NADPH-requiring enzyme acyl-DHAP reductase.

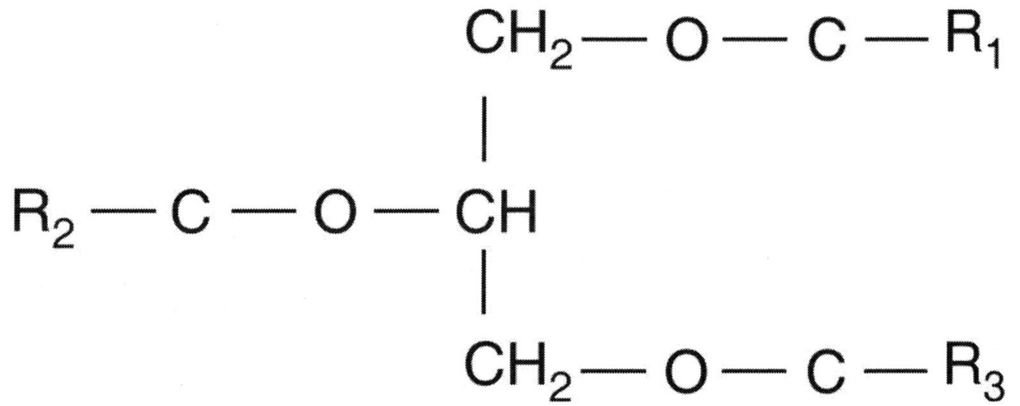

Figure 7. Basic composition of triacylglyceride. Triacylglycerides are composed of a glycerol backbone to which 3 FA are esterified.

The FA incorporated into TAG are activated to acyl-CoAs through the action of acyl-CoA synthetases. Two molecules of acyl-CoA are esterified to glycerol-3-phosphate to yield 1,2-diacylglycerol phosphate (commonly identified as phosphatidic acid). In eukaryotes, phosphatidic acid is then converted to diacylglycerol by phosphatidic acid phosphatase. Triacylglycerol biosynthesis is then catalized by diacylglycerol acyltransferase, an enzyme bound to the cytoplasmic face of the endoplasmic reticulum (Figure 8).

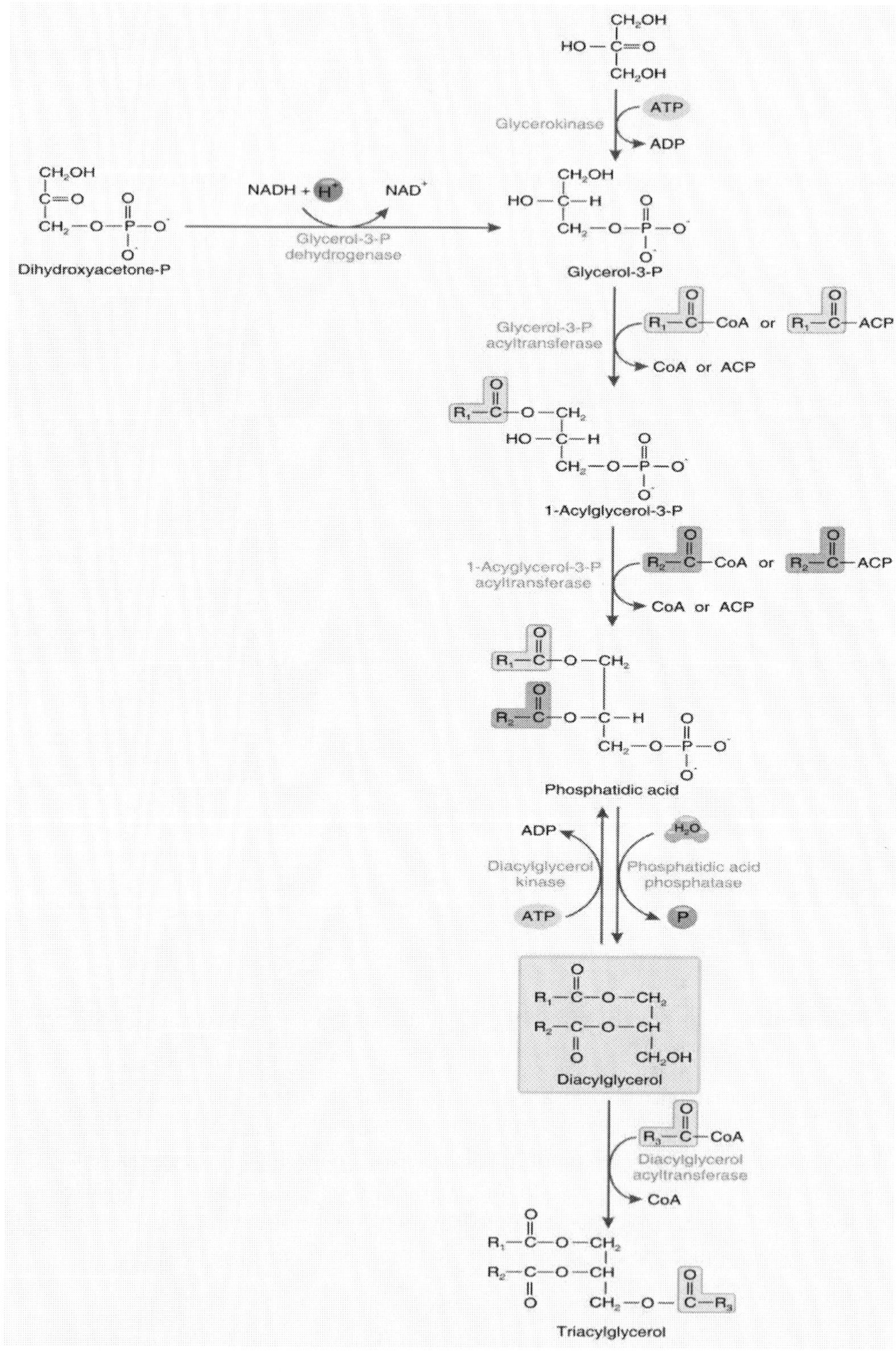

Figure 8. Triacylglycerol synthesis. Triacylglycerols are formed by the action of acyltransferases on mono- and diacylglycerol.

Glycerophospholipids

Phosphatidic acid can also be converted to cytidine diphosphodiacylglycerol (or simply CDP-diacylglycerol). Diacylglycerol and CDP-diacylglycerol are the precursors for all other glycerophospholipids in eukaryotes. Glycerophospholipids, also referred to as phospholipids, are ubiquitous in nature and are key components of the lipid bilayer of cells, as well as being involved in metabolism and signaling. The basic structure of phospolipids is very similar to that of the triacylglycerides except that C–3 (sn-3) of the glycerol backbone is esterified to phosphoric acid (Figure 9). The hydrogen atom of the phosphoric acid can be replaced by several molecules, such as ethanolamine, choline, serine, inositol, among others.

The 'head' of a phospholipid is hydrophilic (attracted to water) whereas the hydrophobic 'tails' repel water. The hydrophillic head contains the negatively charged phosphate group, and may contain other polar groups. The hydrophobic tail usually consists of long FA hydrocarbon chains. When placed in water, phospholipids form a variety of structures depending on the specific properties of the phospholipid. Many phospholipids contain saturated FA in sn-1 position and unsaturated ones in sn-2, but this is not a general rule. In addition to serving as a primary component of cellular membranes and binding sites for intra and intercellular proteins, some glycerophospholipids such as phosphatidylinositols and phosphatidic acids are either precursors of, or are themselves, membrane-derived second messengers, in eukaryotic cells.

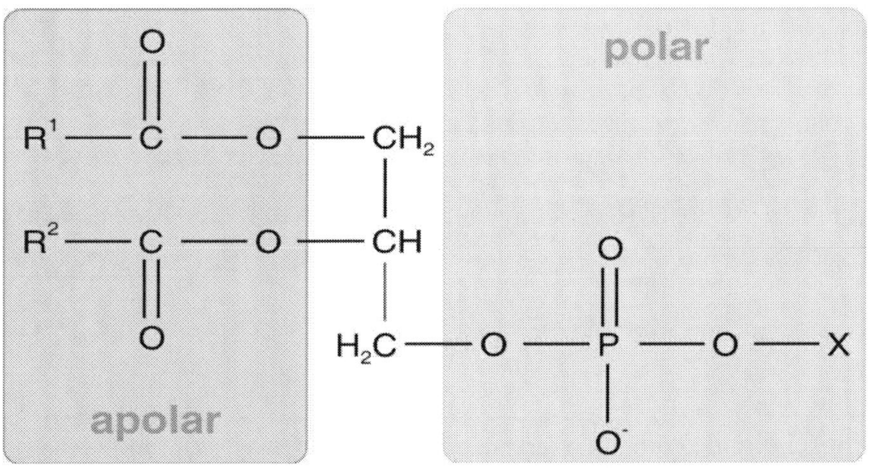

Figure 9. Basic structure of phospholipids. Phospholipids are composed of a glycerol backbone to which 2 FA are esterified and at the sn-3 position is esterified to a phosphoric acid to which different substituents can be linked.

Phosphatidylethanolamine and phosphatidylcholine, the major phospholipids of eukaryotic membranes, are synthesized from DAG (Figure 10). Phosphatidyl-ethanolamine synthesis begins with phosphorylation of ethanolamine to form phosphoethanolamine. The next reaction involves transfer of a cytidylyl group from CTP to form CDP-ethanolamine and pyrophosphate. A specific phosphoethanolamine transferase then links phosphoethanolamine to the diacylglycerol backbone. Biosynthesis of phosphatidylcholine is entirely analogous because animals synthesize it directly. All choline utilized in this pathway must be acquired from the diet.

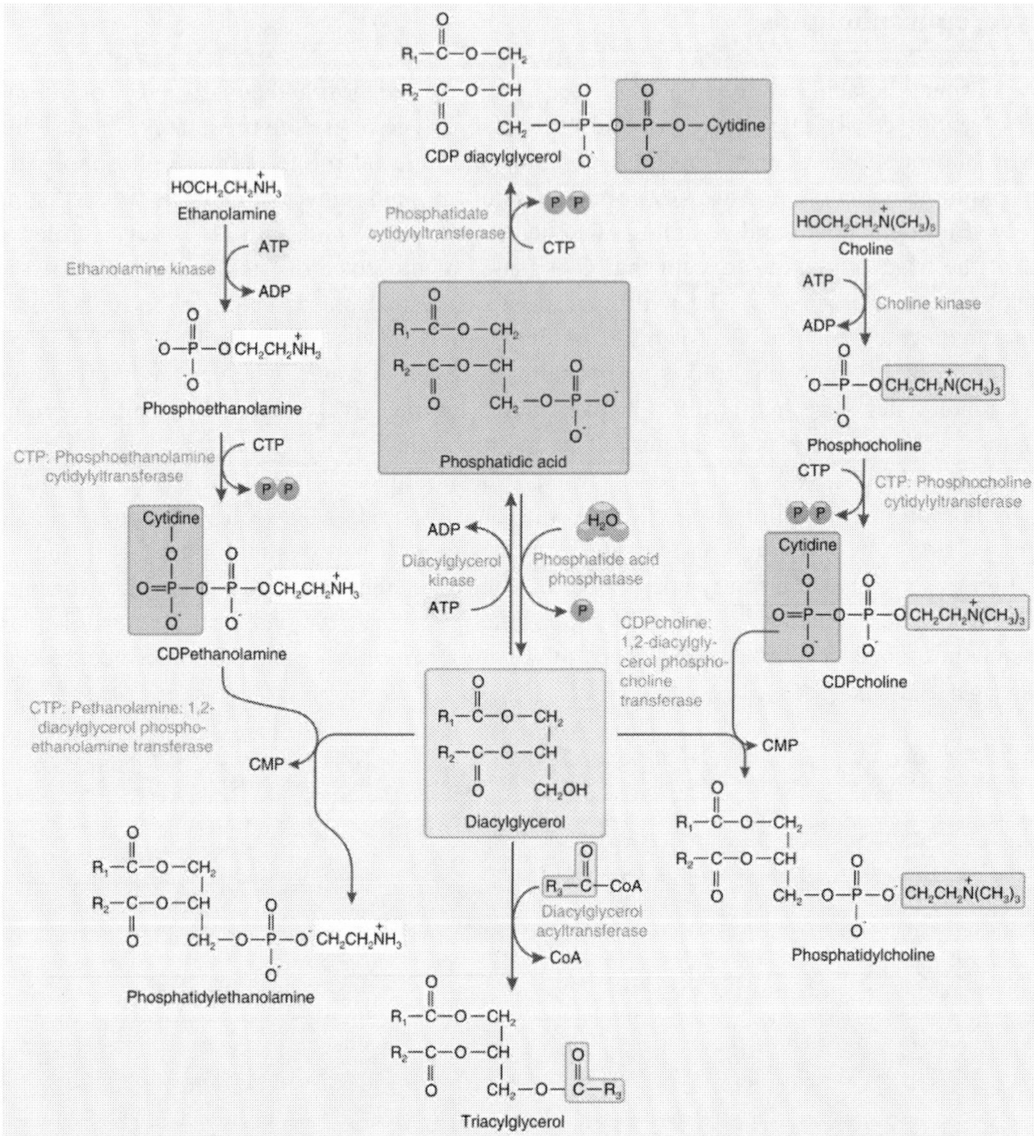

Figure 10. Phospholipid synthesis. Diacylglycerol and CDP-diacylglycerol are the main precursors of glycerolipids in eukaryotes. Phosphatidylethanolamine and phosphatidylcholine are formed by reaction of diacylglycerol with CDP-ethanolamine or CDP-choline, respectively.

Mammals synthesize phosphatidylserine (PS) in a calcium ion-dependent reaction involving aminoalcohol exchange. The enzyme that catalyzes this reaction is associated with the endoplasmic reticulum and will accept phosphatidylethanolamine (PE) and other phospholipid substrates. A mitochondrial PS decarboxylase can subsequently convert PS to PE. No other pathway converting serine to ethanolamine has been found. Eukaryotes also use CDP-diacylglycerol, derived from phosphatidic acid, as a precursor for several other important phospholipids, including phosphatidylinositol (PI), phosphatidylglycerol (PG), and cardiolipin (Figure 11). Phosphatidylinositol accounts for only about 2 to 8% of the lipids in most animal membranes, but breakdown products of PI, including inositol-1,4,5-trisphosphate and DAG, are second messengers in a vast array of cellular signaling processes.

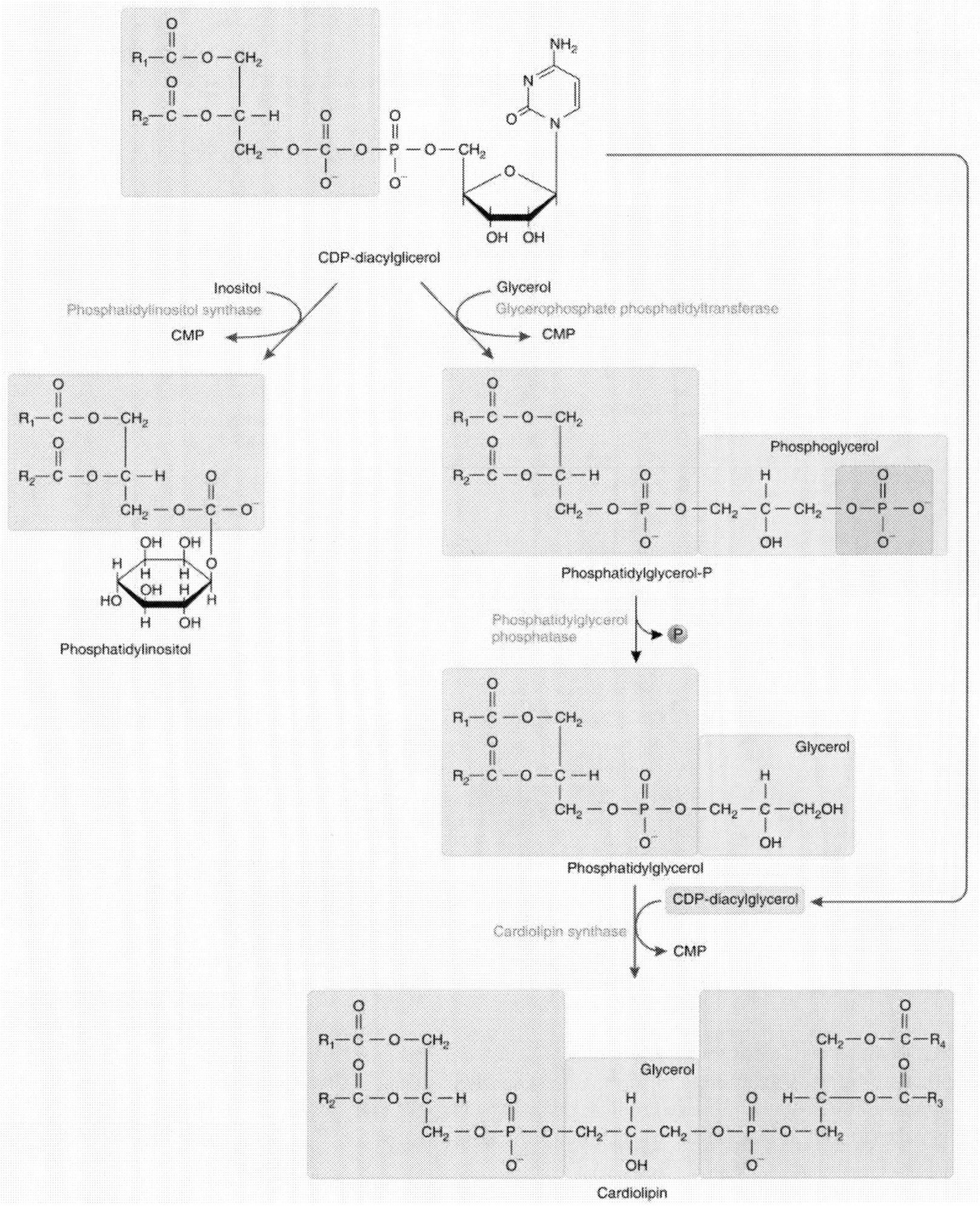

Figure 11. CDP-diacylglycerol is a precursor of phosphatidylinositol, phosphatidylglycerol, and cardiolipin in eukaryotes.

Eicosanoids are signaling molecules produced by oxygenation of twenty-carbon FA that are released after breakdown of selected phospholipids. They exert complex control over many body systems, mainly in inflammation and immunity, and as messengers in the central nervous system. Phospholipase A_2 selectively cleaves FA from the C-2 position of phospholipids. Often these are unsaturated FA that may also be released from phospholipids by the combined actions of phospholipase C (which yields DAG) and diacylglycerol lipase (that releases FA). Animal cells can modify these FA, in processes often involving cyclization and oxygenation, to produce these so-called local hormones that (1) exert their effects at very low concentrations and (2) usually act near their sites of synthesis. These substances include the prostaglandins (PG), as well as prostacyclins, thromboxanes (Tx), leukotrienes, and other hydroxyeicosanoic acids. Two families of enzymes catalyze FA oxygenation to produce the eicosanoids: cyclooxygenase (COX) generates the prostanoids; and lipoxygenase (LOX), in several forms, generates the leukotrienes (Figure 12).

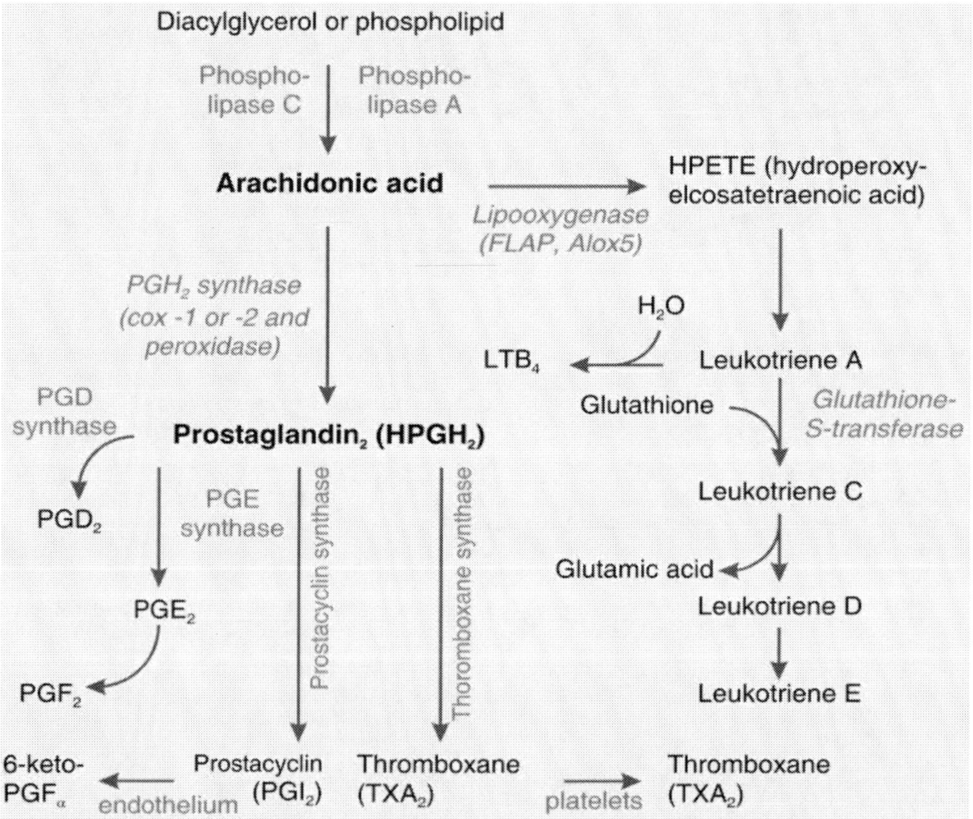

Figure 12. Synthesis of eicosanoids. Arachidonic acid, derived from breakdown of phospholipids (PL), is the precursor of prostaglandins, thromboxanes, and leukotrienes. Two families of enzymes catalyze FA oxygenation to produce the eicosanoids: cyclooxygenase (COX) generates the prostanoids; and lipoxygenase (LOX), in several forms, generates the leukotrienes.

Sphingolipids

Sphingolipids, ubiquitous components of eukaryotic cell membranes, are present at high levels in neural tissues. They are built upon sphingosine backbones rather than glycerol (Figure 13).

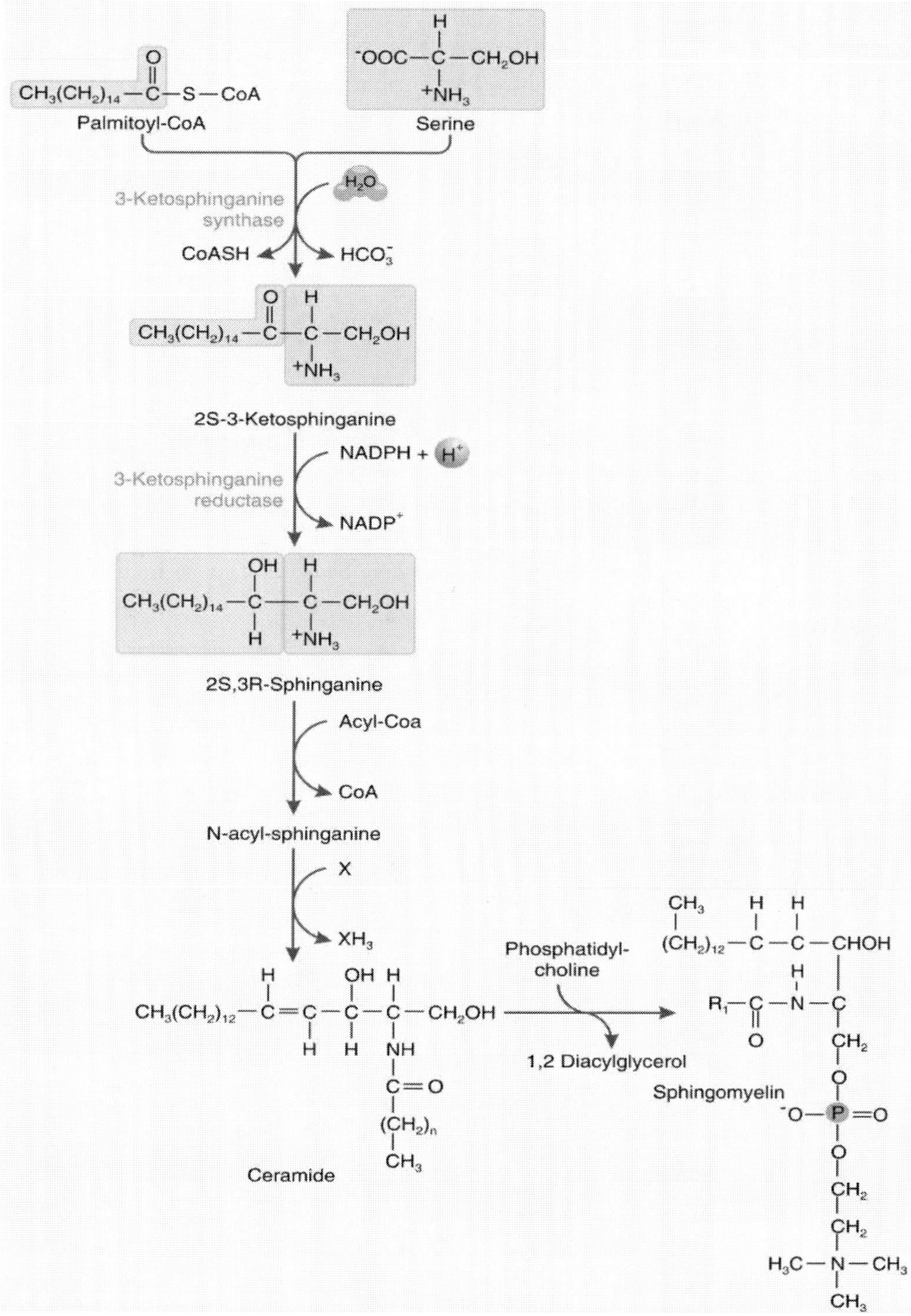

Figure 13. Synthesis of sphingolipids. Biosynthesis of sphingolipids in animals begins with the 3-ketosphinganine synthase reaction, a PLP-dependent condensation of palmitoyl-CoA and serine. Subsequent reduction of the keto group, acylation, and desaturation (via reduction of an electron acceptor, X) form ceramide, the precursor of other sphingolipids.

The initial reaction, which involves condensation of serine and palmitoyl-CoA with release of bicarbonate, is catalyzed by 3-keto-sphinganine synthase. Reduction of the ketone product to form sphinganine is catalyzed by 3-keto-sphinganine reductase, with NADPH as a reactant. In the next step, sphinganine is acylated to form N-acyl sphinganine, which is then desaturated to form ceramide. Sphingosine itself does not appear to be an intermediate in this pathway in mammals. Sphingosine is N-acetylated by a variety of FA generating a family of molecules referred to as ceramides. Ceramide is the building block for all other sphingolipids. Sphingomyelin, for example, is produced by transfer of phosphocholine from phosphatidylcholine.

Saccharolipids

Saccharolipids describe compounds in which FA are linked directly to a sugar backbone, forming structures that are compatible with membrane bilayers. In the saccharolipids, a sugar replaces the glycerol backbone that is present in glycerolipids and glycerophospholipids. The most familiar saccharolipids are the acylated glucosamine precursors of the Lipid A component of the lipopolysaccharides in Gram-negative bacteria. Typical lipid A molecules are disaccharides of glucosamine, which are derivatized with as many as seven fatty-acyl chains (Figure 14).

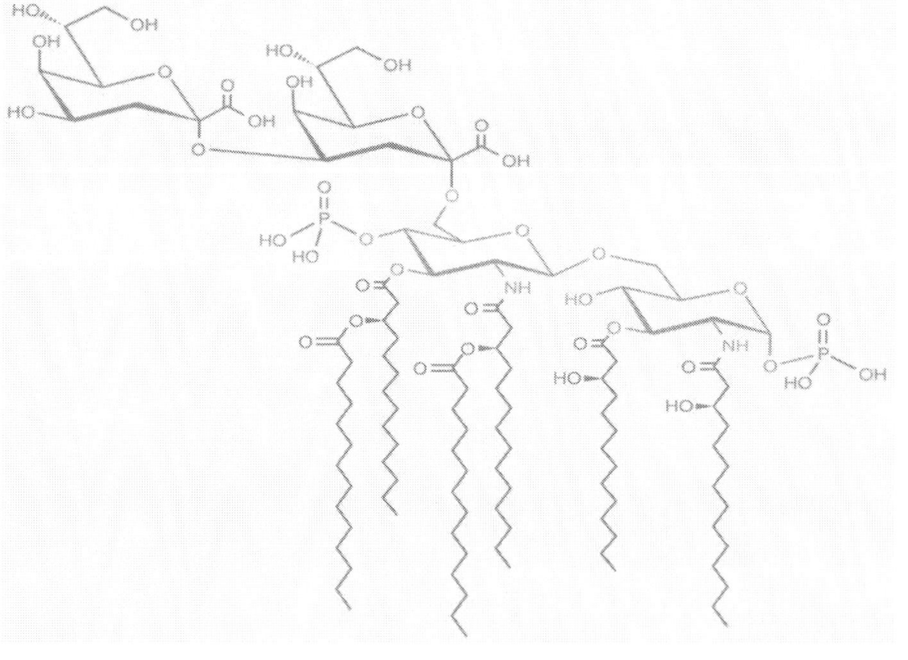

Figure 14. Structure of the saccharolipid Kdo2-Lipid A. Glucosamine residues in blue, Kdo (3-deoxy-D-manno-octulosonic acid) residues in red, acyl chains in black and phosphate groups in green.

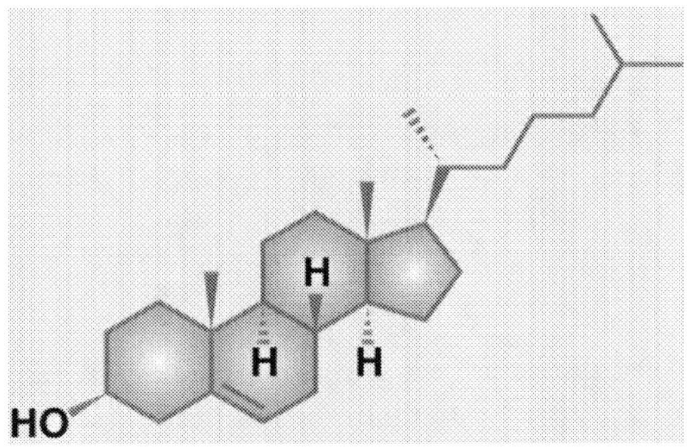

Figure 15. Sterol structure. Sterol lipids are formed by a fused four-ring core structure derived from one molecule of acetyl CoA and one molecule of acetoacetyl-CoA.

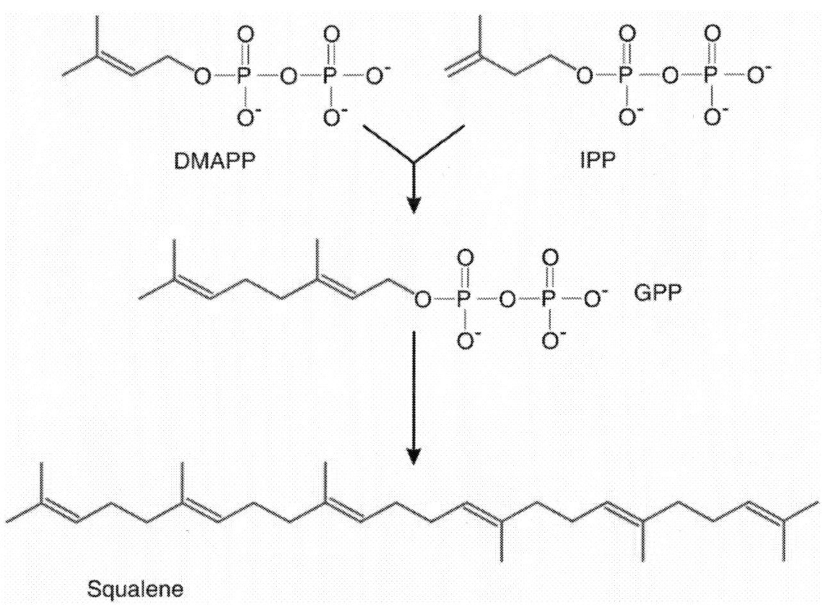

Figure 16. Prenol lipids structure. Prenol lipids are synthesized from the 5-carbon precursors isopentenyl diphosphate (IPP) and dimethylallyl diphosphate (DMAPP) that are produced mainly via the mevalonic acid (MVA) pathway.

Sterols

Sterols are formed by a fused four-ring core structure derived from one molecule of acetyl CoA and one molecule of acetoacetyl-CoA (Figure 15). Sterols of plants are called phytosterols and sterols of animals are called zoosterols. Important zoosterols are cholesterol and some steroid hormones. Cholesterol is an extremely important biological molecule that has roles in membrane structure as well as a precursor for the synthesis of the steroid hormones, vitamin D and bile acids. The cholesterol biosynthetic pathway begins in the cytosol with the synthesis of mevalonate from acetyl-CoA. Squalene is then synthesized from mevalonate, and multiple enzimatic reactions (more than 20) will occur to form cholesterol.

Prenol Lipids

Prenol lipids are synthesized from the 5-carbon precursors isopentenyl diphosphate (IPP) and dimethylallyl diphosphate (DMAPP) that are produced mainly via the mevalonic acid (MVA) pathway (Figure 16). The simple isoprenoids (linear alcohols, diphosphates, etc.) are formed by the successive addition of terpene (C_5H_8) units, and are classified according to the number of these units. Structures containing greater than 40 carbons are known as polyterpenes. Carotenoids are important simple isoprenoids that function as antioxidants and as precursors of vitamin A. Another biologically important class of molecules is the quinones and hydroquinones, which contain an isoprenoid tail attached to a quinonoid core of non-isoprenoid origin. Vitamin E and vitamin K, as well as the ubiquinones, are examples of this class.

Modulation of Intracellular Signalling by Fatty Acids

Polyunsaturated FA (PUFA) may control immune responses indirectly by modification of cell-membrane composition. Omega−3 PUFA incorporated into cell phospholipids could alter the levels of phospholipid-derived second messengers, such as DAG and ceramide. Arachidonic, docosahexaenoic (DHA) and eicosapentaenoic (EPA) acids affect Jurkat cells function (a human T-lymphocyte cell line) by incorporation into DAG and through specific effects of their DAG metabolites on RasGRP (Ras guanyl-releasing protein) and, subsequently, on MAPK (mitogen-activated protein kinase) signalling [11].

MAPK, such as ERK (extracellular-signal-regulated kinase) 1/2, are involved in cell proliferation and differentiation [12]. Denys et al. (2002) have shown that EPA and DHA decrease the activity of ERK1/2 in PMA (phorbol myristate acetate) stimulated Jurkat cells [13]. These two FA also inhibited the PMA-induced plasma membrane translocation of PKC (protein kinase C)-α and -ε and the nuclear translocation of NF-κB (nuclear factor κB) [14]. However, other studies have shown that these effects are mediated by PKC-independent mechanisms [13]. Stulnig et al. (1998) found that JNK (c-Jun N-terminal kinase) phosphorylation and activation is selectively inhibited in Jurkat cells and peripheral blood T-cells treated with EPA, whereas phosphorylation of p38 MAPK remains unaltered [15]. Hii et al. (1998) have shown that arachidonic acid does not enhance p38 MAPK phosphorylation in Jurkat cells, but increases ERK and JNK activities [16]. An increase of ERK activity in Jurkat cells treated with DHA and EPA was also observed, which is in contrast with the study by Denys et al. (2005)[14]. These contradictory results may be partially explained by the concentrations and treatment periods used. For example, Denys et al. (2002) investigated ERK activity after 5 min of treatment with DHA and EPA at concentrations up to 60 μmol/L, whereas Hii et al. (1998) treated cells with up to 20 μmol/L of these FA for 10 min. These studies indicate that PUFA have a selective action on intracellular proteins involved in T-cell activation.

IL-2 plays an essential role in lymphocyte proliferation [17]. Studies have recently shown that DHA, EPA, palmitic and stearic acids decrease the stimulatory effect of IL-2 on human lymphocyte proliferation, increasing the percentage of cells in G1 phase and decreasing the proportion of cells in S and G2/M phases [18]. DHA, EPA, palmitic and stearic acids inhibited the signaling pathway activated by this cytokine after a short period of treatment (1 hour). The IL-2 receptor consists of three subunits: α, β, and γ. The IL-2-induced heterodimerization of β and γ subunits resulted in intermolecular transphosphorylation of their corresponding receptor associated Janus kinase 1 (JAK1) and JAK3 [19, 20]. Members of the STAT (signal transducer and activator of transcription) family of transcription factors, downstream effectors of the JAK, are also regulated by IL-2. IL-2-induced tyrosine phosphorylation of STAT5 allows src homology 2 domain-mediated homodimerization or heterodimerization with a resultant induction of nuclear migration and sequence-specific DNA binding by the STATs [21]. Palmitic and stearic acids, DHA and EPA decreased JAK1, JAK3, STAT5, ERK 1/2 and Akt phosphorylation induced by IL-2. The inhibitory effect of these FA on lymphocyte proliferation was attributed to a decrease in activation of JAK/STAT, ERK, and Akt pathways induced by IL-2. Probably, the inhibition of JAK1, JAK3, STAT5, ERK1/2, and Akt phosphorylation caused by DHA, stearic and palmitic acids is associated with an alteration of CD25 expression at the cell surface. Oleic and linoleic acids stimulated lymphocyte proliferation by increasing ERK1/2 phosphorylation through PKC-ζ activation. Therefore, these FA inhibit lymphocyte proliferation by altering the protein phosphorylation state.

Xue et al (2006) investigated the effects of EPA and DHA on MAPK in the endothelium. The authors measured kinase activities in TNF-α-activated endothelial cells from human umbilical vein (HUVEC) [22]. EPA or DHA significantly reduced the TNF-α-induced activation of p38 and JNK kinases at a concentration of 20 μM but EPA was a more potent inhibitor than DHA. In contrast, both EPA and DHA significantly counteracted the TNF-α-mediated deactivation of ERK1/2 kinases. Meanwhile, both EPA and DHA significantly attenuated the TNF-α-induced expression of p38 and ERK1/2 mRNA. DHA, but not EPA, also reduced the TNF-α-induced JNK mRNA expression.

Chen et al (2005) observed that DHA plays a central role in antagonizing cytokine-induced adhesion molecule expression by attenuating NF-kB signaling in the early steps of inflammation in human vascular endothelial cells [23]. These cells contain the cognate receptors for TNF-α and IL-1β, and the activation of these receptors leads to increased activity of NF-kB. Up regulation of inflammatory cytokines, especially TNF-α [24], IL-1β, [25] and vascular endothelial growth factor (Aiello et al, 1995) along with their corresponding receptors have been well documented in diabetic animal models and in the eyes of diabetic patients. These cytokines induce expression of adhesion molecules in cultured human retinal endothelial cells that is mediated by NF-kB. Therefore, DHA can inhibit adhesion molecule expression through a decrease in NF-kB activity blocking the action of pro-inflammatory cytokines.

Hirafuji et al (2002) showed that DHA increases ·NO production by vascular smoth muscle cells (VSMC) by potentiating inducible nitric oxide synthase (iNOS) expression induced by IL-1β. DHA potentiated the IL-1β-induced phosphorylation of p44/42 MAPK but had no significant effect on p38 MAPK phosphorylation [26].

The modulatory potential of conjugated linoleic acid (CLA) on cytokine-induced eicosanoid production from smooth muscle cells which contributes to the chronic inflammatory response associated with atherosclerosis, has been investigated [27]. *Cis*-9, *trans*-11 CLA and *trans*-10, *cis*-12 CLA were shown to reduce proportions of the eicosanoid precursor arachidonic acid in smooth muscle cells total lipids and to inhibit TNF-α-induced expression of enzymes involved in eicosanoid formation (cPLA2, COX-2, mPGES) and thus, production of the prostaglandins PGE2 and PGI2. These effects were not exclusively related to the decrease in the arachidonic acid content. Conjugated linoleic acid treatment of smooth muscle cells for 24 hours diminished NF-κB DNA-binding activity induced by TNF-α.

Fatty acids also alter TNF-α signaling in fibroblast-like synovial cells. Electrophoretic mobility shift assays revealed that TNF-α activates NF-kB in fibroblast-like synovial cells but this was inhibited by α-linolenic acid at concentration of 1 mM. TNF-α induced IKK mediated phosphorylation of GST-I kB and pre-treatment with linoleic acid inhibited this pathway [28].

Lipotoxicity

The status of intracellular and extracellular lipid homeostasis is determinant for cellular health. In humans, this equilibrium is affected by diet. However, industrialized nations conspire to eliminate the impact of a famine and the population continues to consume extra amount of food. The current state of chronic over-nutrition or positive energy balance has created a metabolic conflict between carbohydrates and lipids [29]. This condition of calorie-overload results in increased storage of FA. The accumulation of fat surpasses the ability of the cell to utilize it, leading to cellular dysfunction [30, 31].

Increased availabity of free FA (FFA) or decreased normal oxidative metabolism in non-adipose tissues including skeletal and cardiac myocytes, hepatocytes, and pancreatic b-cells results in chronic cellular dysfunction and injury in common disease states such as insulin resistance, pancreatic b-cell dysfunction, cardiomyopathy, and steatohepatitis [30]. Accumulation of lipids in skeletal muscle, heart, liver, pancreas, kidneys or blood vessels has the potential to cause organ-specific toxic effects that compromise their normal functionality [29]. After chronic high-energy food consumption, there is an increase in plasma glucose and insulin concentrations that accelerate lipogenesis in adipose tissue, increasing the amount of fat stored as triglycerides. Adipose tissue loses the capacity to respond to insulin, limiting the capacity to store FA in the form of triglycerides shifting the flux of FA to non-adipose tissues mainly to the liver. An accumulation of FA that exceeds the capacity of non-adipose tissues to oxidize them enhances the metabolic flux of FA to other harmful, nonoxidative pathways, producing ceramides in a condition called lipotoxicity [32].

LipotoxicIty is usually accompanied by accumulation of neutral lipids in cells as triglycerides. Most data indicate that the triglycerides themselves serve primarily as energy storage with toxicity deriving mainly from FA and their products such as ceramides and DAG, that accumulate either as a result of failure of esterification or breakdown of the triglycerides. Multiple pathways can be involved in the acute and chronic cellular effects of FA in excess [33].

Normal cellular FA homeostasis reflects in a balance between processes that generate or deliver FA and processes that utilize these molecules. In mammalian cells, free FA are generated through *de novo* synthetic pathway and liberated when triglycerides and phospholipids are hidrolyzed by cellular lipases. Free fatty acids can be imported into mammalian cells by both protein and non-protein mediated mechanisms, either when cellular demand or extracellular free FA concentrations are high [34]. Free fatty acids derived from these processes can be used for membrane biosynthesis, energy production through β-oxidation, generation of lipid signaling molecules, post-translation protein modification, and transcriptional regulation [30].

When cells accumulate more FA than are required for anabolic or catabolic processes, the excess is esterified and stored as triglycerides in lipid droplets. Fatty acids stored may be mobilized through the action of cellular lipases, in a process regulated by hormones and by droplet-associated proteins. Adipocytes have a unique capacity to store large amounts of FA excess in cytosolic lipid droplets. Cells of non-adipose tissues have a limited capacity for lipid storage. When this capacity is exceeded, cellular dysfunction or cell death occur [30].

The cardiac lipid overload that occurs with inherited defects in the mitochondrial FA oxidation pathways is associated with heart failure and sudden death. Increased plasma levels of FA lead to intramyocellular lipid accumulation in humans and this condition plays a critical role in the genesis of insulin resistance and type 2 diabetes. Lipid overload in pancreatic β-cells leads to deregulated insulin secretion with short-term increases and chronic decreases. Triglyceride and FA accumulation in the liver is associated with non-alcoholic steatohepatitis characterized by an inflammatory response with evidence of hepatocyte damage and fibrosis that can progress to cirrhosis. In the progression of renal failure, FA carried by albumin in the proximal tubule are thought to have toxic effects on proximal tubular epithelial cells and contribute to pathological changes of the tubulo interstitium [30].

Another mechanism for lipid accumulation is observed in tissues with high turnover/metabolism of FA, such as the heart. While long-chain FA are the major source of energy in the normal adult mammalian heart, acquired and inherited cardiac disorders are associated with a switch in energy substrate utilization from FA to glucose. High plasma FA and triglyceride levels lead to increased import of these compounds into non-adipose tissues, contributing to intracellular lipid accumulation [30]. Tissues other than adipose tissue, which have an increased accumulation of lipids, often exhibit cellular dysfunction. This has been targeted as a factor contributing to insulin resistance in obese and diabetic subjects [35].

Lifestyle-related diseases such as obesity, diabetes mellitus, hyperlipidemia, hepatic steatosis, and coronary artery disease are associated with chronic consumption of a high-fat or high-carbohydrate diet [36]. Obesity and diabetes mellitus are nutritional disorders that have become major public health concerns in industrialized countries, not only because of their increasing prevalence, having reached epidemic proportions, but also because of their frequent association with major cardiovascular risk factors (dislipidemia, atherosclerosis, and coronary artery disease). Obesity results from disequilibrium between environmental factors including lifestyle habits and genetic predisposition and energy expenditure that lead to weight gain. It is associated with hyperinsulinemia, insulin resistance, abnormalities in lipid metabolism, and nonalcoholic hepatic liver disease [32].

Obesity and its associated disorders is one of the most pressing health issues facing medicine today. Much time and effort have been spent in order to establish the molecular bases concerning the development of obesity in an attempt to identify suitable targets for pharmacological intervention. Lipotoxicity is a major, if not the central, factor linking obesity and cardiovascular outcomes. The regulation of body adipose tissue can be described as linear equation, balancing both food intake and energy expenditure to derive the amount of fat stored. Thus, increases in food intake or decreases in energy expenditure result in elevated fat deposition. Fat may be directed to either oxidative processes in skeletal muscle or brown adipose tissue, or to triglycerides storage in white adipose tissue [37].

White adipose tissue is a key organ for the development of the metabolic syndrome. It is well known that insulin resistance (in obesity) results in two specific effects on white adipose tissue: insulin-stimulated uptake of glucose and FA into this tissue is reduced and there is impaired suppression of FA and glycerol release post-prandially. This increases FA flux towards tissues such as the skeletal muscle, heart, liver and pancreatic β-cell, which may result in lipotoxicity, that may be prevented by either an increased white adipose tissue FA storage capacity or by increased FA oxidation capacity in other tissues [37].

The theme lipotoxicity encompasses toxicity that results not only from lipid overloading induced by excessive delivery versus oxidation of circulating free FA but also from lipids synthesized from an overload of glucose by *de novo* lipogenesis occuring in adipocytes and other cell types [29].

The influence of n-3 PUFA supplementation on different parameters of the diabetic condition such as endothelial dysfunction [38], dyslipidemia [39] and haemoglobin glycosylation [40] has been investigated. There is epidemiological evidence for a protective effect of long-chain n-3 PUFA on type 2 diabetes risk. The prevalence of type 2 diabetes is low in Greenland [41] and Alaskan eskimos [42], populations known to have a very high intake of n-3 PUFA [43].

Oxidative stress has been associated with pancreatic β-cell death in the onset of type 1 and 2 diabetes and also in the complications of the pathological condition. Kesavulu et al. (2002) [44] observed that supplementation with n-3 PUFA (1.08 g EPA plus 720 mg DHA daily) has beneficial effects on serum triglycerides and HDL-cholesterol levels, lipid peroxidation and antioxidant enzyme activities, which may lead to decreased rate of occurrence of vascular complications in diabetes. Mori et al (2003) have recently demonstrated that 6 week supplementation (4 g daily) of pure EPA or DHA does not adversely affect but attenuates oxidative stress in type 2 diabetic patients. The authors did not observe any other effect between the two PUFA on this regard [45]. Following the same study design, Smaouni et al. (2006) found that both erythrocyte Cu/Zn superoxide dismutase (SOD) and glutathione peroxidase (GPx) activities were positively correlated with erythrocyte content of n-3 PUFA in type 2 diabetic patients. In this study, the authors did not find specific correlation between EPA or DHA content and the antioxidant parameters studied.

The first published report examining the contribution of n-3 PUFA intake on human type 1 diabetes was a case-control study from Norway. This study showed that children with diabetes were less likely to have been given cod liver oil during infancy than children without diabetes [46]. This was a large case-control study conducted after the conclusion of the same group that vitamin D or n-3 PUFA (EPA and DHA) or both in the cod liver oil had a protective effect against type I diabetes [47].

Norris et al (2007) suggested that high consumption of total n-3 PUFA is associated with low islet autoimmunity in children at increased genetic risk for type 1 diabetes [48]. Recently, a Net-based clinical trial, called "The Nutritional Intervention for the Prevention of Type 1 Diabetes", has been established the hypothesis that dietary supplementation with anti-inflammatory doses of DHA in uterus and in infancy may block early islet inflammatory events involved in the pathogenesis of type 1 diabetes. If this hypothesis is confirmed, dietary supplementation with n-3 PUFA could be recommended to prevent the development of type 1 diabetes. In spite of the importance of this observation, the effects of EPA and DHA for the development of type I diabetes were not compared yet. Aarnes et al. (2002) investigated the changes in viability of insulin-producing INS-1E cells incubated for 6 days with EPA and DHA at 0.07 mM concentration [49]. There was a significant negative effect of DHA whereas EPA did not cause significant alteration. Association of the n-3 PUFA with IL-1β decreased cell viability, mainly observed for DHA.

We have found that EPA presents low toxicity, compared with other FA even at high concentrations (0.3 and 0.4 mM), to RIN-m5F insulin secreting cells [50]. EPA and DHA did not cause marked effect on viability of these cells (data not published). However, DHA presented very high toxicity by inducing DNA fragmentation in 35% of the cells already at 0.1 mM reaching around 100% at 0.2 mM. EPA presented a much lower cytotoxicity at the same concentrations and time of incubation [50, 51]. The toxicity of FA has been attributed to the inability of the cells to incorporate them into neutral lipid droplets [52]. Omega-3 PUFA are readily incorporated into neutral lipids, when compared with oleic acid in pancreatic islet cells. DHA is more accumulated than EPA after 24 h treatment at 0.1 mM. However, the inverse relationship between cytotoxicity and the intracellular lipid accumulation was not observed for both EPA [50], and DHA, as it has been proposed for oleic and palmitic acids [52]. On the other hand, Suresh and Das (2001), studying the effect of alloxan on RIN cell viability, found a protective effect of various FA, including EPA and DHA [53]. Preincubation with EPA or DHA (15 mg/mL) recovered cell viability back to around 95% and 85%, respectively. Different types of FA appear to have different effects on pancreatic islets isolated from 4 hours starved mice. Polyunsaturated fatty acids (linoleic, linolenic, and arachidonic acids) stimulated insulin secretion, whereas short-chain (butyric acid), medium-chain (hexanoic and octanoic acids), and saturated (palmitic acid) FA inhibited insulin secretion. Pancreatic islets exposed to high concentrations of FA for periods of 24-48 h resulted in enhanced insulin secretion even at low glucose concentrations, decreased insulin synthesis, depleted insulin storage, and impaired response of the β–cell to stimulation by glucose, being characteristic of type 2 diabetes. Lupi et al (2002) showed that FA have a cytotoxic effect on human pancreatic islet cells [54].

Our group has investigated whether the toxic effect of the n-3 PUFA on RINm5F cell is associated with the phosphorylation state of Akt, ERK and protein kinase C (PKC)-δ. The regulation of these kinases was compared in three experimental designs: (a) 4 h-exposure, (b) 4 h-exposure and a subsequent withdrawn of the FA for a 20 h period and (c) 24 h-exposure. Although DHA activated Akt in the short-period treatment, this effect did not persist. In fact, DHA inhibited Akt phosphorylation in 24 h experiment. These findings, led us to postulate that DHA cytotoxicity is probably related to inhibition of Akt and ERK phosphorylation in the 24 h-exposure experiment. EPA promoted a late increase of Akt phosphorylation, as seen in the 4/24 h experiment. The activation of Akt might induce anti-apoptotic effects by phosphorylation of GSK-3 (glycogen synthase kinase-3) [51].

Our group has also observed that Raji cells are more sensitive to the toxic effect of PUFA than Jurkat cell [55]. Linoleic acid, arachidonic acid, DHA and EPA can cause a pronounced effect on death of lymphocytes and neutrophils than monounsaturated or saturated FA [56-58]. Cury-Boaventura et al. (2004) showed that both oleic and linoleic acids lead to accumulation of triglycerides in Jurkat cells [57]. Gorjão et al (2007) showed that stearic and palmitic acids inhibited proliferation and caused lymphocyte death [18]. Martins de Lima et al. (2006) compared the toxicity of FA on a murine macrophage cell line [59]. They showed that all FA tested were toxic to the cells: palmitic acid at 75 μM, stearic acid, arachidonic acid, EPA and DHA at 150 μM. Oleic and linoleic acids were less toxic causing a signal of death at 225 μM and 250 mM, respectively. Cury-Boaventura et al (2006b) tested the toxicity of oleic and linoleic acids on human lymphocytes and did not found toxic effect up to 100 mM. Only at 200 μM loss in lymphocyte plasma membrane occurred [60].

In contrast to *cis* MUFA and *cis* PUFA, there is little information available about the mechanisms that mediate the effects of *trans* FA on human health. The amount of *trans* FA in the diet can reach 20 g in one day, far exceeding the recommended maximum intake of 2 g per day. This is of interest because of the adverse effects of these FA with respect to cardiovascular disease, infant development, diabetes, and inflammation.

The vascular endothelium is always exposed to FA at different levels. Fatty acids released from the adipose tissue by lipolysis during fasting and exercise reach the endothelium and may regulate endothelial cell function. In atherosclerosis, there is endothelial cell death and detachment from the underlying layer. Endothelial cells apoptosis also leads to increased vascular permeability, plaque erosion and plaque rupture [61, 62]. In fact, endothelial cell apoptosis has been postulated as the key event in a variety of pathological conditions including atherosclerosis and hypertension [63].

There are many epidemiological and observational studies that *trans* FA are associated with cardiovascular diseases. Mozaffarian (2006) associated *trans* FA consumption with higher circulating markers of systemic inflammation [64]. Micha and Mozaffarian (2008) described that FA affect not only serum lipid levels, but arrhytmia, hemodynamics, inflammation, endothelial function, insulin sensitivity, and thrombosis; leading to obesity, diabetes mellitus, atheroclerosis, plaque rupture, and sudden death [65]. Mozaffarian and Clarke, (2009) studied an isocaloric replacement of *trans* FA with either PUFA and MUFA or saturated FA. Even 1% energy replacement decreased the total cholesterol/HDL cholesterol ratio. This was mirrored by changes in apolipoprotein levels [66]. The inverse of each of these values indicated the effect of the converse replacement of PUFA and MUFA or saturated FA with *trans* FA. For example, lipoprotein levels were increased by replacement of any of the dietary fats with *trans* FA. Reported effects include alteration in the LDL particle size and in the composition of postprandial lipoproteins [67]. Compared with unsaturated fats, diet with *trans* FA also increases serum triglycerides. *Trans* FA can influence endothelial health; greater *trans* FA intake was associated with high levels of soluble adhesion molecules (intercellular adhesion molecule-1 (ICAM) and vascular cell adhesion molecule-1 (VCAM)). *Trans* FA was associated with plasma E-selectin concentrations, another marker of endothelial dysfunction (Mozaffarian, 2006). Thompson et al (2008) reviewed the positive correlation between the *trans* FA intake with prostate cancer, non-Hodgkin's lymphoma, pancreatic cancer, whereas breast cancer and colorectal cancer did not have any association [68].

THERAPEUTIC OF LIPIDS

The physiological importance of lipids is illustrated by the numerous diseases to which lipid therapy can contribute, including neurological disorders, rheumatoid arthritis, scarring, multiple sclerosis, asthma, psoriasis, cardiovascular diseases and diabetes.

Neurological Disorders

Lipid metabolism may be of particular importance for the central nervous system (CNS). The crucial role of lipids in the therapy of CNS has been demonstrated in many neurological disorders, including Alzheimer's disease, schizophrenia and depression [69].

The cerebral tissue is composed by 36% saturated FA, 20% of monounsaturated FA, 17% n-6 PUFA, 14% n-3 PUFA, 8% n-7 PUFA and 5% corresponding to FA derivatives [70, 71]. The main n-3 PUFA in the brain is the DHA representing 10 to 20% of the total FA, as well as the α-linolenic acid and EPA that correspond to less than 1% of the total composition of FA in the brain [72]. In fact, DHA is the most abundant n-3 PUFA in the brain, which accounts for about 8% of the dry weight of the brain [73].

Alzheimer Disease

Dementia has been recognized as one of the most important medical problems in the elderly. Alzheimer's dementia (AD) is a growing health problem in aging populations in many countries. The prevalence of AD in the American population is expected to increase as the population of individuals over 65 years old grows, and by 2050, it is estimated that there will be approximately 13.2 million cases [74]. Considering that there is no cure for AD, alternative therapies are urgently needed. Recent studies suggest that nutritional intervention may have therapeutic benefits for AD. In particular, research has shown that n-3 PUFA from marine oils, containing EPA (20:5n-3) and DHA (22:6n-3) may have therapeutic usage for the prevention and treatment of AD [75]. Epidemiological studies suggest that increased intake of DHA from fish and marine oils is associated with a reduced risk for AD [75, 76]. Data from animal models support the proposition that DHA may be an effective treatment for AD through anti-amyloid, antioxidant, and neuroprotective mechanisms [77]. Ten of the thirteen epidemiological studies to date report an inverse association between AD risk and n-3 PUFA status or intake [75]. Administration of n-3 PUFA for 12 months in 204 patients with mild to moderate AD showed no delay in the rate of cognitive decline according to the cognitive portion of the Alzheimer Disease Assessment Scale. However, positive effects were observed in a small group of patients with very mild AD [78, 79].

DHA has been shown to decrease proapoptotic protein BAD (Bcl-2 antagonist of cell death) by 47% [80], and the secretion of Aβ by 30–70% [77, 81-83] and increase in PI3-kinase neuroprotective pathway [80], anti-apoptotic protein Bcl-2 [84] and improve cognition in brain regions.

High plasma cholesterol levels have been linked to AD pathology in epidemiological studies whereas higher Aβ production and deposition into amyloid plaques has been observed in animals fed a high cholesterol diet [85]. Cholesterol is reduced in lipid rafts of Aβ-infused rats pre-treated with dietary DHA [81], suggesting that DHA-lowering effects on brain cholesterol is a potential mechanism to explain the therapeutic effect of n-3 PUFA in AD. In addition, the effect of n-3 PUFA observed on AD pathology is related in part to the anti-inflammatory effect of n-3 PUFA through its suppressing effect on cPLA2 and COX-2, both associated to AD [86, 87].

The effect of n-3 PUFA on AD might only be seen after long term treatment, i.e., lifetime consumption. In addition, the type of n-3 PUFA used in epidemiological (fish) versus clinical studies (fish oil/ EPA/DHA) might also explain the discrepancies. Many researchers reported that DHA consumption is inversely associated with AD risk but not EPA [88]. Preclinical studies leading to a better understanding of the mechanism of action by which DHA and other n-3 PUFA affect the progression of AD are still required.

Schizophrenia

Abnormal neurotransmission has been found in those who suffer from schizophrenia [89]. Currently 45 million people worldwide suffer from this illness. Schizophrenic symptoms may be the result of altered neuronal membrane structure and metabolism [90]. In turn, neuronal cell membrane structure and metabolism itself is dependent on plasma levels of certain essential FA and their hydroxy-metabolites [91, 92]. Dietary supplementation of the essential FA may have a direct and positive effect on the symptoms of schizophrenia [93, 94].

Epidemiological studies associated the low consumption of n-3 PUFA from marine foods with an increase in schizophrenia symptoms [93] [95]. Low DHA concentration in plasma and prefrontal cortex has been found in patients with schizophrenia [72, 96, 97]. DHA and EPA treatment seems to improve the schizophrenia symptoms [73, 98, 99]. Premature neonatal with DHA deficit has more risk to show schizophrenia [100, 101] and the duration of the breastfeeding is inversely correlated with the occurrence of this disease. Children breastfed for less than 2 weeks have 1.7 times more chance to show schizophrenia when compared to children who were breast-fed for more than 2 weeks [102].

Depression

A stronger relation has been found between the consumption of fish and depression [103-105], as well as a reduction of n-3 PUFA in the erythrocyte membrane of patients with depression [106-108]. An increase in arachidonic acid:DHA, arachidonic acid:EPA and n-6:n-3 ratios has also been observed in adult [109] and elderly patients with depression [110], and in pregnant women with postpartum depression [111]. Depression is also associated with the levels of arachidonic acid products such as PGE2 e TXB2. Omega-3 PUFA can inhibit the synthesis of these products by competing for cyclooxygenase and lipooxygenase and so reducing the symptoms. Patients supplemented with n-3 PUFA have an improvement in the symptoms of depression and conception of suicide when compared to patients who received placebo [99, 112-115]. Omega-3 PUFA supplementation seems to improve the symptoms of the postpartum depression [116, 117]. Besides, DHA concentration in plasma predicts low concentration of the marker of serotonin turnover in the brain being strongly connected with the risk of depression, suicide, violence and alcoholism [103]. Omega-3 PUFA can decrease the risk of depression by modulation of serotonin and dopamine metabolism [70], signal transduction mediated through fosfatidilinositol, production of cytokines and eicosanoids [118, 119], inhibition of second messengers connected with protein kinase [120] and regulation of the calcium type L channel [121].

Multiple Sclerosis

Multiple sclerosis (MS) is an inflammatory and demyelinating disease of the CNS and is the most important non traumatic cause of neurological disability in young adults, especially in women. Several studies have shown that MS risk is high in countries with a high intake of saturated FA and is low in countries with high intake of PUFA [122-125]. Several mechanisms proposed for the therapeutic use of PUFA, particularly, linoleic, γ-linolenic, EPA and DHA for MS include immunomodulatory and anti-inflammatory actions, effects on microcirculation and erythrocyte aggregation, antioxidant action and also effects related to their role as constituents of the myelin membrane [126, 127].

Dietary supplementation of PUFA seems to have no major effect on disease progression, the main clinical outcome in MS, and it does not substantially affect the risk of clinical relapses over 2 years. However, the data available are not enough to establish the real benefit or harm that might result from PUFA supplementation. Previous study demonstrated that patients with MS with linoleic acid supplementation present shorter and less severe recidive compared to oleic acid supplementation showing significant improvement of the symptoms [128].

A disturbance of n-6 PUFA metabolism is observed in MS [128]. The high dose of γ-linolenic acid-rich oil had a marked clinical effect in relapsing-remitting MS, decreasing the relapse rate and the progression of the disease compared to placebo. In the placebo group, production of TNF-α and IL-1β by mononuclear cells was increased and anti-inflammatory TGF-β markedly decreased with loss of membrane linoleic and arachidonic acids. In contrast, no such changes were observed in the high dose of γ-linolenic acid group. The disturbed n-6 PUFA metabolism in MS gives rise to loss of membrane long chain n-6 PUFA, particularly during the relapse phase, as well as loss of these important FA for CNS structure and function and consequent long term neurological deficit in MS [129].

Rheumatoid Arthritis

Rheumatoid arthritis (RA) is a chronic inflammatory autoimmune disease and the major cause of disability that affects about 1% of the adult population being more common in women than in men [130]. Joint inflammation is manifested by swelling, pain, functional impairment, morning stiffness, osteoporosis and muscle wasting.

The risk for development of RA is low in populations that use a n-3 PUFA rich diet. The Japanese population that has an elevated consumption of foods rich in these FA, when compared with the Western population, presents an incidence of RA three times smaller [131]. Studies with n-3 PUFA, mainly DHA and EPA, has been shown to have beneficial effects in animal models of arthritis and patients. Dietary fish oil reduces the incidence and severity of type II collagen-induced arthritis in mice [132]. In another study, both EPA and DHA were found to suppress streptococcal cell wall-induced arthritis in rats, with EPA being more effective. The benefits of fish oil for patients include reduced duration of morning stiffness, number of tender or swollen joints, joint pain, time to fatigue, use of non-steroidal anti-inflammatory drugs and increased grip strength [133-135].

Studies have reported anti-inflammatory effects of fish oil in patients with RA. Fish oil treatment decreased plasma IL-1β and C-reative protein levels [136, 137], IL-1β and TNF-α production by monocytes [138-140], LTB4 production by neutrophils [133, 134] and monocytes [134], PGE2 production by mononuclear cells [141], and normalized neutrophil chemotactic response [142] in patients with RA. DHA and EPA also decreased the expression of surface and adhesion molecules such as ICAM-1 and LFA-1, which induce monocyte migration to inflamed joints [143, 144]. A recent meta-analysis provided further evidence of the efficacy of n-3 PUFA in RA [145].

Asthma

Asthma is one of the most common chronic conditions and affects both adults and children. Its prevalence has increased significantly in Western nations with overall prevalence rates ranging from 7–15% [146]. It is a chronic inflammatory disease of the airway characterized by increase of the responsiveness and production of mucus that leads to episodes of cough, difficult breathing and deficient air flow [147]. Respiratory benefits might be attributed to changes in the status of inflammation mediators such as leukotrienes and prostaglandins. The hyper sensitivity observed in these patients can be caused by the exaggerated lymphocyte helper response. When exposed to a stimulus, these cells promote secretion of pro-inflammatory agents including LTB4 and PGE2. LTB4 induces bronco-constriction and mucus production by respiratory cells. The potential therapeutic and protective effects of dietary supplementation with the n-3 PUFA have been studied. Several authors have reported the low incidence of asthma in the Northern aboriginal population due to their consumption of large quantities of oily fish rich in n-3 PUFA [148].

Fish oil beneficial effects in the treatment of asthma are based mainly on changes in the production of several inflammatory mediators such eicosanoids and cytokines. Asthmatic patients present elevated concentrations of immunoglobulin E (IgE). EPA reduces the production of IgE through a decrease in the production of PGE2. This effect decreases inflammation and improves the clinical condition of the asthmatic patients. Okamoto et al. (2000) suggested that supplementation of perilla seed oil rich in n-3 PUFA is useful for the treatment of asthma in terms of suppression of LTB4 and LTC4 generation by leukocytes, improving pulmonary function [149]. Deng YM (2007) observed the same in guinea pig and *in vitro* [150]. Other recent study provided evidence that dietary supplementation with n-3 PUFA reduces bronchial inflammation even after low-dose of an allergen challenge [151].

Children with diet rich in fish oil had three times less susceptibility for the development of asthma compared to diet poor in this type of food [152]. Diet supplementation with n-3 PUFA significantly improved asthma control test, pulmonary function tests and pulmonary inflammatory markers in children with moderately persistent bronchial asthma [153]. Intake of n-3 PUFA was also positively correlated with forced expiratory volume [154]. However, in a meta-analysis study no evidence was found that n-3 PUFA are efficient and beneficial for children and adults with asthma [155].

Psoriasis

Psoriasis is an inherited inflammatory skin disease mediated by T-cells and characterized by hyperproliferation and poor differentiation of epidermal keratinocytes that affect 2% of the population. Psoriasis is influenced by environmental factors such as infections and stress. High intake of n-3 PUFA shows beneficial effects for patients with psoriasis [156]. EPA use in psoriasis has been demonstrated in trials using oral, intravenous, and topical preparations, with generally positive outcomes. Depth profile analysis revealed that EPA and its metabolite, 15-HEPE, are deposited in the epidermis, particularly in the metabolically active basal layer. This is considered advantageous in psoriasis therapy [157].

Intravenous infusions of n-3 PUFA containing EPA lead to an increase in LTB5 in psoriatic plaques within 4-7 days of starting treatment, in comparison with the control patients infused with n-6 PUFA. Topic application of fish oil also improved erythema and scaling, reduced plaque thickness and scaling of psoriasis symptoms demonstrating the efficacy of topically applied fish oil in reducing psoriasis symptoms compared to liquid paraffin [158].

Healing Wound or Scarring

The nutritional deficiency of FA impairs the healing process. FA such as linoleic acid, have been employed for the prevention and treatment of pressure ulcers. Essential FA are required to ensure epidermal integrity and to maintain the water barrier in the skin. EPA and γ-linolenic acids intake significantly decrease the occurrence of new pressure ulcers in critically ill patients [159]. The topic application of FA in wound treatment should also be considered as an important therapeutic strategy. FA are commonly used as scarring agents. Oily substances are known to protect against microorganisms, avoid tissue dehydration, keep the body temperature and reduce traumas during dressing changes. Fatty acids are believed to speed up the cicatricial process, stimulating leukocyte chemotaxis, improving angiogenesis, and wound hydration [160].

Previous studies showed that the topic treatment with oleic and linoleic acids in animals presented a significant reduction in the wound area, necrosed cells and •NO production, increased the mass around the wound and reduced the necrosis acting with pro-inflammatory properties, contributing to speed up the repair process [161, 162]. However the treatment with these FA did not affect the vascular permeability and the lipid contact in the cicatricial tissue [162].

In traditional Chinese medicine, extracts from the Chinese mole cricket, that are rich in FA, have been used to treat boils, abscesses and ulcers successfully for over two centuries and are still being used today. Wounds treated with oily extracts from the European mole cricket, rich in linoleic acid methyl ester, epithelialized significantly faster than in the control [163].

Fatty acids are commonly associated to treatment of stomach and intestine ulcers, and in colon wounds. Previous studies showed that small chain FA infusion into the colon furthers the healing process in the injured tissues, improving the colon anastomosis resistance in rats and protects the colonic epithelial cells in rats and in patients submitted to Hartmann surgery [164, 165]. Friedel and Levine (1992) observed an increase in the mucosa DNA content in rats kept in total parenteral nutrition, receiving small chain FA infusions in the proximal colon [166].

Cardiovascular Disease and Diabetes

Omega-3 PUFA have been reported to decrease the risk of cardiovascular disease and to be beneficial for patients with type 2 diabetes although recent trials and a meta-analysis did not confirm this proposition [167-170]. Prospective cohort studies showed that the risk of coronary heart disease is much lower in patients with type 2 diabetes that consume n-3 PUFA [171, 172]. The effects of n-3 PUFA on plasma triglycerides, total cholesterol, HDL-cholesterol and LDL-cholesterol levels have been reviewed systematically in meta-analyses in patients with and without diabetes [173, 174]. Omega-3 PUFA supplementation may reduce triglycerides concentration in patients with type 2 diabetes.

The risk markers for cardiovascular disease include abnormalities in LDL particle size and HDL subfraction concentrations [175], apolipoprotein concentrations [176] and plasma lipase activity [177]. Omega-3 PUFA supplementation has been shown to have both beneficial and adverse effects on lipid metabolism [178]. Previous studies demonstrated that n-3 PUFA supplementation decreases triglycerides, VLDL-cholesterol and VLDL-triglycerides, but may have an adverse effect on LDL-cholesterol [179].

This systematic review and meta-analysis confirm the triglyceride lowering effects of n-3 PUFA, demonstrate the potential dose-response effects and shows improvements in thrombogenesis. Omega-3 PUFA raise LDL levels without concomitant changes in lipid particle size. Changes seen in conventional risk factors are not sufficient to explain the cardiovascular disease risk reductions suggested to occur with n-3 PUFA [179]. Zhao et al. (2009) suggest in a meta-analysis of randomized controlled trials that dietary supplementation with n-3 PUFA reduces the incidence of sudden cardiac death in patients with myocardial infarction but may have adverse effects in angina patients [180].

Lipid Analysis

The determination of the total lipid content in a tissue as well as the amount of a particular group of lipids and their FA profile provide important information to understand the mechanisms of diseases that are connected to lipid metabolism. To conclude this chapter, in the following sessions we will briefly discuss the most common lipid analysis.

Sample Handling and Storage

Prior to the analysis, some cautions must be taken in the collection, handling and storage of the samples from where the lipids will be extracted. Ideally, lipids should be extracted immediately after removal of the tissue from the animal or culture system. Aside from the oxidative damages, the osmotic shock caused by freezing can release lipolytic enzymes that even at – 20 ºC can change lipid molecules, particularly if they are stored in contact with organic solvents. The best way to protect the lipid content of biological samples is to store them at – 80 ºC, under an atmosphere of nitrogen, without adding solvents.

Extracting the Lipids from Samples

Depending on the sample and the aim of the study, it is possible to skip the lipid extraction step prior to the analysis [181]. However, this procedure is very important to most applications. There are several methods to extract lipids from a biological sample, but the extraction with organic solvents might be the best choice in a research laboratory scale. This task is made easier due to the hydrophobicity of the most common lipids. Protocols using different solvents to extract lipids from tissues can be found in the literature and they share a common feature: lipid extraction is usually performed using a mixture of two solvents - one with low and another with high polarity.

The most common protocol for lipid extraction of animal tissues was developed by Folch et al., who in 1957 published an improved version of their protocol to extract lipids from brain [182, 183]. In this protocol, total lipids are extracted from a tissue by homogenizing it with chloroform and methanol (2:1). After the homogenization, water or a saline solution is added in a volume that corresponds to 20% of the volume of chloroform and methanol used. The mixture is shaked and allowed to settle, leading to the formation of two layers that are best defined after a brief centrifugation. Most of the lipids are found in the bottom layer (primarily composed by chloroform). The upper phase is collected and methanol:water (1:1) is added in the volume of one-fourth of the lower phase (chloroform) to re-wash it. Recovering and drying the lower phase is the last step of lipid extraction. Bligh and Dyer revisited this method in 1959, making it more suitable for extraction of lipids from tissues with elevated content of water (80%) [184]. These authors described the extraction of lipids from tissues by homogenizing them with chloroform and methanol 1:2, and not 2:1, as in Folch`s protocol. It was acknowledged that the recovery of non-polar lipids (like triglycerides) might be underestimated, despite satisfactory extraction of phospholipids. The authors advise to re-extract samples with more chloroform when they are particularly rich in non polar lipids.

Despite being the most used solvent, chloroform is very harmful to the environment and human health and it would be better to replace this solvent by others. Solvent systems composed by hexane:isopropanol or hexane:ethanol [185] have the advantage of being less toxic and it is advisable to use these methods whenever is necessary after testing their efficiency.

Lipid Extract - Handling and Storage

Once extracted, lipids are free from the damages induced by lipolytic enzymes but they can still undergo oxidation. The oxidation happens spontaneously when the sample is in contact with air and is accelerated by light, especially if the lipids are composed of unsaturated FA. Oxidized FA can have impaired quantification by chromatographic analysis and caution must be taken to avoid this interference. Lipid samples can naturally contain antioxidants like tocopherols, but it is advisable to add 50 – 100 mg/L of butylated hydroxyl toluene (BHT) in the moment of extraction to protect the sample against oxidation. The BHT does not interfere with analysis due to its low molecular weight, being eluted with the first peaks, in a chromatographic analysis. The best way to store lipid extracts is diluting them in a non polar solvent (like hexane) and store at -20 ºC in glass vials under a nitrogen atmosphere with appropriate addition of antioxidants.

Lipid Analysis – Principles of Chromatography

A wide range of assay kits for quantification of different classes of lipids like cholesterol, free FA, TAG and others are commercially available. These methods are based on simple principles and can be performed even in laboratories were a spectrophotometer is the most sophisticated analytical equipment. When one aims to identify lipid molecules or quantify specific groups of lipids, chromatography is the best choice, due to its versatility and accuracy. The next paragraphs cover the basic principles of chromatographic analysis without intending to cover details or technical questions, which are beyond the scope of this chapter.

In a chromatographic system there are always three components: the mobile phase, the stationary phase and the sample, composed by distinct molecules that have to be separated throughout the process. The sample is loaded in the system and a flow of the mobile phase drags it through the stationary phase. Depending on their polarity, different molecules in a sample will interact differentially with the mobile and stationary phase. If the interaction with the stationary phase is strong enough, these molecules will take longer to pass through it. However, if they have similar polarity to the mobile phase or low affinity with the stationary phase they will be sooner eluted. It is possible to separate different groups of molecules in the sample by varying the polarity of the stationary and mobile phases.

Chromatographic systems can be classified into two categories according to the arrangement of the components: 1 – Column Chromatography, if the stationary phase is packed into a column, and the mobile phase is forced to pass through it; 2 – Thin Layer Chromatography (TLC), when the stationary phase forms a layer on a flat surface and the mobile phase passes through it by capillarity. Column chromatography can also be divided into two categories, depending on the nature of the mobile phase. When it is a liquid solvent we call it Liquid Chromatography that can be of low or high pressure, being the last the most common application. This is known as High Pressure Liquid Chromatography (HPLC). If the mobile phase is a gas, we call the technique of Gas Chromatography (GC).

Independent of the classification, individual components of the sample will interact differently with the particles of the stationary phase. This allows different groups of molecules to be eluted at the end of the stationary phase in different time points in Column Chromatography systems. The time point that a group of molecules is eluted is known as Retention Time and is characteristic of these molecules in given conditions of system operation. Retention Time is used to identify the eluted molecules when compared with a standard. In TLC, the sample molecules are not eluted from the stationary phase, but they are separated in distinct spots on the plate with a particular distance from the point where the sample was loaded. This distance (known as Rf value) is a characteristic of each molecule and used for their identification like the Retention Time.

Thin-Layer Chromatography - TLC

Thin-layer chromatography is the simplest form of chromatography. It is widely used as a preparative step in the analysis of lipids, but it can also give quantitative results. In this technique, the samples are applied as spots approximately 15 mm from the edge of a plate (usually a glass plate) covered with the stationary phase (that can be silica). The plate is then placed into a tank, which contains a small volume of the mobile phase with the sample spots positioned in the bottom of the plate. The mobile phase will migrate by capillarity through the stationary phase, dragging the sample with it. As discussed above, the components of the sample will follow the flow of mobile phase according to their affinity to the stationary and mobile phase. When the solvent front reaches the upper side of the plate, it is removed from the tank and allowed to dry. Then the different components of the sample can be visualized in different regions of the plate, based on their differential mobility.

Depending on the purpose of the separation, different combination of solvents and stationary phases can be used. Thin-layer chromatography can be used to separate single lipid groups from a total lipid extract, or to separate the components of a group of lipids. For example, phospholipids can be separated in a band that can be scratched from the plate, re-extracted from the silica and re-chromatographed, using a different solvent system to separate the phospholipids species. There is also the possibility of running a two-dimensional TLC. In this case, the plate is developed to separate wider groups of lipids, and then is rotated 90 ° and developed on a perpendicular direction to the first development to separate the components of the lipid groups previously separated. Examples of this application can be found in Wuthier et al [186].

The detection and quantification of lipids in a TLC plate can be accomplished by incubating or spraying the plate with an agent that react with the lipid bands, indicating their location. The lipids can be quantified by measuring the area of the bands with a photo-densitometer. Alternatively, the bands of separated lipid groups can be scratched from the plate and the lipids eluted from the silica and quantified. Phospholipids can be quantified by phosphorus determination, for example [187].

Column Chromatography

High Pressure Liquid Chromatography - HPLC

HPLC is probably the most common application of chromatography, being used to several classes of molecules, including lipids. The stationary phase (usually a C_{18} or a C_8 hydrocarbon) is chemically bounded to small support particles that are packed into a cylindrical column. As the column is filled with particles that have the stationary phase bound, the flow of mobile phase happens under a high pressure, giving the name of the technique. An arrangement of pumps is responsible to keep the necessary pressure to ensure the flow of the mobile phase through the column. Different patterns of separation of the molecules can be achieved by varying the polarity of the mobile phase using different combinations of solvents.

The column outlet is equipped with a detector (there are several types of detectors) that generates a signal with proportional amplitude to the amount of the molecule that is passing through it. This signal is transduced as a peak, which the area is proportional to the amount of the molecule that generated it. The system can be equipped with fraction collectors programmed to collect a fraction of the mobile phase in a determined time frame - when the interest molecules elutes from the column. This allows the system to separate, quantify, identify and recover a group of molecules from a mixture!

HPLC can be used to quantify specific lipid groups, or to determine the FA composition of a lipid sample. Protocols to separate the most common lipid classes can be found in the book Lipid Analysis (Christie, W.W). It is also possible to separate FA by HPLC. The procedures described in Abushufa et al. (1994) and Nishiyama-Naruke et al. (1998) using fluorescence detector are recommended, but there are many protocols available using different detectors [188, 189].

Gas Chromatography - GC

In the GC technique, a mixture of volatile components are carried by a stream of gas which flows through a column packed with particles coated with the stationary phase. Capillary columns are more common for analytical purposes; they are very thin and long usually with less than 0.2 mm of internal diameter and longer than 100 m. In the GC, the separation of the compounds is influenced by their volatility. The more volatile molecules are the first to elute at the end of the column. Non-volatile samples must be volatilized (e.g. FA are converted into methyl esthers) prior to their injection into the system. The volatility of the molecules through the column can be changed by alterations in the temperature. For this reason, the column of a gas chromatographer is placed into an oven, where the temperature can be programmed to best perform the separation of the molecules.

The most common detector in GC applications is the Fire Ionization Detector (FID) that consists of two platinum electrodes across a flame of hydrogen. The gas that leaves the column is introduced into the flame, and the molecules that elute from the column are ionized in the flame. This generates a current across the electrodes that is transduced as a peak, with proportional area to the concentration of the molecule that originated the current. The protocol described by Hartman and Lago in 1973 is a very good reference to convert FA into methyl esters, allowing their analysis by GC.

The highest standard of reliability in qualitative and quantitative lipid analysis is achieved when techniques like GC or HPLC are coupled with mass spectrometry. In mass spectrometry, molecules that leave the column are bombarded by a stream of high-energy electrons, being fragmented and generating smaller ionized entities. These are then separated in a magnetic or electrostatic field according to their mass and charge. As molecules tend to split in specific regions it is possible to determine the structure of the compound based on the nature of the fragments produced. This characteristic of mass spectrometry analysis minimize the possibility of mistakes in the identification of the separated molecules that can happen when it relies only on the comparison of the retention times of the separated molecules with those of standards.

Relevance of Lipid Analysis

Lipidomics is the result of the advances in lipid analysis and statistical tools, leading to a comprehensive analysis of lipid species in cells and tissues. Following the advent of gene and protein profiling, called genomics and proteomics, lipidomics emerged as a new way to study metabolism. It has the advantage of being a phenotypic analysis, detecting differences that result from expression of genes, activities of enzymes and environmental influences. As a detailed discussion on this topic is out of the scope of this chapter, we suggest the reviews by Chunxiu et al., (2009) and Watson (2006) as sources of information about analytical techniques and statistical tools necessary to the development of lipidomics [190].

This phenotypic analysis can be used to identify specific markers of diseases, helping in the prediction and diagnosis of several disorders. Considering at least 40 FA and 15 lipid classes, there is a repertory of at least 600 lipid species that can be easily investigated [191]. The probability that some of these species could be useful markers of specific conditions is very high and should be explored.

REFERENCES

[1] Lefkowisch, J.I., *Chemical Technology and Analysis of Oils, Fats and Waxes*. 5 ed. Vol. 1. 1913, London: Macmillan and Co.

[2] van Dorp, D.A., Essential fatty acid metabolism. *Proc. Nutr. Soc.*, 1975. 34: p. 279 - 286.

[3] Evans, H.M., Burr, G. O., A new dietary deficiency with highly purified diets. *Proc. Soc. Exp. Biol. Med.*, 1927. 24: p. 1.

[4] Burr, G.O.B.a.M.M., A new deficiency disease produced by the rigid exclusion of fat from thc dict. *J. Biol. Chem*, 1929. 32: p. 345 - 367.

[5] Burr, G.O.B.a.M.M., On the nature and role of the effect of the fatty acids essential in nutrition. *J. Biol. Chem.*, 1930. 86: p. 587 - 621.

[6] Mead, J.F., Slater, R. B. A., Howton, D. R., Popjak, G., *Lipids. Chemistry, biochemistry and nutrition.* 1986, New York: Plenum Press. 132-147.

[7] Holman, R.T., *George O.* Burr and the discovery of essential fatty acids. *J. Nutr*, 1988. 118(5): p. 535-40.

[8] Bergstrom, S., Prostaglandins from bedside observation to a family of drugs. *Prog. Lipid Res,* 1981. 20: p. 7-12.

[9] Gurr, M.I., Harwood, J. L., *Lipid Biichemistry, an introduction.* 4 ed. 1991, London: Chapman and Hall. 99 - 118.

[10] Lefkowith, J.B., et al., *Essential fatty acid deficiency: a new look at an old problem.* Prostaglandins Leukot Med, 1986. 23(2-3): p. 123-7.

[11] Madani, S., et al., Implication of acyl chain of diacylglycerols in activation of different isoforms of protein kinase C. *FASEB J,* 2001. 15(14): p. 2595-601.

[12] Pages, G., et al., Defective thymocyte maturation in p44 MAP kinase (Erk 1) knockout mice. *Science*, 1999. 286(5443): p. 1374-7.

[13] Denys, A., A. Hichami, and N.A. Khan, Eicosapentaenoic acid and docosahexaenoic acid modulate MAP kinase enzyme activity in human T-cells. *Mol. Cell Biochem*, 2002. 232(1-2): p. 143-8.

[14] Denys, A., A. Hichami, and N.A. Khan, n-3 PUFAs modulate T-cell activation via protein kinase C-alpha and -epsilon and the NF-kappaB signaling pathway. *J. Lipid Res,* 2005. 46(4): p. 752-8.

[15] Stulnig, T.M., et al., Polyunsaturated fatty acids inhibit T cell signal transduction by modification of detergent-insoluble membrane domains. *J. Cell Biol*, 1998. 143(3): p. 637-44.

[16] Hii, C.S., et al., Stimulation of p38 phosphorylation and activity by arachidonic acid in HeLa cells, HL60 promyelocytic leukemic cells, and human neutrophils. Evidence for cell type-specific activation of mitogen-activated protein kinases. *J. Biol. Chem*, 1998. 273(30): p. 19277-82.

[17] Wang, G., G.R. Krueger, and L.M. Buja, A continuous model studying T cell differentiation and lymphomagenesis and its distinction with discrete models. *Anticancer Res*, 2004. 24(3a): p. 1813-6.

[18] Gorjao, R., et al., Regulation of human lymphocyte proliferation by fatty acids. *Cell Biochem. Funct*, 2007. 25(3): p. 305-15.

[19] Kirken, R.A., et al., Activation of JAK3, but not JAK1, is critical for IL-2-induced proliferation and STAT5 recruitment by a COOH-terminal region of the IL-2 receptor beta-chain. *Cytokine*, 1995. 7(7): p. 689-700.

[20] Russell, S.M., et al., Interaction of IL-2R beta and gamma c chains with Jak1 and Jak3: implications for XSCID and XCID. *Science*, 1994. 266(5187): p. 1042-5.

[21] Darnell, J.E., Jr., I.M. Kerr, and G.R. Stark, Jak-STAT pathways and transcriptional activation in response to IFNs and other extracellular signaling proteins. *Science,* 1994. 264(5164): p. 1415-21.

[22] Xue, H., et al., Eicosapentaenoic acid and docosahexaenoic acid modulate mitogen-activated protein kinase activity in endothelium. *Vascul Pharmacol,* 2006. 44(6): p. 434-9.

[23] Chen, W., et al., Anti-inflammatory effect of docosahexaenoic acid on cytokine-induced adhesion molecule expression in human retinal vascular endothelial cells. *Invest Ophthalmol. Vis. Sci,* 2005. 46(11): p. 4342-7.

[24] Limb, G.A., et al., Evidence for control of tumour necrosis factor-alpha (TNF-alpha) activity by TNF receptors in patients with proliferative diabetic retinopathy. *Clin. Exp. Immunol,* 1999. 115(3): p. 409-14.

[25] Kowluru, R.A., S.N. Abbas, and S. Odenbach, Reversal of hyperglycemia and diabetic nephropathy: effect of reinstitution of good metabolic control on oxidative stress in the kidney of diabetic rats. *J. Diabetes Complications,* 2004. 18(5): p. 282-8.

[26] Hirafuji, M., et al., Docosahexaenoic acid potentiates interleukin-1beta induction of nitric oxide synthase through mechanism involving p44/42 MAPK activation in rat vascular smooth muscle cells. *Br. J. Pharmacol*, 2002. 136(4): p. 613-9.

[27] Ringseis, R., et al., *CLA* isomers inhibit TNFalpha-induced eicosanoid release from human vascular smooth muscle cells via a PPARgamma ligand-like action. *Biochim Biophys Acta,* 2006. 1760(2): p. 290-300.

[28] Lee, C.K., et al., Alpha-lipoic acid inhibits TNF-alpha induced NF-kappa B activation through blocking of MEKK1-MKK4-IKK signaling cascades. *Int. Immunopharmacol,* 2008. 8(2): p. 362-70.

[29] Slawik, M. and A.J. Vidal-Puig, Lipotoxicity, overnutrition and energy metabolism in aging. *Ageing Res. Rev*, 2006. 5(2): p. 144-64.

[30] Schaffer, J.E., Lipotoxicity: when tissues overeat. *Curr. Opin. Lipidol*, 2003. 14(3): p. 281-7.

[31] Unger, R.H., Lipotoxic diseases. *Annu. Rev. Med*, 2002. 53: p. 319-36.

[32] Torres, N. and A.R. Tovar, The role of dietary protein on lipotoxicity. *Nutr. Rev*, 2007. 65(6 Pt 2): p. S64-8.

[33] Weinberg, J.M., Lipotoxicity. *Kidney Int*, 2006. 70(9): p. 1560-6.

[34] Schaffer, J.E., Fatty acid transport: the roads taken. *Am. J. Physiol. Endocrinol. Metab*, 2002. 282(2): p. E239-46.

[35] Russell, A.P., Lipotoxicity: the obese and endurance-trained paradox. *Int. J. Obes. Relat. Metab. Disord*, 2004. 28 Suppl 4: p. S66-71.

[36] Carmiel-Haggai, M., A.I. Cederbaum, and N. Nieto, A high-fat diet leads to the progression of non-alcoholic fatty liver disease in obese rats. *Faseb J,* 2005. 19(1): p. 136-8.

[37] Lelliott, C. and A.J. Vidal-Puig, Lipotoxicity, an imbalance between lipogenesis de novo and fatty acid oxidation. *Int. J. Obes. Relat. Metab. Disord*, 2004. 28 Suppl 4: p. S22-8.

[38] Das, U.N., *Beneficial effect(s) of n-3 fatty acids in cardiovascular diseases: but, why and how?* Prostaglandins Leukot Essent Fatty Acids, 2000. 63(6): p. 351-62.

[39] Barre, D.E., The role of consumption of alpha-linolenic, eicosapentaenoic and docosahexaenoic acids in human metabolic syndrome and type 2 diabetes--a mini-review. *J. Oleo Sci*, 2007. 56(7): p. 319-25.

[40] Sullivan, D.R., et al., The effects of dietary n - 3 fatty acid in animal models of type 1 and type 2 diabetes. *Diabetes Res. Clin. Pract,* 1990. 9(3): p. 225-30.

[41] Martinsen, N., et al., Predictions of type 2 diabetes and complications in Greenland in 2014. *Int. J. Circumpolar Health,* 2006. 65(3): p. 243-52.

[42] Schraer, C.D., et al., *Prevalence of diabetes mellitus in Alaskan Eskimos, Indians, and Aleuts. Diabetes Care,* 1988. 11(9): p. 693-700.

[43] De Caterina, R., et al., n-3 fatty acids in the treatment of diabetic patients: biological rationale and clinical data. *Diabetes Care*, 2007. 30(4): p. 1012-26.

[44] Kesavulu, M.M., et al., Effect of omega-3 fatty acids on lipid peroxidation and antioxidant enzyme status in type 2 diabetic patients. *Diabetes Metab*, 2002. 28(1): p. 20-6.

[45] Mori, T.A., et al., Effect of eicosapentaenoic acid and docosahexaenoic acid on oxidative stress and inflammatory markers in treated-hypertensive type 2 diabetic subjects. *Free Radic. Biol. Med*, 2003. 35(7): p. 772-81.

[46] Stene, L.C. and G. Joner, Use of cod liver oil during the first year of life is associated with lower risk of childhood-onset type 1 diabetes: a large, population-based, case-control study. *Am. J. Clin. Nutr*, 2003. 78(6): p. 1128-34.

[47] Stene, L.C., et al., Use of cod liver oil during pregnancy associated with lower risk of Type I diabetes in the offspring. *Diabetologia,* 2000. 43(9): p. 1093-8.

[48] Norris, J.M., et al., Omega-3 polyunsaturated fatty acid intake and islet autoimmunity in children at increased risk for type 1 diabetes. *Jama*, 2007. 298(12): p. 1420-8.

[49] Aarnes, M., S. Schonberg, and V. Grill, Fatty acids potentiate interleukin-1beta toxicity in the beta-cell line INS-1E. *Biochem. Biophys. Res. Commun*, 2002. 296(1): p. 189-93.

[50] Azevedo-Martins, A.K., et al., Fatty acid-induced toxicity and neutral lipid accumulation in insulin-producing RINm5F cells. *Toxicol. In Vitro*, 2006. 20(7): p. 1106-13.

[51] Simon, M.N., et al., Persistent activation of Akt or ERK prevents the toxicity induced by saturated and polyunsaturated fatty acids in RINm5F beta-cells. *Toxicol. In Vitro,* 2008. 22(4): p. 1018-24.

[52] Cnop, M., et al., Inverse relationship between cytotoxicity of free fatty acids in pancreatic islet cells and cellular triglyceride accumulation. *Diabetes*, 2001. 50(8): p. 1771-7.

[53] Suresh, Y. and U.N. Das, Protective action of arachidonic acid against alloxan-induced cytotoxicity and diabetes mellitus. *Prostaglandins Leukot Essent Fatty Acids*, 2001. 64(1): p. 37-52.

[54] Lupi, R., et al., Prolonged exposure to free fatty acids has cytostatic and pro-apoptotic effects on human pancreatic islets: evidence that beta-cell death is caspase mediated, partially dependent on ceramide pathway, and Bcl-2 regulated. *Diabetes,* 2002. 51(5): p. 1437-42.

[55] Lima, T.M., et al., Ranking the toxicity of fatty acids on Jurkat and Raji cells by flow cytometric analysis. *Toxicol. In Vitro*, 2002. 16(6): p. 741-7.

[56] Cury-Boaventura, M.F. and R. Curi, Regulation of reactive oxygen species (ROS) production by C18 fatty acids in Jurkat and Raji cells. *Clin. Sci.* (Lond), 2005. 108(3): p. 245-53.

[57] Cury-Boaventura, M.F., C. Pompeia, and R. Curi, Comparative toxicity of oleic acid and linoleic acid on Jurkat cells. *Clin. Nutr,* 2004. 23(4): p. 721-32.

[58] Cury-Boaventura, M.F., C. Pompeia, and R. Curi, Comparative toxicity of oleic acid and linoleic acid on Raji cells. *Nutrition*, 2005. 21(3): p. 395-405.

[59] Martins de Lima, T., et al., Comparative toxicity of fatty acids on a macrophage cell line (J774). *Clin. Sci.* (Lond), 2006. 111(5): p. 307-17.

[60] Cury-Boaventura, M.F., et al., Comparative toxicity of oleic and linoleic acid on human lymphocytes. *Life Sci,* 2006. 78(13): p. 1448-56.

[61] Rossig, L., S. Dimmeler, and A.M. Zeiher, Apoptosis in the vascular wall and atherosclerosis. *Basic Res. Cardiol,* 2001. 96(1): p. 11-22.

[62] Choy, J.C., et al., Endothelial cell apoptosis: biochemical characteristics and potential implications for atherosclerosis. *J. Mol. Cell Cardiol*, 2001. 33(9): p. 1673-90.

[63] Mallat, Z. and A. Tedgui, Apoptosis in the vasculature: mechanisms and functional importance. *Br. J. Pharmacol,* 2000. 130(5): p. 947-62.

[64] Mozaffarian, D., et al., Trans fatty acids and cardiovascular disease. *N. Engl. J. Med*, 2006. 354(15): p. 1601-13.

[65] Micha, R. and D. Mozaffarian, Trans fatty acids: effects on cardiometabolic health and implications for policy. *Prostaglandins Leukot Essent Fatty Acids*, 2008. 79(3-5): p. 147-52.

[66] Mozaffarian, D. and R. Clarke, Quantitative effects on cardiovascular risk factors and coronary heart disease risk of replacing partially hydrogenated vegetable oils with other fats and oils. *Eur. J. Clin. Nutr,* 2009. 63 Suppl 2: p. S22-33.

[67] Ascherio, A., Trans fatty acids and blood lipids. *Atheroscler Suppl*, 2006. 7(2): p. 25-7.

[68] Thompson, A.K., et al., Trans-fatty acids and cancer: the evidence reviewed. *Nutr. Res. Rev*, 2008. 21(2): p. 174-88.

[69] Adibhatla, R.M., J.F. Hatcher, and R.J. Dempsey, Lipids and lipidomics in brain injury and diseases. *Aaps J*, 2006. 8(2): p. E314-21.

[70] Innis, S.M., The role of dietary n-6 and n-3 fatty acids in the developing brain. *Dev. Neurosci*, 2000. 22(5-6): p. 474-80.

[71] McNamara, R.K., The emerging role of omega-3 fatty acids in psychiatry. *Prostaglandins Leukot Essent Fatty Acids*, 2006. 75(4-5): p. 223-5.

[72] McNamara, R.K. and S.E. Carlson, Role of omega-3 fatty acids in brain development and function: Potential implications for the pathogenesis and prevention of psychopathology. *Prostaglandins Leukot Essent Fatty Acids*, 2006. 75(4-5): p. 329-49.

[73] Muskiet, F.A. and R.F. Kemperman, Folate and long-chain polyunsaturated fatty acids in psychiatric disease. *J. Nutr. Biochem,* 2006. 17(11): p. 717-27.

[74] Hebert, L.E., et al., Alzheimer disease in the US population: prevalence estimates using the 2000 census. *Arch. Neurol*, 2003. 60(8): p. 1119-22.

[75] Boudrault, C., R.P. Bazinet, and D.W. Ma, Experimental models and mechanisms underlying the protective effects of n-3 polyunsaturated fatty acids in Alzheimer's disease. *J. Nutr. Biochem*, 2009. 20(1): p. 1-10.

[76] Kalmijn, S., et al., Dietary intake of fatty acids and fish in relation to cognitive performance at middle age. *Neurology*, 2004. 62(2): p. 275-80.

[77] Lim, G.P., et al., A diet enriched with the omega-3 fatty acid docosahexaenoic acid reduces amyloid burden in an aged Alzheimer mouse model. *J. Neurosci,* 2005. 25(12): p. 3032-40.

[78] Freund-Levi, Y., et al., Omega-3 fatty acid treatment in 174 patients with mild to moderate Alzheimer disease: OmegAD study: a randomized double-blind trial. *Arch Neurol,* 2006. 63(10): p. 1402-8.

[79] van Marum, R.J., Current and future therapy in Alzheimer's disease. *Fundam. Clin. Pharmacol,* 2008. 22(3): p. 265-74.

[80] Calon, F., et al., Docosahexaenoic acid protects from dendritic pathology in an Alzheimer's disease mouse model. *Neuron*, 2004. 43(5): p. 633-45.

[81] Hashimoto, M., et al., Docosahexaenoic acid-induced amelioration on impairment of memory learning in amyloid beta-infused rats relates to the decreases of amyloid beta and cholesterol levels in detergent-insoluble membrane fractions. *Biochim. Biophys. Acta*, 2005. 1738(1-3): p. 91-8.

[82] Oksman, M., et al., *Impact of different saturated fatty acid, polyunsaturated fatty acid and cholesterol containing diets on beta-amyloid accumulation in APP/PS1 transgenic mice. Neurobiol. Dis*, 2006. 23(3): p. 563-72.

[83] Green, K.N., et al., Dietary docosahexaenoic acid and docosapentaenoic acid ameliorate amyloid-beta and tau pathology via a mechanism involving presenilin 1 levels. *J. Neurosci,* 2007. 27(16): p. 4385-95.

[84] Lukiw, W.J., et al., A role for docosahexaenoic acid-derived neuroprotectin D1 in neural cell survival and Alzheimer disease. *J. Clin. Invest*, 2005. 115(10): p. 2774-83.

[85] Puglielli, L., R.E. Tanzi, and D.M. Kovacs, Alzheimer's disease: the cholesterol connection. *Nat. Neurosci*, 2003. 6(4): p. 345-51.

[86] Bazan, N.G., The onset of brain injury and neurodegeneration triggers the synthesis of docosanoid neuroprotective signaling. *Cell Mol. Neurobiol,* 2006. 26(4-6): p. 901-13.

[87] Rao, J.S., et al., Dietary n-3 PUFA deprivation alters expression of enzymes of the arachidonic and docosahexaenoic acid cascades in rat frontal cortex. *Mol. Psychiatry,* 2007. 12(2): p. 151-7.

[88] Morris, D.H., Methodologic challenges in designing clinical studies to measure differences in the bioequivalence of n-3 fatty acids. *Mol. Cell Biochem*, 2003. 246(1-2): p. 83-90.

[89] Heresco-Levy, U. and D.C. Javitt, The role of N-methyl-D-aspartate (NMDA) receptor-mediated neurotransmission in the pathophysiology and therapeutics of psychiatric syndromes. *Eur. Neuropsychopharmacol*, 1998. 8(2): p. 141-52.

[90] Horrobin, D.F., A.I. Glen, and K. Vaddadi, The membrane hypothesis of schizophrenia. Schizophr. *Res*, 1994. 13(3): p. 195-207.

[91] Di Marzo, V. and D. Piomelli, Participation of prostaglandin E2 in dopamine D2 receptor-dependent potentiation of arachidonic acid release. *J. Neurochem*, 1992. 59(1): p. 379-82.

[92] Nunez, E.A., Preface--fatty acids and cell signalling. *Prostaglandins Leukot Essent Fatty Acids,* 1993. 48(1): p. 1-4.

[93] Mellor, J.E., J.D. Laugharne, and M. Peet, Schizophrenic symptoms and dietary intake of n-3 fatty acids. *Schizophr. Res*, 1995. 18(1): p. 85-6.

[94] Laugharne, J.D., J.E. Mellor, and M. Peet, Fatty acids and schizophrenia. *Lipids,* 1996. 31 Suppl: p. S163-5.

[95] Stokes, C. and M. Peet, Dietary sugar and polyunsaturated fatty acid consumption as predictors of severity of schizophrenia symptoms. *Nutr. Neurosci*, 2004. 7(4): p. 247-9.

[96] Reddy, R.D., M.S. Keshavan, and J.K. Yao, Reduced red blood cell membrane essential polyunsaturated fatty acids in first episode schizophrenia at neuroleptic-naive baseline. *Schizophr. Bull*, 2004. 30(4): p. 901-11.

[97] Muskiet, F.A., et al., Long-chain polyunsaturated fatty acids in maternal and infant nutrition. *Prostaglandins Leukot Essent Fatty Acids*, 2006. 75(3): p. 135-44.

[98] Peet, M., et al., Two double-blind placebo-controlled pilot studies of eicosapentaenoic acid in the treatment of schizophrenia. *Schizophr. Res*, 2001. 49(3): p. 243-51.

[99] Peet, M. and D.F. Horrobin, A dose-ranging exploratory study of the effects of ethyl-eicosapentaenoate in patients with persistent schizophrenic symptoms. *J. Psychiatr. Res*, 2002. 36(1): p. 7-18.

[100] Jones, P.B., et al., Schizophrenia as a long-term outcome of pregnancy, delivery, and perinatal complications: a 28-year follow-up of the 1966 north Finland general population birth cohort. *Am. J. Psychiatry*, 1998. 155(3): p. 355-64.

[101] Ichiki, M., et al., Intra-uterine physical growth in schizophrenia: evidence confirming excess of premature birth. *Psychol. Med,* 2000. 30(3): p. 597-604.

[102] Sorensen, H.J., et al., Breastfeeding and risk of schizophrenia in the Copenhagen Perinatal Cohort. *Acta Psychiatr. Scand*, 2005. 112(1): p. 26-9.

[103] Hibbeln, J.R., et al., *Essential fatty acids predict metabolites of serotonin and dopamine in cerebrospinal fluid among healthy control subjects, and early- and late-onset alcoholics. Biol. Psychiatry*, 1998. 44(4): p. 235-42.

[104] Hibbeln, J.R., et al., Do plasma polyunsaturates predict hostility and depression? *World Rev. Nutr. Diet*, 1997. 82: p. 175-86.

[105] Tanskanen, A., et al., Fish consumption, depression, and suicidality in a general population. *Arch Gen. Psychiatry*, 2001. 58(5): p. 512-3.

[106] Adams, P.B., et al., Arachidonic acid to eicosapentaenoic acid ratio in blood correlates positively with clinical symptoms of depression. *Lipids*, 1996. 31 Suppl: p. S157-61.

[107] Assies, J., et al., Fatty acids and homocysteine levels in patients with recurrent depression: an explorative pilot study. *Prostaglandins Leukot Essent Fatty Acids*, 2004. 70(4): p. 349-56.

[108] Peet, M., et al., *Depletion of omega-3 fatty acid levels in red blood cell membranes of depressive patients. Biol. Psychiatry*, 1998. 43(5): p. 315-9.

[109] Frasure-Smith, N., F. Lesperance, and P. Julien, Major depression is associated with lower omega-3 fatty acid levels in patients with recent acute coronary syndromes. *Biol. Psychiatry*, 2004. 55(9): p. 891-6.

[110] Tiemeier, H., et al., Plasma fatty acid composition and depression are associated in the elderly: the Rotterdam Study. *Am. J. Clin. Nutr*, 2003. 78(1): p. 40-6.

[111] Otto, S.J., R.H. de Groot, and G. Hornstra, Increased risk of postpartum depressive symptoms is associated with slower normalization after pregnancy of the functional docosahexaenoic acid status. *Prostaglandins Leukot Essent Fatty Acids*, 2003. 69(4): p. 237-43.

[112] Puri, B.K., et al., Eicosapentaenoic acid in treatment-resistant depression associated with symptom remission, structural brain changes and reduced neuronal phospholipid turnover. *Int. J. Clin. Pract,* 2001. 55(8): p. 560-3.

[113] Nemets, B., Z. Stahl, and R.H. Belmaker, Addition of omega-3 fatty acid to maintenance medication treatment for recurrent unipolar depressive disorder. *Am. J. Psychiatry,* 2002. 159(3): p. 477-9.

[114] Su, K.P., et al., Omega-3 fatty acids in major depressive disorder. A preliminary double-blind, placebo-controlled trial. *Eur. Neuropsychopharmacol*, 2003. 13(4): p. 267-71.

[115] Williams, A.L., et al., Do essential fatty acids have a role in the treatment of depression? *J. Affect. Disord,* 2006. 93(1-3): p. 117-23.

[116] Chiu, C.C., et al., Omega-3 fatty acids for depression in pregnancy. *Am. J. Psychiatry*, 2003. 160(2): p. 385.

[117] Freeman, M.P., Omega-3 fatty acids and perinatal depression: A review of the literature and recommendations for future research. *Prostaglandins Leukot Essent Fatty Acids,* 2006. 75(4-5): p. 291-7.

[118] Mamalakis, G., et al., Depression and adipose polyunsaturated fatty acids in an adolescent group. *Prostaglandins Leukot Essent Fatty Acids*, 2004. 71(5): p. 289-94.

[119] Murck, H., et al., Ethyl-eicosapentaenoate and dexamethasone resistance in therapy-refractory depression. *Int. J. Neuropsychopharmacol*, 2004. 7(3): p. 341-9.

[120] Mirnikjoo, B., et al., Protein kinase inhibition by omega-3 fatty acids. *J. Biol. Chem,* 2001. 276(14): p. 10888-96.

[121] Cott, J. and J.R. Hibbeln, Lack of seasonal mood change in Icelanders. *Am. J. Psychiatry*, 2001. 158(2): p. 328.

[122] Lauer, K., The food pattern in geographical relation to the risk of multiple sclerosis in the Mediterranean and Near East region. *J. Epidemiol. Community Health*, 1991. 45(3): p. 251-2.

[123] Lauer, K., The risk of multiple sclerosis in the U.S.A. in relation to sociogeographic features: a factor-analytic study. *J. Clin. Epidemiol*, 1994. 47(1): p. 43-8.

[124] Lauer, K., Environmental associations with the risk of multiple sclerosis: the contribution of ecological studies. *Acta Neurol. Scand. Suppl*, 1995. 161: p. 77-88.

[125] Esparza, M.L., S. Sasaki, and H. Kesteloot, Nutrition, latitude, and multiple sclerosis mortality: an ecologic study. *Am. J. Epidemiol*, 1995. 142(7): p. 733-7.

[126] Swank, R.L. and J.W. Goodwin, How saturated fats may be a causative factor in multiple sclerosis and other diseases. *Nutrition*, 2003. 19(5): p. 478.

[127] Schwarz, S. and H. Leweling, Multiple sclerosis and nutrition. *Mult. Scler,* 2005. 11(1): p. 24-32.

[128] Namazi, M.R., The beneficial and detrimental effects of linoleic acid on autoimmune disorders. *Autoimmunity,* 2004. 37(1): p. 73-5.

[129] Harbige, L.S. and M.K. Sharief, Polyunsaturated fatty acids in the pathogenesis and treatment of multiple sclerosis. *Br. J. Nutr*, 2007. 98 Suppl 1: p. S46-53.

[130] Firestein, G.S., Evolving concepts of rheumatoid arthritis. *Nature,* 2003. 423(6937): p. 356-61.

[131] Shapiro, J.A., et al., Diet and rheumatoid arthritis in women: a possible protective effect of fish consumption. *Epidemiology,* 1996. 7(3): p. 256-63.

[132] Leslie, C.A., et al., Dietary fish oil modulates macrophage fatty acids and decreases arthritis susceptibility in mice. *J. Exp. Med,* 1985. 162(4): p. 1336-49.

[133] Kremer, J.M., *n-3* fatty acid supplements in rheumatoid arthritis. *Am. J. Clin. Nutr,* 2000. 71(1 Suppl): p. 349S-51S.

[134] Cleland, L.G., M.J. James, and S.M. Proudman, Omega-6/omega-3 fatty acids and arthritis. *World Rev. Nutr. Diet*, 2003. 92: p. 152-68.

[135] Calder, P.C., Session 3: Joint Nutrition Society and Irish Nutrition and Dietetic Institute Symposium on 'Nutrition and autoimmune disease' PUFA, inflammatory processes and rheumatoid arthritis. *Proc. Nutr. Soc,* 2008. 67(4): p. 409-18.

[136] Espersen, G.T., et al., Decreased interleukin-1 beta levels in plasma from rheumatoid arthritis patients after dietary supplementation with n-3 polyunsaturated fatty acids. *Clin. Rheumatol,* 1992. 11(3): p. 393-5.

[137] Sundrarjun, T., et al., Effects of n-3 fatty acids on serum interleukin-6, tumour necrosis factor-alpha and soluble tumour necrosis factor receptor p55 in active rheumatoid arthritis. *J. Int. Med. Res*, 2004. 32(5): p. 443-54.

[138] Kremer, J.M., et al., Dietary fish oil and olive oil supplementation in patients with rheumatoid arthritis. *Clinical and immunologic effects. Arthritis Rheum*, 1990. 33(6): p. 810-20.

[139] Arend, W.P. and J.M. Dayer, Inhibition of the production and effects of interleukin-1 and tumor necrosis factor alpha in rheumatoid arthritis. *Arthritis Rheum*, 1995. 38(2): p. 151-60.

[140] Badolato, R. and J.J. Oppenheim, Role of cytokines, acute-phase proteins, and chemokines in the progression of rheumatoid arthritis. *Semin. Arthritis Rheum*, 1996. 26(2): p. 526-38.

[141] Cleland, L.G., et al., Reduction of cardiovascular risk factors with longterm fish oil treatment in early rheumatoid arthritis. *J. Rheumatol*, 2006. 33(10): p. 1973-9.
[142] Sperling, R.I., et al., Effects of dietary supplementation with marine fish oil on leukocyte lipid mediator generation and function in rheumatoid arthritis. *Arthritis Rheum,* 1987. 30(9): p. 988-97.
[143] Hughes, D.A. and A.C. Pinder, Influence of n-3 polyunsaturated fatty acids (PUFA) on the antigen-presenting function of human monocytes. *Biochem. Soc. Trans*, 1996. 24(3): p. 389S.
[144] Hughes, D.A. and A.C. Pinder, n-3 polyunsaturated fatty acids inhibit the antigen-presenting function of human monocytes. *Am. J. Clin. Nutr*, 2000. 71(1 Suppl): p. 357S-60S.
[145] Goldberg, R.J. and J. Katz, A meta-analysis of the analgesic effects of omega-3 polyunsaturated fatty acid supplementation for inflammatory joint pain. *Pain*, 2007. 129(1-2): p. 210-23.
[146] McKeever, T.M. and J. Britton, Diet and asthma. *Am. J. Respir. Crit. Care Med*, 2004. 170(7): p. 725-9.
[147] Wong, K.W., Clinical efficacy of n-3 fatty acid supplementation in patients with asthma. *J. Am. Diet Assoc,* 2005. 105(1): p. 98-105.
[148] Horrobin, D.F., Low prevalences of coronary heart disease (CHD), psoriasis, asthma and rheumatoid arthritis in Eskimos: are they caused by high dietary intake of eicosapentaenoic acid (EPA), a genetic variation of essential fatty acid (EFA) metabolism or a combination of both? *Med. Hypotheses*, 1987. 22(4): p. 421-8.
[149] Okamoto, M., et al., Effects of perilla seed oil supplementation on leukotriene generation by leucocytes in patients with asthma associated with lipometabolism. *Int. Arch. Allergy Immunol*, 2000. 122(2): p. 137-42.
[150] Deng, Y.M., et al., Anti-asthmatic effects of Perilla seed oil in the guinea pig in vitro and in vivo. *Planta Med*, 2007. 73(1): p. 53-8.
[151] Schubert, R., et al., Effect of n-3 polyunsaturated fatty acids in asthma after low-dose allergen challenge. *Int. Arch Allergy Immunol*, 2009. 148(4): p. 321-9.
[152] Hodge, L., et al., Consumption of oily fish and childhood asthma risk. *Med. J. Aust,* 1996. 164(3): p. 137-40.
[153] Biltagi, M.A., et al., Omega-3 fatty acids, vitamin C and Zn supplementation in asthmatic children: a randomized self-controlled study. *Acta Paediatr*, 2009. 98(4): p. 737-42.
[154] de Luis, D.A., et al., Dietary intake in patients with asthma: a case control study. *Nutrition*, 2005. 21(3): p. 320-4.
[155] Reisman, J., et al., Treating asthma with omega-3 fatty acids: where is the evidence? A systematic review. *BMC Complement Altern. Med*, 2006. 6: p. 26.
[156] Wolters, M., Diet and psoriasis: experimental data and clinical evidence. *Br. J. Dermatol*, 2005. 153(4): p. 706-14.
[157] Zulfakar, M.H., M. Edwards, and C.M. Heard, Is there a role for topically delivered eicosapentaenoic acid in the treatment of psoriasis? *Eur. J. Dermatol*, 2007. 17(4): p. 284-91.
[158] Escobar, S.O., et al., Topical fish oil in psoriasis--a controlled and blind study. *Clin. Exp. Dermatol*, 1992. 17(3): p. 159-62.

[159] Theilla, M., et al., A diet enriched in eicosapentanoic acid, gamma-linolenic acid and antioxidants in the prevention of new pressure ulcer formation in critically ill patients with acute lung injury: A randomized, prospective, controlled study. *Clin. Nutr*, 2007. 26(6): p. 752-7.

[160] Rodrigues, K.L., et al., Cicatrizing and antimicrobial properties of an ozonised oil from sunflower seeds. *Inflammopharmacology*, 2004. 12(3): p. 261-70.

[161] Cardoso, C.R., et al., Influence of topical administration of n-3 and n-6 essential and n-9 nonessential fatty acids on the healing of cutaneous wounds. *Wound Repair Regen*, 2004. 12(2): p. 235-43.

[162] Pereira, L.M., et al., Effect of oleic and linoleic acids on the inflammatory phase of wound healing in rats. *Cell Biochem. Funct*, 2008. 26(2): p. 197-204.

[163] Zimmer, M.M., et al., Effect of extracts from the Chinese and European mole cricket on wound epithelialization and neovascularization: in vivo studies in the hairless mouse ear wound model. *Wound Repair Regen*, 2006. 14(2): p. 142-51.

[164] Rolandelli, R.H., et al., Effects of intraluminal infusion of short-chain fatty acids on the healing of colonic anastomosis in the rat. *Surgery,* 1986. 100(2): p. 198-204.

[165] Kissmeyer-Nielsen, P., et al., Transmural trophic effect of short chain fatty acid infusions on atrophic, defunctioned rat colon. *Dis. Colon Rectum*, 1995. 38(9): p. 946-51.

[166] Friedel, D. and G.M. Levine, Effect of short-chain fatty acids on colonic function and structure. *JPEN J. Parenter Enteral Nutr,* 1992. 16(1): p. 1-4.

[167] Bucher, H.C., et al., N-3 polyunsaturated fatty acids in coronary heart disease: a meta-analysis of randomized controlled trials. *Am. J. Med*, 2002. 112(4): p. 298-304.

[168] Burr, M.L., et al., Lack of benefit of dietary advice to men with angina: results of a controlled trial. *Eur. J. Clin. Nutr*, 2003. 57(2): p. 193-200.

[169] Raitt, M.H., et al., Fish oil supplementation and risk of ventricular tachycardia and ventricular fibrillation in patients with implantable defibrillators: a randomized controlled trial. *Jama*, 2005. 293(23): p. 2884-91.

[170] Hooper, L., et al., Risks and benefits of omega 3 fats for mortality, cardiovascular disease, and cancer: systematic review. *Bmj*, 2006. 332(7544): p. 752-60.

[171] Hu, F.B., et al., Fish and long-chain omega-3 fatty acid intake and risk of coronary heart disease and total mortality in diabetic women. *Circulation*, 2003. 107(14): p. 1852-7.

[172] Erkkila, A.T., et al., Fish intake is associated with a reduced progression of coronary artery atherosclerosis in postmenopausal women with coronary artery disease. *Am. J. Clin. Nutr*, 2004. 80(3): p. 626-32.

[173] Balk, E., et al., Effects of omega-3 fatty acids on cardiovascular risk factors and intermediate markers of cardiovascular disease. *Evid. Rep. Technol. Assess.* (Summ), 2004(93): p. 1-6.

[174] Lewis, A., et al., Treatment of hypertriglyceridemia with omega-3 fatty acids: a systematic review. *J. Am. Acad. Nurse Pract*, 2004. 16(9): p. 384-95.

[175] Mori, T.A., et al., Purified eicosapentaenoic and docosahexaenoic acids have differential effects on serum lipids and lipoproteins, LDL particle size, glucose, and insulin in mildly hyperlipidemic men. *Am. J. Clin. Nutr*, 2000. 71(5): p. 1085-94.

[176] Mori, T.A., R. Vandongen, and J.R. Masarei, Fish oil-induced changes in apolipoproteins in IDDM subjects. *Diabetes Care*, 1990. 13(7): p. 725-32.

[177] Harris, W.S., Fish oils and plasma lipid and lipoprotein metabolism in humans: a critical review. *J. Lipid. Res*, 1989. 30(6): p. 785-807.

[178] Leigh-Firbank, E.C., et al., Eicosapentaenoic acid and docosahexaenoic acid from fish oils: differential associations with lipid responses. *Br. J. Nutr*, 2002. 87(5): p. 435-45.

[179] Hartweg, J., et al., Potential impact of omega-3 treatment on cardiovascular disease in type 2 diabetes. *Curr. Opin. Lipidol,* 2009. 20(1): p. 30-8.

[180] Zhao, Y.T., et al., Prevention of sudden cardiac death with omega-3 fatty acids in patients with coronary heart disease: a meta-analysis of randomized controlled trials. *Ann. Med*, 2009. 41(4): p. 301-10.

[181] Lepage, G. and C.C. Roy, Direct transesterification of all classes of lipids in a one-step reaction. *J. Lipid. Res*, 1986. 27(1): p. 114-20.

[182] Folch, J., et al., Preparation of lipide extracts from brain tissue. *J. Biol. Chem*, 1951. 191(2): p. 833-41.

[183] Folch, J., M. Lees, and G.H. Sloane Stanley, A simple method for the isolation and purification of total lipides from animal tissues. *J. Biol. Chem*, 1957. 226(1): p. 497-509.

[184] Bligh, E.G. and W.J. Dyer, A rapid method of total lipid extraction and purification. *Can. J. Biochem. Physiol,* 1959. 37(8): p. 911-7.

[185] Hara, A. and N.S. Radin, Lipid extraction of tissues with a low-toxicity solvent. *Anal. Biochem*, 1978. 90(1): p. 420-6.

[186] Wuthier, R.E., Two-dimensional chromatography on silica gel-loaded paper for the microanalysis of polar lipids. *J. Lipid. Res*, 1966. 7(4): p. 544-50.

[187] Rosenthal, A.F. and S.C. Han, Phosphorus determination in phosphoglycerides from thin-layer chromatograms. *J. Lipid. Res*, 1969. 10(2): p. 243-5.

[188] Nishiyama-Naruke, A.d.S.J.A.A.C.s.F., M.; Curi, R., HPLC Determination of Underizatized Fatty Acids Saponified at Low Temperature Analysis of Fatty Acids in Oils and Tissues. *Analytical Letters*, 1998. 31: p. 2565 - 76.

[189] Abushufa, R., P. Reed, and C. Weinkove, Fatty acids in erythrocytes measured by isocratic HPLC. *Clin. Chem*, 1994. 40(9): p. 1707-12.

[190] Watson, A.D., Thematic review series: systems biology approaches to metabolic and cardiovascular disorders. Lipidomics: a global approach to lipid analysis in biological systems. *J. Lipid. Res*, 2006. 47(10): p. 2101-11.

[191] Watkins, S.M., Comprehensive lipid analysis: a powerful metanomic tool for predictive and diagnostic medicine. *Isr. Med. Assoc. J*, 2000. 2(9): p. 722-4.

In: Biochemistry Research Updates
Editor: Simon J. Baginski
ISBN 978-1-61209-700-8

Chapter 2

IMMUNE RESPONSE MODULATION BY TARGETED COMPLEXES BASED ON STREPTAVIDIN

Zsuzsanna Szekeres[2], Melinda Herbáth and József Prechl[2]
[1]Department of Immunology, ELTE, Pazmany P. s. 1/C Budapest, Hungary
[2]Immunology Research Group, MTA Pazmany P. s. 1/C Budapest, Hungary

ABSTRACT

Several immunomodulatory constructs based on streptavidin-biotin binding have been described that exploit the modularity and flexibility of this system. In general, a streptavidin core serves as the carrier of targeting and effector molecules. Targeting agents, such as whole antibodies or their fragments and natural ligands, deliver antigen to specific receptors on leukocytes. This targeted delivery can both enhance and fine-tune immune responses, which are directed against streptavidin itself or its cargo. Due to its tetrameric structure, streptavidin-based targeting complexes can crosslink cell surface receptors. This property enables such constructs to initiate signaling events in the target cells. Thus, in addition to improving antigen uptake, activation of APCs can be achieved. By using liposome- or nanobead-bound streptavidin, one can both boost the crosslinking effect and target different cell populations, specialized in the uptake of particulate antigens. To generate streptavidin-based constructs, various molecular biological and chemical methods have been applied. In a preferred setup, monobiotinylated components are used, in order to ensure the formation of complexes with controllable stoichiometry. This strategy also allows the combination of distinct targeting and antigenic moieties in the same complex. Here we review various approaches of streptavidin-based immunotargeting and their effects on immune responses.

INTRODUCTION

Targeted delivery of bioactive molecules, such as antigens, drugs or toxins, specifically to appropriate cells in the body is an attractive approach in disease therapy and prevention. Targeting has several benefits: the primary effect induced by the given molecule remains localized and therefore concentrated, non-specific effects exerted on bystander cells are reduced and smaller amount of the molecule is needed to elicit the same response compared to administering the molecule alone. Targeted delivery of antigens aims to enhance and fine-tune immunomodulatory strategies, e.g. vaccination. On the other hand, highly specific affinity reagents of the immune system, antibodies, are often used for the delivery of therapeutic compounds, a strategy called immuno-targeting or antibody-mediated targeting. In addition to target-specific monoclonal antibodies, natural ligands of the targeted receptor can also be applied as targeting agents.

Targeting a molecule requires at least two components: the targeting unit and the molecular cargo. These two components can be chemically or physically joined together, in a manner that is stable during the period of generation, storage and administration into the body, until the cargo reaches its target. This coupling can be either direct or indirect, the latter permitting some flexibility of the system. Indirect coupling, for example, allows the design of modular complexes where different targeting units and cargos can be joined by a core coupling module.

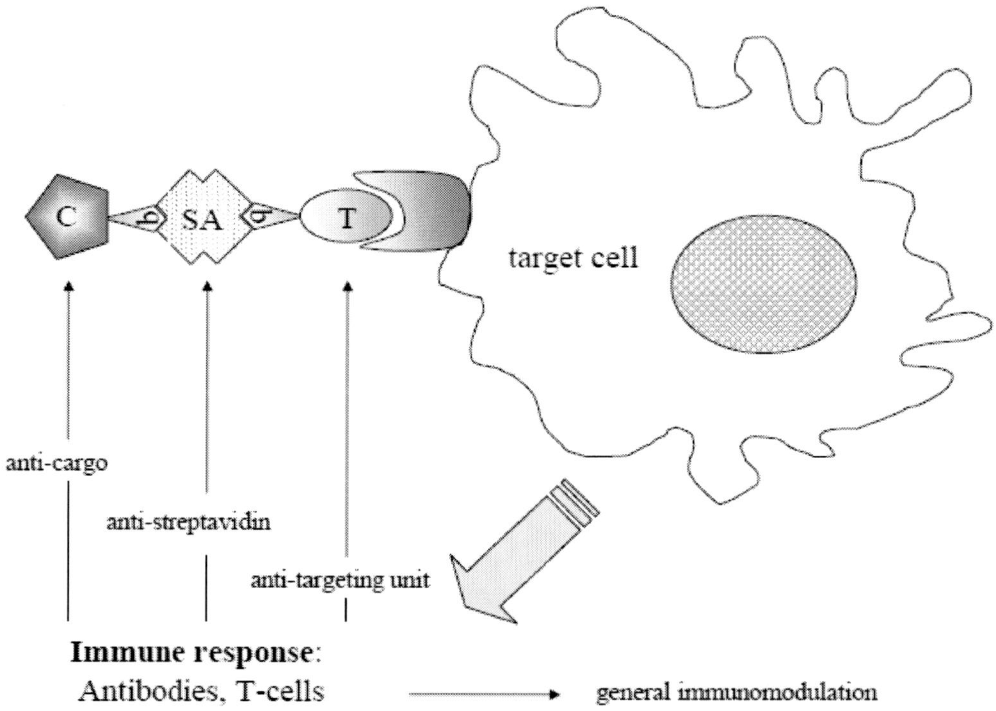

Figure 1. General scheme of immunomodulation by streptavidin-based targeting contructs. Theoretically, immune response against any component of the targeting complex can be generated; in practice, immune responses against the cargo, streptavidin and sometimes against the targeting unit itself are monitored. C, cargo; b, biotin; SA, streptavidin; T, targeting unit.

Again, indirect coupling also needs to be stable and preferably of controllable stoichiometry. Since the binding of biotin and streptavidin (SA) is practically irreversible and SA has four binding sites, SA is an ideal coupling module in these systems. In view of the fact that avidin has properties very similar to SA, we have also included avidin-based targeting approaches in this review, especially where the two are replaceable, but also tried to point out differences between the two proteins.

Avidin is a cationic glycoprotein isolated from egg white, while SA is a non-glycosylated neutral protein, produced by *Streptomyces avidinii.* Both proteins are tetrameric and have similar and extremely high biotin-binding affinities [1]. The biotin binding sites of avidin and SA possess different, highly conserved amino acid sequences, but their secondary, tertiary and quaternary structures have huge similarities that make them functionally analogous [2,3]. Both avidin and SA are stable at low and at high pH [1], resistant against heat [4], denaturants [5] and against proteolytic cleavage [6]. This stability is further enhanced after biotin binding. Some studies examined their mutant variants with the purpose to extend their mechano- or termostability [7,8], to weaken the quaternary structure and prevent aggregation [9] or to change biotin binding affinity [10,11]. This stability and high affinity protein-ligand interaction of avidin and SA make them attractive tools in a number of biological applications, such as diagnostic assays, imaging or pre-targeting in cancer radioimmunotherapy. SA and avidin are also ideal units of immunotargeting complexes due to their ability to bind multiple biotinylated molecules, besides the properties mentioned above, which allows the combination of different components.

Figure 1 shows a prototypical targeting strategy, where a biotinylated antigenic cargo and a biotinylated targeting unit are coupled by a SA molecule. This complex reaches the target cell, delivering the cargo and thereby exerting immunomodulatory effects. This results in the induction of adaptive immune responses, antibody production or T cell responses. It is important to point out that these responses can be aimed against any component of the targeting complex. Indeed, SA itself is also used as an antigenic target, omitting the cargo, especially in experimental immunology.

In this chapter, we review the utilization of SA-based targeting complexes in immunological research, the structure and the effects of these complexes on the immune system. All reviewed targeting strategies are summarized in the tables at the end of this chapter.

ON THE IMMUNOGENICITY OF SA

SA – and also avidin – is obviously non-self protein, whether we talk about humans or mice, and as such is potentially immunogenic. Working with SA containing complexes *in vivo*, i. e. administering them systematically, an anti-SA immune response can be evoked. While this is often a detrimental side-effect in experiments and a potentially dangerous effect in humans, this very response may also be the aim of experiments investigating various targeting strategies in mice. Whichever is the case, this response needs to be taken into consideration. Interestingly, although several studies utilized SA or avidin as components of targeting complexes, surprisingly little is known about the immunogenic properties of these proteins.

A molecular property that can influence the immunogenicity of a protein is the glycosylation status. Carbohydrate side-chains that are different from the host glycosylation pattern are strongly immunogenic, resulting in antibody responses against the molecule bearing these carbohydrates. Avidin has ten asparagine residues in each subunit, and one of them, Asn17, is glycosylated [12]. Analyses have shown that the carbohydrate units of avidin are composed of an average of four to five mannose and three N-acetylglucosamine residues per subunit [13,14] . Chemical modification of avidin by a variety of reagents [15] has shown that biotin-binding activity is dependent on factors associated with the protein portion of the molecule [16]. Martilla et al. demonstrated that removal of the oligosaccharide moiety from avidin has little effect on the thermal stability and no effects in biotin-binding ability, but mutants were more sensitive to limited proteolysis with proteinase K and has clearly increased non-specific binding to different cell types, such as platelets, hepatocytes and lymphocytes [17]. This group also investigated the non-specific binding characteristics of modified avidin, in which positively charged residues were substituted for acidic residues. The wild-type avidin bound strongly to DNA due its positively charged residues, since there was a clear correlation between lowering the isoelectric point and reduced binding to DNA. Reducing the positive charge of avidin drastically lowered the binding to hepatocytes and lymphocytes, but the most decreased binding to cells and to DNA was observed in the case of using non-glycosylated and acidic variants of avidin. This study suggests that the presence of positive charge, rather than carbohydrates chains, in avidin has a main role in its immunogenicity, causing non-specific binding to extraneous material. Schechter et al. have also verified that carbohydrate units could not cause aspecific interaction with other cell surface proteins or carbohydrate chains in organs [18]. They measured the tissue distribution and blood clearance of avidin and SA and detected that avidin has a rapid blood clearance and low tissue levels, even in non-glycosylated form, while SA showed prolonged blood and organ retention. Notwithstanding, when avidin was deglycosylated and neutralized in another study, circulation time of modified avidin was augmented, with the slowest clearance resulting from the combination of deglycosylation and neutralization [19]. It was assumed that sugar side chains on avidin may be recognized by sugar receptors present throughout the reticuloendothelial system, and can thus contribute to the fast blood clearance of avidin. Moreover, cationic proteins can interact with the globular basement membrane. In contrast, galactose moieties attached to SA increased its blood clearance while its liver accumulation correlated with the amount of bound galactose.

SA is a non-glycosylated protein, suggesting that it is not as immunogenic as avidin. SA showed no detectable binding to lymphocytes, indicating that it is less "sticky" than avidin [17]. Evidence for its antigenic properties comes from pretargeting radioimmuntherapy studies, when SA–antibody conjugates were injected into the body and used as capture for biotin derivatives containing radionuclides. Breitz et al. investigated the humoral immune response against a targeting mouse antibody, its SA conjugates and SA itself, during a clinical pretargeting study, when patients with adenocarcinoma were treated with different doses and forms of anti-carcinoma epitope antibody-SA conjugates [20]. Using ELISA analysis of pooled serum from patients, they detected antibodies against members of the conjugates and against the whole conjugate, within 10-14 days after treatment, and this response was primarily directed toward the SA portion of the conjugates. Immunogenic properties of SA were also investigated by Subramanian and Adiga [21]. They identified six linear epitopes in SA and avidin, using the pepscan method and rabbit antisera, and they showed that antibodies against these epitopes are cross-reactive between these two proteins. One of the reported linear epitopes was also detected by Meyer et al., who tested a series of synthetized peptide derived from the loop 1-7 region of SA with human anti-SA sera [22]. Based on the crystal srtucture of SA, they hypothesized that antigenic determinants can be found on the loops between the β-strands in the buried section of the surface of SA, which are also exposed to the solvent. One linear epitope was found to be localized in loop 6 and contained a charged residue, E101. This peptide failed to block the recognition of SA by anti-SA immune serum, thus the main dominant epitopes of SA remain unclear. These data also suggest that important epitopes are conformational, rather then linear in SA. Meyer et al. have also investigated whether mutation of the founded linear epitope reduces antigenicity of SA. Using site directed mutagenesis, they substituted the E101 charged residue in the linear epitope and other charged, aromatic and hyhrophobic residues in its surroundings, to smaller neutral amino acids. The generated series of 37 mutants were screened for biotin dissociation rate and proper folding of tetrameric structure. One of the mutants was proved to be only 20% as antigenic as SA; it elicited weaker IgM and IgG antibody response in rabbit, compared to the original protein. Our group also recently reported work with SA, as basic coupling component of our targeting complexes and as a model antigen at the same time [23]. When mice were immunized with SA-antigenic peptide complexes (5 μg SA/mouse) subcutaneously in incomplete Freund adjuvant, then boosted with same amount of conjugates, strong IgG1 and moderate IgG2a anti-SA antibody response was detected by ELISA. When we used SA-coated microspheres as the coupling unit of our construct, in place of soluble SA, enhanced levels of IgG1 and the appearance of IgG3 anti-SA antibodies was measured, compared to using the soluble SA complexes. The high level of IgG3 is attributable to the T-independent type 2 antigenic profile of the microspheres. Laitinen et al. have also demonstrated that changing the extent of aggregation and size of SA or avidin containing complexes can influence their immunogenicity. They produced monomeric avidin, in which Trp-110 was converted to lysine and Asn-54 was mutated to alanine. The resultant mutant binds biotin specifically in a reversible manner and remains in the monomeric state even upon binding to biotin. Monoavidin was compared immunologically to native avidin by ELISA using a polyclonal and two monoclonal anti-avidin preparations. The polyclonal rabbit sera recognized monoavidin only partially and weaker than native avidin and additionally the monoclonal antibodies failed to recognize the mutant [24].

Conjugation to polyethyleneglycol is often used to modify pharmacokinetics and pharmacodynamics of therapeutic proteins. Salmaso et al. has shown that PEGylation of avidin prolongs its permanence in the circulation, reduces its disposition in the liver and kidneys and prevents the protein cell uptake. Affinity of PEGylated avidin for biotin and biotinylated molecules depends on the length of the PEG chains applied and the application of protective agents. Antigenicity and immunogenicity of PEGylated avidin was found to be reduced compared to native avidin [25].

T cell response against SA is even less studied than the anti-SA humoral immune response, but undoubtedly it would be important for the better understanding of the anti-streptavdin immune response. In our experiments we failed to reactivate SA specific T cells in vitro (unpublished data). However, the appereance of anti-SA IgG isotypes during immunization with SA conjugates indicates the occurrence of isotype switch, which depends on T cell activation. Thus, SA bears epitopes not only for B cells, but for T cells, as well; the immunogenicity of these epitopes depends on the circumstances of immunization.

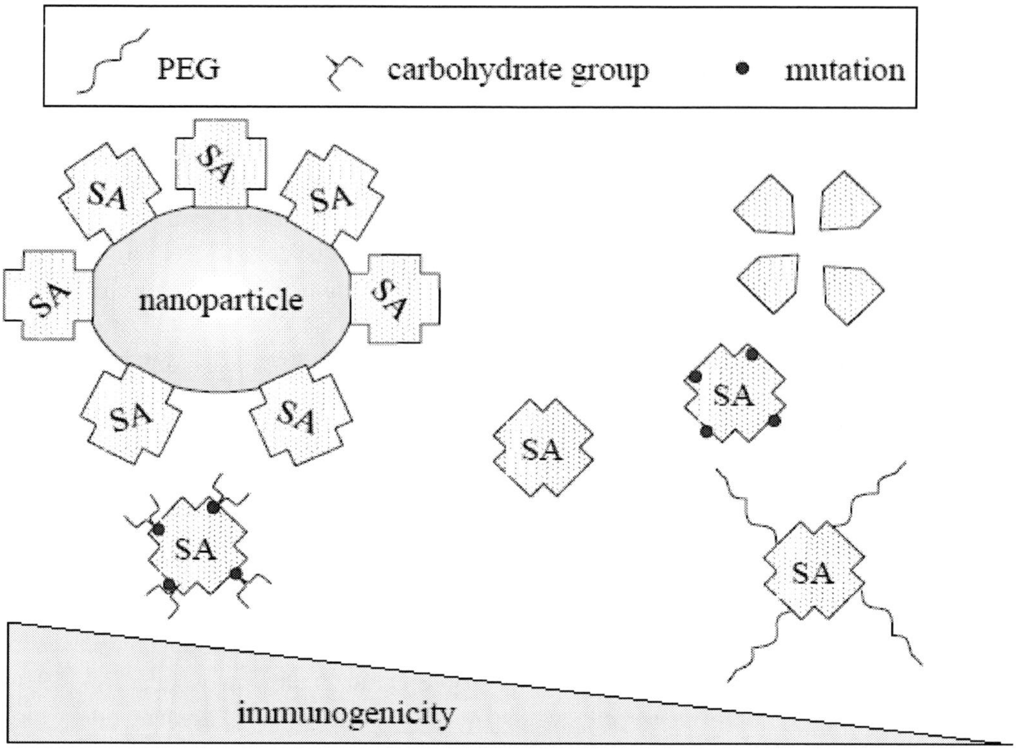

Figure 2. Immunogenicity of streptavidin. Compared to native streptavidin, in the center, glycosylated and particle-bound forms are more immunogenic, whereas PEGylation, certain point mutations and disassemly of the teramer can decrease its immunogenicity.

In addition to SA, its various modified recombinant forms are also used for targeting. Neutravidin is a recombinant modified, deglycosylated avidin with neutral isoelectronic pont. This variant was mainly generated to produce avidin–like protein with reduced aspecific binding to cells and weaker immunogenecity, but with the original high biotin-binding affinity.

Because SA binds biotinylated ligands more stably than avidin [3], and is less antigenic, it is preferred to avidin in targeting devices and biological tools. Figure 2 summarizes the effects of various modifications on immunogenicity.

Assembly of SA-Based Targeting Complexes

For the experimental immunologist, the main advantage of the SA-based targeting schemes is the flexibility of the system, due to its modularity. Different biotinylated targeting units can be coupled to the same SA-cargo pair, or – alternatively - different biotinylated cargos can be coupled with the same SA-targeting unit pair. The way biotin is conjugated to the relevant module is of key importance. The commonly used technique for the biotinylation of proteins employs chemically reactive forms of biotin, mostly primary amine reactive compounds. This technique is only partially predictable: –NH2 groups of the amino terminal end of the polypeptide and exposed lysines bind biotin but the exact number and the position of the incorporated biotin molecules cannot be readily determined. Having more than one biotin per module can result in the formation of high molecular weight lattices with decreased targeting efficiency. Sequence-specific incorporation of biotin molecules can be achieved by synthesis or enzymatically driven processes. Thus, peptides and oligonucleotides are synthesized with a pre-designed biotin at the required position. The enzymatic approach is applicable to recombinant proteins. A recognition site for the bacterial biotin ligase (BirA) enzyme is fused to the N- or C-terminal end of the protein; the enzyme catalyzes the covalent attachment of biotin to the lysine residue within the recognition site. Monobiotinylated molecules can be mixed with SA in a desired molecular ratio to achieve partial or full saturation of the biotin binding sites. It is important to remove unbound biotin and unbound biotinylated modules using SDS-PAGE, dialyzis or size-exclusion chromatography at the appropriate steps, otherwise conjugate formation or targeting efficiency is reduced. As SA has four biotin binding sites, theoretically tetrameric conjugates can be generated. However, in practice, because of the spatial constraints, often only a mixture of trimers, di- or monomers is formed.

In addition to biotin-mediated binding, genetic fusion to SA monomers is also used for the generation of targeting complexes. Either the cargo or the targeting unit can be fused to SA; the assembly of SA into the tetrameric form will result in the tetramerization of the fused component, as well. Biotinylation and fusion are also used in various combinations, as shown in figure 3.

The molecular ratio of the components is an important question in targeting; biochemical characterization of the complexes and titration of the components can help fine-tune the production process. Under certain conditions, one might like to avoid crosslinking of the biotinylated components, e.i. one biotinylated antibody represented on one SA is the favorable composition.

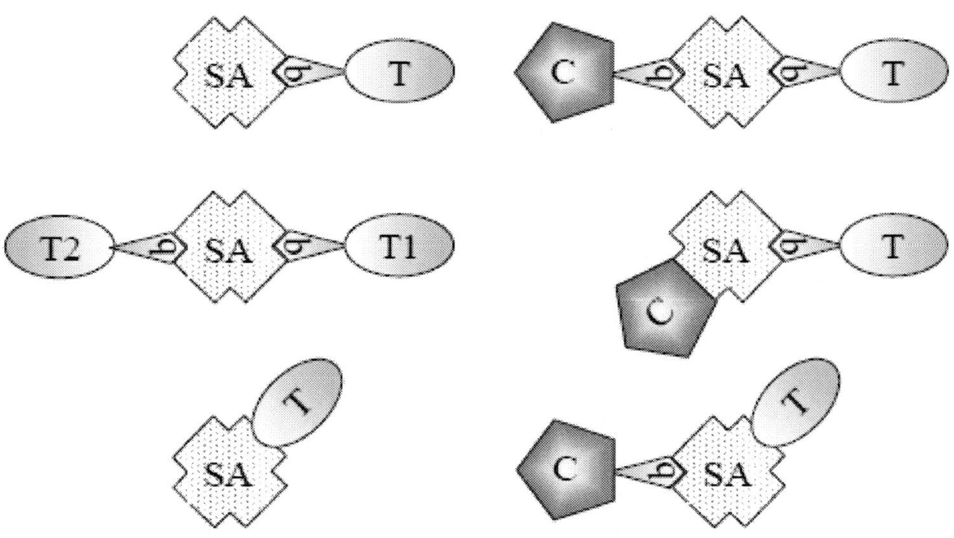

Figure 3. Composition of targeting complexes based on monobiotinylated components and fusion proteins. Note that SA molecules are shown with biotin binding sites unsaturated, by using appropriate molar ratios saturation with one or more biotinylated molecules can be achieved.

This reduction in valency can be achieved by adding titrated amount of free biotin before conjugating the biotinylated antibody [26]. Monovalent recombinant avidin or SA can also be produced when the amino acids responsible for the interaction between the subunits are substituted [24]. In cases when reversible binding is desired, monomeric avidin or SA can be used [17]. Most often, however, crosslinking of the receptors on the target cell surface improves targeting efficiency both by increasing the avidity of the targeting complex towards the target and by triggering stronger signals via the targeted receptors. Additionally, monobiotinylated targeting units with different specificities can be combined on single SA molecules in a stoichiometric manner [23]. Such strategies can achieve the co-crosslinking of different receptors on a particular cell or the targeting of multiple cell types.

Affinity Reagent-Mediated Targeting

Specific delivery of the targeting complexes can be accomplished by affinity reagents recognizing our target or by natural ligands of the target. In addition to the most commonly used affinity reagents, monoclonal antibodies, aptamers have also been exploited in SA-based targeting approaches.

In the next two sections we review studies using monoclonal antibodies, their recombinant and engineered variants and fragments, and the functionally antibody-like nucleic acid aptamers. Reports, in which immunological effects of targeting were not investigated, for example imaging or drug delivery studies are not discussed here. We examine both the immunological results against SA, often the model antigen in these studies, against the molecular cargo, where applicable, and also additional immunogical changes in the host caused by targeting. We focused primarily on SA-based complexes, but several works using avidin-based targeting constructs are also reviewed since avidin was widely used in earlier immunotargeting experiments.

Use of Whole Antibodies as Targeting Devices

Rat monoclonal antibodies against murine targets and murine monoclonal antibodies against human targets were the devices of choice following the appearance of hybridoma technology. These antibodies usually belonged to the IgG antibody class and biotinylation was accomplished by chemical coupling.

One of the earlier studies using avidin-based immuno-targeting was reported by Carayanniotis and Barber [27]. They demonstrated that immunization of mice with avidin-major histocompability complex II (MHCII) specific antibody in saline induced significant secondary IgG anti-avidin response compared to injection of avidin alone in saline. Moreover, no response was observed when the same amount of avidin conjugated to isotype control antibody was injected or when the delivering antibody was of a different MHCII-allele specificity than the MHCII expressed by the experimental mouse strain. Continuing this work, Carayanniotis et al. investigated whether synthetic peptides derived from herpes simplex virus could also be immunogenic when delivered to MHCII positive cells. The applied immuno-targeting complex was assembled by mixing biotinylated peptides and biotinylated anti-MHCII antibody with avidin. This conjugate elicited peptide-specific IgG antibody response in mice bearing the proper MHCII allele, while the same avidin-peptide conjugates were non-immunogenic when coupled to various controls [28]. Other works also showed that peptides and proteins, such as a fragment of influenza hemagglutinin [29] or luteinizing hormone releasing hormone (LHRH) [30], conjugated to anti-MHCII antibody, via avidin-biotin bridge, proved to be immunogenic. In the latter case, targeting the luteinizing hormone releasing hormone (LHRH)-avidin conjugate to MHCII resulted in a twofold augmentation of titers relative to controls. These antibodies were almost exclusively of the IgG1 isotype. Similarly, monoclonal antibody targeting in sheep resulted in a significantly enhanced immune response (reciprocal titers of 15 000 and 20 000 for LHRH and avidin, respectively), which was comparable to that achieved by immunizing with Quil A as adjuvant. Both the monoclonal antibody targeting and Quil A treatment tended to favour the production of antibody isotype IgG1 over IgG2. In another experiment, Skea et al. immunized ferret and rabbit with a hemagglutinin fragment-avidin-anti-MHCII antibody complex. High levels of anti-peptide IgG antibody were observed, with or even without the presence of alum adjuvant. Furthermore, the induced antibodies possessed stronger blocking property against hemagglutination compared to adding alum. These findings indicated that targeting protein or peptide antigens to MHCII has significant potential to induce antibody response against an antigen in an adjuvant free manner. The suggested explanation was that MHCII is expressed mainly on professional antigen presenting cells (APC) which have central role in the initiation of both humoral and cellular immune response. Thus, targeting APC proved to be successful strategy for the modulation of immune responses.

Skea et al. interested, whether delivering avidin to other receptors on APCs, or to non-APC cells, such as T cells, could also elicit significant anti-avidin antibody response [26,31]. They targeted CD3, CD4, CD8, CD11b, CD11c, CD18, CD22, CD45, CD205, MHCII, MAC-2 antigen on macrophages and dendritic cell (DC) inhibitory receptor 2 (DCIR2, the specific antibody is 33D1) by biotinylated antibody-avidin conjugates. To avoid the extensive cross-linking and aggregation of molecules during conjugation, which could influence the immunogenicity of such complexes, monovalent avidin was used and the size of conjugates was controlled by HPLC analysis. The targeting complexes against CD3, CD4, MHCII, CD45 and DCIR2 were able to enhance the level of anti-avidin IgG responses in mice, in adjuvant-independent manner, compared to avidin injected alone. These data showed that T cell-targeting by transporting antigen to CD3 or to CD4, as well as APC targeting, could be successful. Yet the two most effective immuno-targeting complexes were the DC specific 33D1 antibody-avidin and the anti-MHCII-avidin complexes, verifying that APC-targeting was still the most effective method. DCs are known to be most important APCs during the initiation of a primary immune response. In the case of these DC-specific complexes, they determined the isotype distribution: most of the induced antibody was lgGl (65-75%), indicating the dominance of Th2 response, but a significant proportion of IgG2a (20-30 %) was also found, showing the presence of Th1 cytokines, as well. Furthermore, memory response was also established using these two immuno-targeting complexes. Targeting CD11c, CD18, CD205, MAC-1 and MAC-2, CD22 and CD8 was less effective than targeting MHCII or DC by 33D1 antibody or the antibody response did not differ from that of mice injected with avidin in PBS.

To find other candidates for efficient APC targeting, Frleta et al. compared the efficacy of avidin targeted to CD40, MHCII and CD11c, and tested the hypothesis that stimulation of the

CD40 receptor could result in enhanced serological response [32]. They found that avidin-anti CD40 antibody conjugates elicited a weak humoral immune response against avidin, the avidin-anti-CD11c antibody complexes did not induce a detectable antibody response, while immunization with avidin-anti MHCII antibody enhanced the primary and the secondary immune response to avidin. Moreover, neither the combination of avidin-anti-CD40 and avidin-anti-MHCII, nor combination of three targeting approaches did significantly increase the anti-avidin humoral response compared to targeting antigen to MHCII alone. This group also described that when avidin (25μg) was only mixed with targeting antibody without coupling, there was no serologic response, indicating that avidin is non-immunogenic alone, injected in saline, at the doses used in the experiment.

As we have seen so far, antibody-mediated delivery strategies tended to target antigen to APCs, but because of its efficacy, other cell types were also proposed to be attractive targets. Kuroda et al. worked on targeting prostate cells, using biotinylated, humanized monoclonal antibody (hJ591) against prostate-specific membrane antigen (PSMA), which was conjugated to the saporin toxin-SA fusion protein [33]. This immunoconjugate was evaluated for antitumor activity against prostate cancer cell lines. Results showed that after 72 hours of treatment with the conjugates, the percentage of apoptotic cells was 60.29% and 40.73% in the two PSMA pozitiv cell line, respectively, compared to 4.70% in PSMA negative cells. Moreover, the conjugate also had anticancer activity in a prostate cancer xenograft model.

In another study, Muzykantov et al. tested antibodies to platelet-endothelial adhesion molecule 1 (PECAM-1, a stably expressed endothelial antigen) as targeting unit for vascular immunotargeting [34]. Interestingly, they found that endothelial cells could poorly internalize anti-PECAM-1 antibody but conjugation of SA to the antibody could stimulate internalization of anti-PECAM-1 in vitro, as well as in vivo, in perfused rat lungs or in lungs of intact animals. Facilitation of the targeting of anti-PECAM was specific, because SA reduced pulmonary uptake of isotype control antibody. If they conjugated biotinylated antioxidant enzyme catalase to the PECAM-1-SA complex, they detected the protection of rat lung against H_2O_2-induced injury. Thus, modification of a poor carrier antibody with biotin and SA provides an approach for the facilitation of antibody-mediated intracellular drug targeting. The finding that SA could specifically facilitate targeting of biotinylated anti-PECAM, suggested that the size of conjugates affected this mechanism. Diverse anti-PECAM conjugates ranging from 80 to 5000 nm in diameter were generated and it was found that the conjugates smaller than 350 nm were preferentially internalized by human umbilical vein endothelial cells (HUVECs) [35]. In addition, small (200 to 250 nm) and large (600 to 700 nm) conjugates of glucose oxidase (GOX) (producing H_2O_2 from glucose) with rat antibodies to murine PECAMs or with rat anti-thrombomodulin antibody were synthesized and injected into mice intravenously [36]. Only small conjugates had high pulmonary uptake and they caused a profound oxidative injury in the pulmonary vasculature. The size of these conjugates was controlled by the extent of biotinylation and molar ratio between the conjugate components.

A specialized form of T-cell targeting has been described by Rusakiewicz et al. [37]. They combined T-cell receptor ligand and costimulatory molecules in the targeting complex by using soluble MHCI-peptide monomers coupled to anti-CD27 or anti-CD28 or anti-CD40L antibodies on a SA core. This soluble antigen presenting complex was generated by mixing enzymatically biotinylated MHCI-peptide with biotinylated antibodies in 3:1 molar ratio and adding this mix to SA in 4:1 molar ratio. When constructs containing MHCI-cytomegalovirus peptide along with anti-CD27 and/or anti-CD28 were applied for stimulation of peripheral blood mononuclear cells, derived from cytomegalovirus seropositive donors, they detected efficient triggering and activation of virus-specific T-cell costimulation. Lower or no response at all was detected in seronegative donors. When anti-CD40L antibody was also included in the complex, enhanced T cell proliferation was measured compared to the effect of anti-CD27/anti-CD28 costimulation alone. They also examined the effect of MHCI-peptide tetramer with soluble costimulatory molecules, that it was also capable of inducing the expansion of CMV-specific memory T cells. However, the conjugation of MHCI-peptide with costimulatory molecules on SA core was mandatory for the successful induction of primary anti-virus responses.

Use of Antibody Fragments as Targeting Devices

Fragments of whole antibodies can be obtained either by enzymatic fragmentation or by molecular engineering of recombinant immunoglobulins. Recombinant technology has the advantage that by introducing a biotin ligase recognition site, biotin can be coupled to the protein enzymatically, in a sequence-specific fashion.

Some forms of antibody fragments can only be produced by engineering. One such fragment is the single-chain antibody (scFv) that has been used in several targeting experiments. This recombinant protein is the smallest fragment of an antibody with a full antigen recognition site, composed of the variable region of one heavy and one immunoglobulin light chain connected by a serine-glycine rich linker region. This monovalent fragment is preferably applied in tumor therapy and imaging, because the lack of the Fc-portion prevents binding to Fc-receptors (FcR) and activation of the complement system. Moreover, scFv can deeply infiltrate tissues and has faster clearance than a whole antibody. To crosslink receptors and activate cells, monovalent scFvs need to be multimerized, wherein application of the SA-biotin oligomerization technology can also help. ScFvs can be biotinylated enzimatically using the bacterial biotin ligase enzyme BirA, once a recognition site is inserted to either end of the the single chain antibody sequence. The resulting monobiotinylated scFvs can easily be combined with SA to form di-, tri- or tetravalent forms. Another method for the multimerization of scFvs using SA, is the genetic fusion of SA to scFv, where oligomerization occurs during protein production [38,39]. Multimerization increases both valency and avidity of single-chain antibodies (scFv). Moreover, combinations of different antigens and targeting agents can be coupled to a single SA molecule.

There are several papers describing modulation of immune response by antigen targeting to Fc-receptors for IgG (FcγR) and complement receptors (CR) on APC using scFv. These receptors are widely expressed on various APCs and exert strong modulatory effects on these cells, thereby also polarizing the immune response. Adamova et al. demonstrated enhanced antibody production to hFcyRI-targeted SA in vivo, using transgenic mice, in both primary and secondary immune responses. They generated a recombinant anti-hFcyRI-SA-fusion protein where a single subunit of SA was fused to a duplicated H22 scFv specific for human FcγRI. The SA-specific antibody response consisted mainly of IgG, the dominant subclass was IgG1. Furthermore, cytokine responses following immunization with hFcγRI-SA-biotinylated dextran complex was also examined and suggested enhancement of both Th1 and Th2 responses [40].

Our group investigated the effect of combined targeting of SA to low affinity FcγRII and III and/or to complement receptor 1 and 2 (CR1/2) in a murine model. As targeting units, we used monobiotinylated 7G6 scFv specific for CR1/2 or monobiotinylated 2.4G2 scFv specific for FcγRII and III, which were conjugated to SA individually or in combination. Mice were injected with these conjugates and humoral immune response against SA, and myc-tag and hexahistidin-tag fused to scFv were simultaneously measured. Results have shown that targeting SA complexes to mFcγRII/III induces significantly stronger IgG1 antibody response than targeting to mCR1/2, yet both strategies enhanced the antibody response compared to the control group immunized with non-targeted peptide-SA complexes. Combined targeting of CR1/2 and FcγRII/III receptors did not result in cumulative enhancement of the antigen specific immune response [23].

Angyal et. al have also verified that targeting low affinity FcγRs is a potent method of enhancing antigen specific antibody production [41]. They designed a monobiotinylated scFv from the rat anti-mouse FcγRII and III clone 2.4G2, linked to avidin–FITC, and tested its effect on the FITC–hapten-specific T-independent type 2 (TI-2) and T-dependent (TD) immune responses. When 2.4G2 scFv-avidin-FITC construct was applied as a booster, following primary immunization with TI-2 (FITC–dextran) or TD (FITC–keyhole limpet haemocyanin) antigens, elevated numbers of FITC -specific IgM/IgG-producing B cells were detected.

In vivo DC targeting in mice was evaluated by Wang et al., who generated scFv that recognized the CD205 multilectin receptor fused with a truncated core-SA domain [42]. Immune response in mice was induced with a variety of different cathegories of biotinylated antigens, such as ovalbumin, epithelial mucin 1 (MUC-1) peptide, DNA (derived from SARS-CoV membrane or spike, EBOV and WEEV) or gangliosides (GM2 and GM3). Results showed that in the presence of CD205 scFv-SA and costimulatory anti-CD40 mAb, both humoral and cell-mediated responses can be augmented against each targeted antigen. In the multiple antigens targeting strategy, they also achieved humoral and cell-mediated responses to OVA, SARS-CoV spike RBD, MUC-1 and anthrax PA.

Delivery of immunogenic peptides coupled to major histocompatibility complex class I (MHCI) to the surface of tumor cells allows activation of cytotoxic T lymphocytes which then recognize and kill the tumor cell. Savage et al. used a two-step targeting system: recombinant biotinylated MHC-class I/peptides were attached to the surface of B cells in vitro via the anti-CD20 B9E9 scFv-SA fusion protein [43]. The first step is the delivery of the anti-CD20 B9E9 scFv-SA fusion protein, the second is the delivery of recombinant biotinylated MHC-class I/ peptide. This method is able to amplify and selectively expand peptide specific CTLs from unselected populations of peripheral blood mononuclear cells. Continuing this idea, mice were immunized with ovalbumin and survival of EL4Hu20 lymphoma cells targeted with MHCI/ovalbumin peptide complexes and control MHC complexes were compared. Data showed that MHCI/ovalbumin-targeted lymphoma cells in immunized mice represented only 12% of the control cells [44].

Robert et al. used a similar approach for triggering cytotoxic T cells [45]. In the targeting complexes they used tumor specific Fab antibody fragments as targeting agent, which were chemically coupled to SA by a thioether bond between their reduced C terminal cysteine residues and the maleimide groups randomly distributed on lysine residues of SA. These Fab-

SA conjugates were subsequently used for tetramerization of biotinylated MHCI-influenza matrix peptide or MHCI-β2-microglobulin complex. Binding of Fab-SA-MHC tetramers to tumor cell lines resulted in their lysis in vitro.

Fusion proteins of scFv-SA [46,47,48,49] and immunoglobulin-SA [50,51] have also been used for pre-targeting experiments, where these proteins are first injected into body to recognize tumor cells and then serve as capture agents for the later administered radiolabeled biotin. In these pre-targeting experiments immunological effects were not reported, however.

Nucleic acid aptamers possess many properties similar to antibodies, thus they can be applied as targeting units. Nucleic acids are less immunogenic than proteins, therefore anti-targeting unit responses can be minimized with their use. Synthetic production of oligonucleotides also permits the designed incorporation of biotin and subsequent generation of SA-based complexes with defined stoichiometry. Two high-affinity DNA aptamers, developed previously by Ferreira et al. [52] against MUC1 antigen on MCF7 breast cancer cells, were linked to the first component of complement (C1q) via a biotin–SA system. These constructs induced significant complement mediated lysis of MCF7 cells in vitro [53].

Targeting with Particulate Complexes Containing Antibodies

Phagocytic cells of the immune system, that is monocytes, macrophages, DCs and neutrophil granulocytes, are capable of engulfing cell-sized particles. Some size-preference has been observed: whereas macrophages tend to take up micrometer sized particles, DCs accumulate smaller particles with diameter in the range of a few hundred nanometers. Non-phagocytic APCs, B cells, are also capable of internalizing virus-sized particles. Conjugation of our targeting constructs to particles therefore introduces another element into the targeting strategy. In fact, particle-mediated delivery is quite potent itself without the use of molecular targeting agents. Besides the increase in size, conjugation to particles can also increase the valency of the constructs. A nanobead with a diameter of 40 nanometers can carry several SA molecules; on a 400 nanometer bead hundreds to thousands of SA molecules can be displayed with a proportional increase in biotin-binding ability. This enormous valency can significantly amplify signals delivered via targeted receptors. Furthermore, lots of different molecules can be conjugated to these particles giving way to sophisticated and complex targeting strategies. Some examples of particle-mediated delivery strategies are shown in Figure 4.

Walsh et al. targeted FcγRI on APC with a two-component modular strategy [54] . A single subunit of SA was fused to duplicated H22 scFv specific for human FcγRI (SA-scFvH22) and this fusion protein was conjugated with the biotinylated antigen adsorbed onto microspheres. They used tetanus toxoid (TT) as model antigen, which was applied in equimolar ratio for conjugation to SA-scFvH222. In vitro, using macrophages, the (SA-scFvH222)-microsphere-TT complexes significantly enhanced the proliferation of specific T cell compared to non-targeted-antigen.

We have also demonstrated that conjugation of antigen to microspheres significantly increases the antigen specific antibody response. We studied nanoparticle-mediated targeting of SA to FcyRII and III using 2.4G2 scFv and/or to CR1/2 using 7G6 scFv molecules [23]. 500 nm polystyrene microspheres coated with SA were used, conjugated with either one or both monobiotinylated scFv. The scFvs contained a myc-hexahistidine-tag peptide which also served as model antigen, beside SA. Immune response for both was analyzed, using SA-biotin-myc-hexahistidine peptide complex for the immune assay as antigen. After immunization, only FcyRII and III targeting complexes induced high levels of IgG1 antibody and the effect of CR1/2 targeting complex did not differ from the control. Combined targeting of the two receptors did not further increase the immune response. Interestingly, anti-SA specific IgG3 was elicited in all groups, even in mice injected with non-targeting complexes, whereas IgG3 levels were not affected in any of the groups receiving soluble SA in our previous experiments.

In addition to various polymer nanobeads, liposomes can also serve as particulate units of the targeting complexes. Klegerman et al. developed bifunctionally targeted liposomes, conjugated to antibodies specific for both a stem cell marker CD34 and an adhesion molecule CD54, expressed by inflammatory endothelium [55]. These complexes, built for bridging stem cells and atheroma, and for increasing the permeability of endothelial cell monolayers by NO delivered within the liposome, could ameliorate pathological processes in the atheroma.

The combination of various targeting agents and antigenic cargo on a particle can resemble an antigen presenting cell and be suitable for triggering T cells. Such an artificial APC can prime and/or expand of T cells of low frequency and often low avidity. Liebeth et al. and Schilbach et al. immobilized MHCI-peptide complexes, along with costimulatory antibodies or recombinant proteins including anti-CD28 antibody, recombinant CD80 and CD54 onto SA-coated microbeads [56,57]. These artificial antigen-presenting cells were shown to be capable of rapidly generating higher yields of cytotoxic $CD8^+$ T cells than that achieved using conventional autologous DCs as APCs.

LIGAND-MEDIATED TARGETING

Soluble Ligand-Based Targeting Complexes

First we look at strategies employing SA or avidin as a model antigen, without an antigenic cargo, than discuss reports utilizing biotin-coupled or fused cargo for immune-modulation.

Recombinant ovine (Rov) IL-lα and IL-lβ have been studied extensively as adjuvants for use in sheep, with many of the initial experiments utilizing the model protein antigen avidin [58,59]. Rov IL-lα and avidin was emulsified with alum. After the secondary immunization there was a significant eight-fold increase in antibody titer compared to the control group that received avidin in alum without rovIL-1α, and also an enhanced anti-avidin cellular response as measured by delayed type hypersensitivity assay [58]. Similar adjuvant activities were found for rovIL-1β, and additional experiments showed that rovIL-1 and avidin must be administered in a site drained by the same lymph node for the adjuvant effect of rovIL-1β to be observed [59]. Lofthouse et al. demonstrated the activity of recombinant ovine IL-2 as an adjuvant using the model protein avidin, and found 4-fold enhancement in secondary antibody titer. Multiple administration of IL-2 resulted in a further significant increase in antibody titers. As in the case of IL-1, the combination of alum and IL-2 was more effective for induction of antibody than the use of these adjuvants individually. In contrast to IL-1 however, IL-2 alone or in combination with alum did not appear to enhance delayed type hypersensitivity responses (reviewed in [60]).

Vaccine immunogenicity can be enhanced by conjugation of adjuvant to the antigen, because adjuvant and antigen will contact the very same antigen-presenting cell.

Heath et al. used biotinylated recombinant IFN-γ as an adjuvant, conjugated to the model antigen avidin. The conjugation resulted in significantly enhanced delayed-type hypersensitivity responses to avidin, and caused a slight, but insignificant increase in secondary antibody responses compared to unconjugated controls [61]. Similarly, the adjuvant-like effect of CpG oligonucleotides is improved, when their proximity to antigen is maintained. Klinman et al. used SA to link biotinylated CpG oligonucleotides and the biotinylated form of the model antigen ovalbumin (OVA), then immunized mice with these complexes. Co-localizing CpG oligos with OVA boosted IgG anti-OVA responses by fourfold (tenfold, when compared to OVA alone), and lowered the IgG1:IgG2a ratio compared to OVA and CpG mixture [62,63].

Tedder et al. targeted type 2 complement receptor (CD21) using its natural ligand, the C3dg fragment of complement protein 3. C3dg-SA tetramers were assembled by the coupling of four monobiotinylated C3dg to SA [64,65]. Immunizations with the C3dg-SA tetramers in wild type mice induced significant SA-specific IgM and IgG responses, compared to injection of SA alone, but the level of elicited streptavdidn–specific antibody titer was shown to be inversely dose dependent. C3dg functioned as an adjuvant with low-dose SA in wild-type mice, but did not enhance SA-specific immune responses to the same extent when SA and C3dg were used at 10-fold higher concentrations. However, immunization of CD21/CD35$^{-/-}$ mice with either low-dose SA-C3dg or high-dose SA induced measurable Ab responses, indicating that C3dg can function in an CD21 independent pathway, as well. Measuring the B-cell antigen receptor-induced [Ca^{2+}] responses in spleen B cells in vitro, low SA-C3dg concentrations augmented it, whereas higher SA-C3dg concentrations were inhibitory. They concluded that physiologically relevant concentrations of C3d-antigen complexes may have the potential to positively or negatively regulate B cell signaling and in vivo immune responses through both CD19/CD21 complex-dependent and independent pathways. In an other study, the same C3dg-SA tetramers augmented B cell [Ca^{2+}] responses and specifically bound CD21 on wild-type, but not on CD21$^{-/-}$ B cells [66].

Costimulatory molecule 4-1BB ligand (4-1BBL) is a promising candidate as component of vaccines against cancer or chronic infections. Elpek et al. generated SA-4-1BBL by fusing the extracellular domains of 4-1BBL to the C terminus of a modified form of core SA [67]. SA-mediated oligomerization of 4-1BBL proteins induces the cross-linking of 4-1BB receptors. This study demonstrated that regulatory T cells had no suppressive function in the physical presence of SA-4-1BBL for effector T cells in vivo, and this effect was completely reversible upon removal of the chimeric protein. Vaccination with SA-4-1BBL fusion protein activated DCs, thus contributed to enhanced antigen uptake and induced CD8 T cell and memory response [68]. Immunization with human papillomavirus 16 E7 oncoprotein mixed with SA-4-1BBL [68] or biotinylated E7 and survivin in conjugation with SA-4-1BBL [69] resulted in increased antigen uptake and cross-presentation by DCs, leading to the generation of effective CD4 T and CD8 T-cell effector and memory responses against cervical and lung carcinoma tumors respectively, and to a higher intratumoral CD8 T effector/regulatory T cell ratio.

Chimeric molecules consisting of the extracellular domains of human and mouse CD40L and a modified form of core SA were generated by Kilinc et al. and tested for activation of antigen-presenting cells. These conjugates upregulated MHCII on macrophages, as well as on B cells, and stimulated macrophages for inflammatory cytokine production, such as IL-1β and IL-6 in vitro. Moreover, the human CD40L chimeric construct induced production of iNOS in IFNγ-primed macrophages [70].

Subunit vaccines based on tumor-associated antigens (TAA) represent an attractive approach for cancer treatment, however, the poor immunogenicity of TAAs requires potent adjuvants for therapeutic efficacy. Cell surface display of desired proteins (for example costimulator ligand expression on tumor cells) is a promising method for immunomodulation. Transduction of tumor cells for the expression of costimulatory molecules has proven to be an efficient way to enhance tumor immunogenicity. Moro et al. introduced a three step approach that allows the easier decoration of tumor cells with these molecules. This method is based on the sequential incubation of cultured tumor cells with biotinylated antibodies specific for a membrane TAA, neutravidin, and biotinylated soluble B7-1 or B7-2 fused with the IgG constant regions (bio-B7-IgG). Targeting mouse tumor cells with biotin-B7-IgG in vitro was efficient in expanding tumor-specific CTLs for adoptive immunotherapy and generating therapeutic nonreplicating whole cell vaccines. In vivo treatment of tumor-bearing animals with biotin-B7-IgG induced therapeutic T-cell immunity [71].

Poor uptake of macromolecules, into the endocytic compartment of targeted cells can be a hindrance to achieve effective DNA delivery, tumor therapy, and stimulation of immune responses by transfection of the target cell or by antigen uptake and presentation by APCs. Scientists developed different methods to overcome this problem. Hussey et al. employed a biotinylated cholesterylamine derivative, termed Streptaphage, which inserts into lipid rafts and the biotin moiety captures SA complexes on the biotinylated cell. When SA crosslinks the membrane-anchored biotins, lipid raft aggregation occurs and the SA (that can carry additional molecules) is taken up by clathrin mediated endocytosis [72].

Targeting with Particles Carrying Receptor Ligands

As discussed earlier in this chapter, some vaccine formulations take advantage of the size and other adjuvant activities of particles in place of or in addition to the molecular targeting of specific receptors. Various artificial particles, like liposomes, polymers and dendrimers, and also natural particulate elements, such as virions or erythrocytes have been employed for this purpose.

Immunostimulating complexes (ISCOMs), are potent adjuvants with a spherical cage-like structure, generated from cholesterol, phospholipids and saponins. One of their limitations is that the hydrophilic proteins are associated poorly to the ISCOM matrix (ISCOM particles without any antigen). Wikman et al. reviewed recombinant strategies applied by their group to achieve efficient ISCOM incorporation of soluble antigens [73]. They applied biotin-SA bridges to couple recombinant immunogens to the ISCOM matrix. In the first concept (indirect coupling), a hexahistidine-tagged SA fusion protein (His6-SA) was bound to Ni^{2+}-loaded ISCOM matrix. Biotinylated immunogens (malaria antigen M5 or *Neospora* antigen SRS2') were thereafter associated to the SA coated ISCOMs.

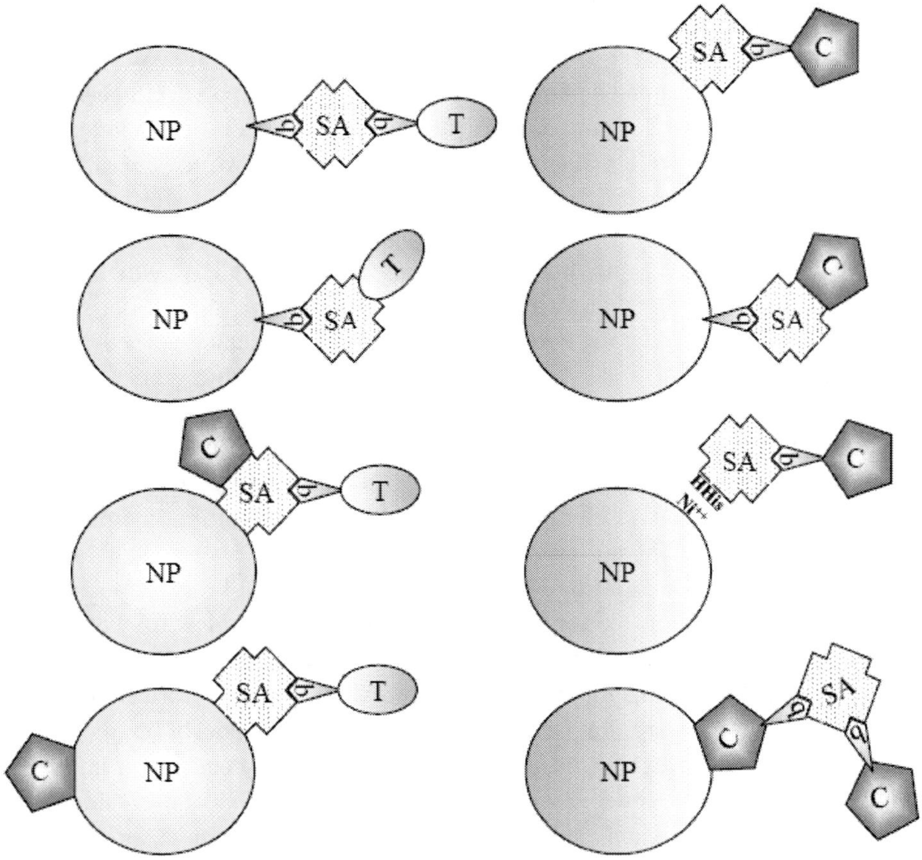

Figure 4. Examples of particle-mediated delivery and its combinations with molecular targeting. Note that for the sake of clarity and simplicity, nanoparticles are shown with one module of a kind attached, SA molecules are shown with one or two biotin binding sites occupied. Both the density of nanoparticle-bound molecules and saturation of biotin binding sites can be achieved at will. Consequently, particles can be decorated with several different targeting units or cargos, resulting in complex delivery strategies. Particles themselves can serve as targeting units, this is emphasized by the gradient fill of the nanoparticle symbols. NP, nanoparticle.

In the second concept (direct coupling), recombinant immunogens were expressed as SA fusion proteins (His6-SA-M5 and His6-SA-SRS2'), and were directly coupled to the biotinylated ISCOM matrix. These preparations induced high-titer antigen-specific antibody responses in mice. Antibodies induced by SRS2' constructs demonstrated reactivity to the native antigen NcSRS2 [74]. Mice immunized with recombinant ISCOMs prepared according to the first concept, were partially protected against challenge infection with *Neurospora caninum* [75]. Since it was not evident from these initial studies, which of these two concepts was preferable, Pinitkiatisakul et al. introduced a third strategy for further comparative studies. In this third concept, the His6-SA-SRS2' immunogen was conjugated to the Ni^{2+} loaded ISCOM matrix via its His6 tag. Immunization studies showed that all three preparations induced *N. caninum*-specific IgG1 and IgG2a antibody responses, proliferation of spleen cells and high levels of IFN-γ and IL-4. It was the direct coupling method that induced the strongest cellular responses. The potential advantage of the Ni^{2+} loaded ISCOM matrix coated with His6-SA would be that it could be used together with any biotinylated protein, thus antigen does not have to be produced as a fusion protein to SA [76].

Controlled release of antigen and adjuvant from the administered vaccine can enhance the ensuing immune responses. Lofthouse et al. designed two types of silicone implants as vaccine delivery vehicles, which change the antigen release profile and thus can allow reduced antigen and adjuvant dosage and a single administration. The silicone implants consisted of the model antigen avidin (or alternatively *Clostridium* toxoids), IL-1β as adjuvant, along with mannitol and/or sodium citrate additives, all of which were solved, mixed and lyophilized, then blended with silicone. Rods of 1.6 mm in diameter were extruded and were cut to 10 mm length to produce the matrix type implants. The covered rod type implants were covered with an outer layer of silicone that exposed the inner matrix only on the cut surfaces thus providing slower and prolonged antigen and adjuvant release. Immunization of sheep with the matrix type implant resulted in antibody responses equivalent to those induced by alum, predominated by IgG1 isotype. The covered rods elicited markedly enhanced antibody responses over alum, and resulted in sustained antibody titers. In addition, the covered rod implants elicited elevated levels of both IgG1 and IgG2a antibodies. Interestingly, the lowest dose of antigen induced the highest antibody response that could have resulted from the slow but continuous release and the relatively high concentration of released antigen in the body fluids [77].

Deposition of liposomes in lymph nodes may prove useful in a variety of applications such as vaccine antigen delivery, but conventional liposomes are poorly retained in draining lymph nodes (<2% injected dose). Phillips et al. achieved great enhancement in the retention of liposomes in the lymph nodes by injecting biotin-coated liposomes subcutaneously. followed by an adjacent proximal subcutaneous injection of avidin. This method causes aggregation of biotin-coated liposomes migrating through the lymphatic vessels and the aggregated liposomes become trapped in the next encountered lymph node. Injecting rabbits subcutaneously with biotin-coated liposomes followed by an adjacent injection of avidin in both hind feet resulted in retention of 13.7% of the injected liposomes in popliteal nodes and 2.3% in iliac nodes, whereas in control rabbits (injected with biotin-coated liposomes without the avidin injection) retention was only 1.7% of the liposomes in popliteal nodes and 0.3% in iliac nodes after 24 hours. Blood and liver uptake of the biotin-coated liposomes was greatly decreased in the experimental rabbits receiving avidin [78].

Tobacco mosaic virus (TMV) has proven to be an effective carrier for the display of foreign peptides fused to the virus coat protein, and capable of eliciting a neutralizing immune response against numerous pathogens (reviewed by Pogue et al. [79]) via its ability to display antigens in an ordered, repetitive way. Genetic fusion of antigenic peptides to the coat protein however has limitations concerning the size and amino-acid composition of the epitope. To overcome these limitations Smith et al. modified the TMV coat protein to contain a reactive lysine at the amino terminus that allows for biotinylation of the virus capsid, and subsequently displayed the model antigen GFP-SA fusion protein on the biotinylated recombinant TMV particle. The fragment of Canine oral papillomavirus L2 protein (COPV L261-171) fused to SA was also conjugated to biotinylated TMV particles and immunogenicity of both kinds of decorated virions were evaluated. Mice and guinea pigs immunized with the GFP-SA-TMV particles developed an augmented total IgG response (4- to 20-fold) compared to animals either administered GFP-SA mixed with non-biotinylated virus particles or GFP-SA fusion protein alone. Similar responses were detected against the COPV L2 antigen fragment when it was displayed on TMV particles in mice [80].

For the efficient induction of anti-viral responses it is important to ensure access of exogenous antigens to the MHC I processing pathway and thus induce strong CD8 T cell responses. Particulate antigens have been demonstrated to elicit CD4 and CD8 T-cell responses and are therefore good candidates for inducing protection from viral infection. Corinti et al. used biotinylated erythrocytes to carry biotinylated HIV Tat protein via avidin bridges as antigen for boosting the autologous T-cell stimulating effects of IFN-γ-treated human DCs. Cells incubated with soluble Tat could induce strong CD4 and CD8 T-cell proliferation, but red blood cell-delivered Tat (RBC-Tat) was similarly effective in 1250-fold smaller doses. Using RBCs as vaccine delivery vehicles can be advantageous compared to DCs transduced with Tat DNA or RNA, since these cells will produce this otherwise immunosuppressive protein unrestrictedly [81].

Strategies Employing Particles and Molecular Targeting in Combination

Maus et al. investigated methods for achieving effective activation of CD4 T cells for adoptive immunotherapy. They used artificial APCs; magnetic beads coated with HLA-flu hemagglutinin (HA) tetramers on SA, prepared as described by Novak et al. [82] and anti-CD28 antibodies for costimulation. From the various forms of conjugations, the indirect coating method, when the antigen-HLA tetramers were conjugated to the beads via an anti-HLA class II or an anti-SA antibody, was found to be effective in stimulating antigen-specific CD4 T cells. This effect was probably caused by the specific, directional alignment and flexibility of the tetramers on the surface of the beads, in contrast to direct covalent conjugation of the HLA tetramers onto the rigid surface. Additional advantages of indirect coating are the mild physical conditions that can help to preserve the HLA-peptide tetramer structure [83].

In certain instances effective uptake of targeted materials by macrophages without any interference with the immune system is the aim of targeting. Broz et. al. applied "stealth" nanocontainers that were formed from a biotin-functionalized (poly(2-methyloxazoline)–b-poly(dimethylsiloxane)–b-poly(2-methyloxazoline), shortly PMOXA–PDMS–PMOXA triblock copolymer, that have very low unspecific adherence to proteins and cells and can target enclosed diagnostic or therapeutic agents to the targeted cells via biotinylated targeting ligands, coupled to the particles via SA bridges. Broz et al. targeted nanocontainers loaded with fluorescent markers against the macrophage scavenger receptor A1 (SRA1) that is associated with the endocytotic pathway and is responsible for binding and uptake of a number of polyanionic substances. Biotinylated polyguanylic acid (polyG) were used to functionalize the nanocontainers. In human and transgenic cell lines and in mixed cultures receptor-specific binding and vesicular uptake was observed in the absence of cytotoxicity [84].

Presentation of the antigen can be achieved by introducing an expression vector, instead of the protein itself, into the APC. DCs are primary targets of these strategies that often exploit viral entry and infection mechanisms for introducing foreign DNA Transduction efficacy is an important feature of the viral vector systems, which affects the ability to elicit strong immune responses to the encoded antigen. In order to enhance this feature of the recombinant adenovirus (rAd) transduction system, Maguire et al. targeted rAd type 5 particles to alternate cell surface receptors on DC, using Neutravidin (NA) as a linker between targeting units and the biotinylated virus particle. They took advantage of a genetically-modified rAd5 vector (previously described by Parrott et al. [85]) that contains a biotin acceptor-protein (BAP) motif on the C-terminus of the fiber protein, allowing the virus to be metabolically biotinylated and subsequently bound to any ligand of interest via a biotin–avidin bridge. The chosen target receptors on DCs were αVβ3 integrin, ChemR23, DC-SIGN, because of their abundant cell surface expression, ability to undergo endocytosis and involvement in antigen processing/cellular activation. The receptors were targeted using biotinylated high affinity binding peptides (in the case of αvβ3 integrin and Chem R23) or a monoclonal antibody (in the case of DC-SIGN). All targeting complexes increased the transduction efficiency of the modified rAd5 vectors and increased the expression of cell surface markers of DC maturation, as compared to unmodified control vectors. When a rAd5 vector that encoded the immunodominant human cytomegalovirus (HCMV) pp65 protein was targeted to DC via each of the receptors noted above, enhancement of the ability of vector-transduced DC to stimulate the antigen-specific activation of autologous human memory CD8 T cells was observed [86].

Because of the safety concerns and other hurdles faced by DNA-based gene therapy, an alternative method to display proteins on the cell surface was developed and reviewed in [87]. They biotinylated biological surfaces, such as cells, tissues or organs chemically, with biotin compounds binding to NH_2 groups, and used recombinant chimeric molecules composed of the extracellular domains of immunological ligands of interest and core SA (a truncated form of SA) for durable display of the chimeric ligands. The persistence of the chimeric molecules on the surface of cells varied from days to weeks depending on the protein and the cell type displaying it [88,89]. Immunological synapse formation and signal transduction requires minutes to hours, so this transient persistence was sufficient for eliciting immunological effects. Administration of biotinylated and core SA-FasL conjugated donor splenocytes could induce tolerance to cardiac and pancreatic islet allografts by blocking primary and secondary alloreactive responses, by causing apoptosis of alloreactive T cells and induction or expansion of regulatory T cells [89,90,91,92,93]. Even whole organs could be engineered this way: hearts perfused with biotin followed by SA-FasL overcame acute rejection when transplanted into allogeneic hosts [94].

Biotinylated tumor cells decorated with the extracellular portion of CD80 costimulatory molecule fused to core SA (CD80-SA) served as APCs for the generation of antitumor responses ex vivo, and vaccination with the engineered tumor cells prevented tumor growth in a mouse lymphoma model (tumor specific CTLs and long-term immune memory was generated) [88]. In a preclinical study, tumor cells were resected from lung, colon and breast cancer patients and were engineered with CD80-SA. These cells induced proliferation of autologous T cells and specific CTL responses against autologous tumors ex vivo [95]. Co-display of costimulatory molecules on biotinylated tumor cells could also be achieved for cancer vaccine applications. Sharma et al. engineered papillomavirus E7 expressing TC-1 cells to co-display SA-4-1BBL and SA-LIGHT on their surface. Vaccination with irradiated TC-1 cells co-displaying both chimeric proteins showed better efficacy in a prophylactic tumor setting than vaccination with TC-1 cells engineered with either protein alone. Vaccine efficacy was associated with the generation of primary and memory T-cell and antibody responses against the tumor without detectable signs of autoimmunity [96].

Table I. Targeting strategies using whole IgG antibodies

Target receptor	Targeting unit	Link 1	Center	Link 2	Cargo	Effect	Species	Ref.
MHCII	IgG	Biotin (chem.)	-	-	A	high anti-avidin IgG, mainly IgG1 response, induce memory	mouse (in vivo)	[26,27, 31,32]
MHCII	IgG	Biotin (chem.)	A	photocoupled	HSV peptides	anti-HSV peptide IgG response	mouse (in vivo)	[28]
MHCII	gG	Biotin (chem.)	A	Biotin (chem.)	BHA peptide	high anti-BHA peptide IgG response, which can block hemagglutination	rabbit, ferret (in vivo)	[29]
MHCII	IgG	Biotin (chem.)	A	Biotin (chem.)	LHRH	Enhanced level of anti-LHRH and anti-avidin IgG, mainly IgG1	mouse, sheep (in vivo)	[30]
CD3	IgG	Biotin (chem.)	-	-	A	high titer of anti-avidin IgG	mouse (in vivo)	[26]
CD4	IgG	Biotin (chem.)	-	-	A	high titer of anti-avidin IgG	mouse (in vivo)	[26]
CD45	IgG	Biotin (chem.)	-	-	A	high titer of anti-avidin IgG	mouse (in vivo)	[26]
33D1 receptor	IgG	Biotin (chem.)	-	-	A	high titer of anti-avidin IgG, mainly IgG1, induced memory	mouse (in vivo)	[26,31]
CD11c	IgG	Biotin (chem.)	-	-	A	similar respose compared to avidin in PBS or no detectable humoral response	mouse (in vivo)	[26,32]
CD18	IgG	Biotin (chem.)	-	-	A	similar respose compared to avidin in PBS	mouse (in vivo)	[26]
CD205	IgG	Biotin (chem.)	-	-	A	similar respose compared to avidin in PBS	mouse (in vivo)	[26]
MAC-1	IgG	Biotin (chem.)	-	-	A	similar respose compared to avidin in PBS	mouse (in vivo)	[26,31]
MAC-2	IgG	Biotin (chem.)	-	-	A	similar respose compared to avidin in PBS	mouse (in vivo)	[26,31]
CD22	IgG	Biotin (chem.)	-	-	A	similar respose compared to avidin in PBS	mouse (in vivo)	[26]

Table I. (Continued)

Target receptor	Targeting unit	Link 1	Center	Link 2	Cargo	Effect	Species	Ref.
CD8	IgG	Biotin (chem.)	-	-	A	similar respose compared to avidin in PBS	mouse (in vivo)	[26]
CD40	IgG	Biotin (chem.)	-	-	A	weak humoral immune response	mouse (in vivo)	[32]
MHCII and CD40	IgG	Biotin (chem.)	-	-	A	Similar anti-avidin antibody titer, as avidin-MHCII alone	mouse (in vivo)	[32]
MHCII and CD11c	IgG	Biotin (chem.)	-	-	A	Similar anti-avidin antibody titer, as avidin-MHCII alone	mouse (in vivo)	[32]
CD40 and CD11c	IgG	Biotin (chem.)	-	-	A	no detectable humoral immune response	mouse (in vivo)	[32]
MHCII and CD40 and CD11c	IgG	Biotin (chem.)	-	-	A	Similar anti-avidin antibody titer, as avidin-MHCII alone	mouse (in vivo)	[32]
Prostate cancer antigen	IgG	Biotin (chem.)	SA	fusion	saporin	decrease tumor cell viability, anticancer activity	mouse (in vitro and in vivo)	[33]
PECAM-1	IgG	Biotin (chem.)	SA	Biotin (chem.)	catalase	Protection cells from H2O2-caused injury	mouse, rat (in vivo)	[34]
CD27	IgG	Biotin (chem.)	cSA	biotin (BirA.)	HLA-A2/CMV peptide	Expansion of CMV specific CD8+ T cells	human (in vitro)	[37]
CD28	IgG	Biotin (chem.)	cSA	biotin (BirA.)	HLA-A2/CMV peptide	Expansion of CMV specific CD8+ T cells	human (in vitro)	[37]
CD27/CD28	IgG	Biotin (chem.)	cSA	biotin (BirA.)	HLA-A2/CMV peptide	Expansion of CMV specific CD8+ T cells	human (in vitro)	[37]
CD27/CD28/CD40L	IgG	Biotin (chem.)	cSA	biotin (BirA.)	HLA-A2/CMV peptide	higher T cell proliferation compared to the effect of anti-CD27/anti-CD28 costimulation.	human (in vitro)	[37]

Table II. Targeting strategies using affinity reagent fragments

Target receptor	Targeting unit	Link 1.	Center	Link 2.	Cargo	Effect	Species	Ref.
hFcγRI	scFv	fusion	SA	-	-	Enhanced anti-streptavidin IgG, mainly IgG1	mouse (in vivo)	[40]
hFcγRI	scFv	fusion	SA	Biotin (chem.)	dextran	Enhanced anti-streptavidin IgG, and citokin response (IL-2 , IL-4, IL-5, IFNγ)	mouse (in vivo)	[40]
mCR1/2	scFv	Biotin (chem.)	SA	-	-	Moderate level of anti-streptavidin IgG response	mouse (in vivo)	[23]
mFcγRII/ III	scFv	Biotin (chem.)	SA	-	-	Enhanced anti-streptavidin IgG1	mouse (in vivo))	[23]
mFcγRII/ III and CR1/2	scFv	Biotin (chem.)	SA	-	-	Similar anti-SA humoral immune response, than individual targeting of mFcγRII and III	mouse (in vivo)	[23]
mFcγRII/ III	scFv	Biotin (chem.)	A	chemical	FITC	Enhanced anti-FITC antidody response both in TI and TD manner	mouse (in vivo)	[41]
CD-205	scFv	fusion	cSA	biotin (chem. and synth.)	DNA, ovalbumin, MUC-1 peptide, gangliozid	Augmented humoral and cell-mediated responses in the presence of anti-CD40	mouse (in vivo)	[42]
CD20	scFv	fusion	SA	biotin (chem.)	HLA-classI/peptide	Induce proliferation and expansion of antigen specific effector T cells	human (in vitro) mouse (in vivo)	[43, 44]
CD20, Erb-2, CEA	Fab	Chemic.	SA	biotin (synth.)	HLA-A2/Flu	Lysis tumor cells	human (in vitro)	[45]
MUC-1 antigen	aptamer	biotin (synth)	SA	chemical	C1q	Significant complement lysis of target cells	human (in vitro)	[53]

Table III. Targeting strategies using receptor ligands

Target receptor	Targeting unit	Link 1.	Center	Link 2.	Cargo	Effect	Species	Ref.
IFN-γR	IFN-γ	biotin (chem.)	A	-	-	increased cellular response, DTH	mouse (in vivo)	[61]
TLR9	CpG	biotin (chem.)	SA	biotin (chem.)	OVA	Th1: IFN-γ, OVA-specific IgG, lowered IgG2a:IgG1 ratio	mouse (in vivo)	[62]
CD21	C3dg	biotin (BirA)	SA	-	-	positively or negatively regulate B cell signaling and SA-specific immune responses	mice (in vitro, in vivo)	[64,65]
CD21	C3dg	biotin (BirA)	SA	biotin (chem.)	IgM	augmented B cell $[Ca^{2+}]$ responses	human (in vitro)	[66]
4-1BB	4-1BBL	fusion	cSA	physical	HPV-16 E7 peptide	enhaned antigen uptake by DC CD8 T cell and memory response.	mouse (in vitro, in vivo)	[67,68,69]
4-1BB	4-1BBL	fusion	cSA	biotin (BirA)	HPV-16 E7 peptide and survivin	and CD4 and CD8 enhanced antitumor response	mouse (in vivo)	[69]
CD40	CD40L	fusion	cSA	-	-	upregulation of MHCII on Mf and B cells IL-1β, IL-6 production by Mf, production of iNOS in IFNγ-primed Mf	human, mouse (in vitro)	[70]
CD28, CD152	B7-Fc	biotin (chem.)	NA	biotin (chem.)	anti-TAA antibody	expansion of tumor-specific CTLs *in vitro* T-cell immunity *in vivo*	mouse (in vitro, in vivo)	[71]
lipid raft	*N*-alkyl 3*β*-cholesterylamine	biotin (chem.)	SA	-	-	clathrin-mediated endocytosis of SA	human and hamster (in vitro)	[72]

Table IV. Targeting strategies based on particles

Target receptor	Targeting unit	Conjugation 1	Center	Conjugation 2	Cargo	Effect	Species	Ref.
Particles without targeting units								
-	iscom (Ni^{2+} - loaded)	His6 (fused to SA)	SA	biotin (BirA)	Biotin-SRS2' (biotinylated *Neospora caninum* antigen SRS2')	humoral and cellular responses, IgG1 and IgG2a, IFNγ, IL-4 protective immunity	mouse (in vivo)	[74,75,76]
-	iscom (Ni^{2+} - loaded)	His6 (fused to SA)	SA	biotin (BirA)	Biotin-M5 (biotinylated *Plasmodium falciparum* antigen M5 peptide)	recombinant M5-specific ab response	mouse (in vivo)	[74,76]
-	iscom (biotinylated)	Biotin (chem)	SA	fusion	His6-SA-SRS2' (*Neospora caninum* antigen SRS2' fused to SA and a HHis peptide)	humoral and strong cellular responses, IgG1 and IgG2a, IFNγ, IL-4 protective immunity	mouse (in vivo)	[75,76]
-	iscom (Ni^{2+} - loaded)	His6 (fused to SA)	SA	fusion	His6-SA-SRS2' (*Neospora caninum* antigen SRS2' fused to SA and a HHis peptide)	humoral and cellular responses, IgG1 and IgG2a, IFNγ, IL-4 protective immunity	mouse (in vivo)	[76]
-	liposome	biotin (chem.)	A	-	-	enhanced liposome retention in draining lymph nodes	mouse	[78]
-	TMV (Tobacco mosaic virus) particles	biotin (chem.)	SA	fusion	GFP	enhanced total IgG response	mouse, guineapig (in vivo)	[80]

Table IV. (Continued)

Target receptor	Targeting unit	Conjugation 1	Center	Conjugation 2	Cargo	Effect	Species	Ref.
Particles without targeting units								
-	TMV (Tobacco mosaic virus) particles	biotin (chem.)	SA	fusion	N-terminal fragment of Canine oral papillomavirus L2 protein (COPV $L2_{61\text{-}171}$)-SA fusion protein	enhanced total IgG response	mouse (in vivo)	[80]
-	RBC (red blood cell)	biotin (chem.)	A	biotin (chem.)	HIV virus Tat protein	effective and rapid internalization of RBC-Tat by DCs, strong CD4+ and CD8+ T-cell proliferation	human (in vitro)	[81]
-	artificial APCs (magnetic beads)	anti-Class II Ab	SA	biotin (BirA)	HLA DR-Flu HA peptide	strong $CD4^+$ T cell stimulation	human (in vitro)	[83]
-	artificial APCs (magnetic beads)	anti-SA Ab	SA	biotin (BirA)	HLA DR-Flu HA peptide	strong $CD4^+$ T cell stimulation	human (in vitro)	[83]
-	microspheres	chem.	SA	Biotin (chem.)	Myc-His6 peptide	Enhanced IgG1, IgG3 antibody response	Mouse (in vivo)	[23]
Particles with antibodies								
hFcγRI	scFv	fusion	SA	biotin (chem.)	Tetanus toxoid, or tetanus toxoid C-fragment, adsorbed onto microspheres	Enhance T cell activation	human (in vitro)	[54]

Table IV. (Continued)

Target receptor	Targeting unit	Conjugation 1	Center	Conjugation 2	Cargo	Effect	Species	Ref.
Particles with antibodies								
mCR1/2	scFv	Biotin (chem.)	SA-coated polystyrene bead	-	-	Similar IgG1response and moderat IgG3 compared to non-targting group	mouse (in vivo)	[23]
mFcyRII/III	scFv	Biotin (chem.)	SA-coated polystyrene bead	-	-	High level of anti-streptavidin IgG1, moderat IgG3, compared to non targting group	mouse (in vivo)	[23]
mFcγRII/III and CR1/2	scFv	Biotin (chem.)	SA-coated polystyrene bead	-	-	Similar anti-SA humoral immune response, than individual targeting of mFcγRII and III	mouse (in vivo)	[23]
CD34 and CD54	mIgG on liposome	Biotin (chem.)	SA	biotin (chem.)	NO in liposome	Enhanced monocytes migration	human (in vitro)	[55]
CD28	IgG	Biotin (chem.)	SA-coated polystyrene bead	biotin (BirA)	HLA-A2/HA-1 peptide	induction of cytotoxic, expandable, and long-lasting HA-1–specific $CD8^{+}$ effector memory T	human (in vitro)	[56]
Particles with receptor ligands								
CD28, CD152	CD80	chemical	SA coated latex bead	biotin (BirA)	HLA-A2/HA-1 and HLA-A2/HA-2	Enhanced IFNγ and expansion and increased cytotoxicity of CTL	human (in vitro)	[57]
β_2 integrin	CD54	chemical	SA coated latex bead	biotin (BirA)	HLA-A2/HA-1 and HLA-A2/HA-2	Enhanced IFNγ and expansion and increased cytotoxicity of CTL	human (in vitro)	[57]
SRA1	polyG	biotin (chem)	SA	biotin (chem.)	ABA triblock copolymer based nanocontainer	targeting, stealth (very low aspecific binding and uptake)	human (in vitro)	[84]

Table IV. (Continued)

Target receptor	Targeting unit	Conjugation 1	Center	Conjugation 2	Cargo	Effect	Species	Ref.
Particles with receptor ligands								
ChemR23	chemerin-peptid	biotin (synth.)	NA	biotin (BirA)	recomb. Adenovirus (rAd) vector, that encodes Ag (immunodominant human cytomegalovirus (HCMV) pp65 protein) / reporter (GFP)	transfection (+enhance maturation, APC of ag of DC & DC stim. cytokine production of CMV.specific mem CD8+ T in vitro)	human (in vitro)	[86]
$\alpha_v\beta_3$ integrin	ligand peptide of $\alpha_v\beta_3$ integrin	biotin (synth.)	NA	biotin (BirA)	recomb. Adenovirus (rAd) vector, that encodes Ag (immunodominant human cytomegalovirus (HCMV) pp65 protein) / reporter (GFP)	transfection (+enhance maturation, APC of ag of DC & DC stim. cytokine production of CMV.specific mem CD8+ T in vitro)	human (in vitro)	[86]
DC-SIGN	IgG	biotin (chem)	NA	biotin (BirA)	recomb. Adenovirus (rAd) vector, that encodes Ag (immunodominant human cytomegalovirus (HCMV) pp65 protein) / reporter (GFP)	transfection (+enhance maturation, APC of ag of DC & DC stim. cytokine production of CMV.specific mem CD8+ T in vitro)	human (in vitro)	[86]
Fas	FasL	fusion	cSA	biotin (chem.)	splenocytes or heart allograft	depletion of Fas+ cells in vitro, depletion of alloreactive lymphocytes, induction of Treg cells, prevention of allograft rejection in vivo	mouse, rat (in vitro, in vivo) human (in vitro)	[89,94]

Table IV. (Continued)

Target receptor	Targeting unit	Conjugation 1	Center	Conjugation 2	Cargo	Effect	Species	Ref.
Particles with receptor ligands								
CD28, CD152	CD80	fusion	cSA	biotin (chem.)	tumor cells	CD8 Tc antitumor response long term protective mem.	murine (in vivo) human (in vitro)	[88,95]
4-1BB and	4-1BBL and LIGHT	fusion	cSA	biotin (chem.)	HPV E7 expressing TC-1 tumor cells	primary and memory T-cell and antibody responses	mouse	[96]

The method of coupling of the targeting units is indicated in the columns Link 1 and Link 2. Chemically conjugated biotin (chem), enzymatically introduced biotin (BirA) or synthetically added biotin (synth) was used in the studies.

Streptavidin, SA; avidin, A; neutravidin, NA; core streptavidin, cSA; dendritic cell, DC; antibody, Ab.

Conclusion

SA is ideal for the generation of modular immune-targeting complexes. Current strategies employ recombinant fusion proteins and enzymatically or synthetically biotinylated modules, which allow the construction of complexes with strictly controlled stoichiometry. Combination of this technology with various forms of nanoparticle-based experimental and clinical immunological approaches opens the way for building highly sophisticated targeting agents.

Acknowledgments

This work was supported by the grants NKTH-OTKA K68617, TÁMOP 421/B and by the Hungarian Academy of Sciences.

References

[1] Green M. N. , 1978, "Avidin and streptavidin," *Methods in enzymology*, 184, pp. 51-67.

[2] Wilchek M., Bayer E. A., and Livnah O. , 2006, "Essential of biorecognition: the (strept)avidin-biotin system as a model for protein-protein and protein-ligand interaction," *Immunology letters*, 103(1), pp. 27-31.

[3] Laitinen O. H., Nordlund H. R., and Kulomaa M. S. , 2006, "Genetically engineered avidins and streptavidins," *Cellular and Molecular Life Sciences*, 63, pp. 2992-3017.

[4] Gonzalez M., Argarana C., and Fidelio G. , 1999, "Extremely high thermal stability of streptavidin and avidin upon biotin binding," *Biomolecular Engineering*, 16(1-4), pp. 67-72.

[5] Rybak J., Scheurer S., Neli D., and Elia G. , 2004, "Purification of biotinylated proteins on streptavidin resin: a protocol for cantitative elution," *Proteomics*, 4(8), pp. 2296-2299.

[6] Ellison K., Hinton J., Hubbard S., and Beynon R. , 1995, "Limited proteolysisi of native proteins: interaction between avidin and proteinase K," *Protein science*, 4(7), pp. 1337-1345.

[7] Laitinen O. H., Nordlund H. R., Hytönen V. P., and Kulomaa M. S. , 2007, "Brave new (strept)avidins in biotechnology.," *Trends in biotechnology*, 25(6), pp. 269-277.

[8] Nordlund H. R., Laitinen O. H., Uotila S. T., Nyholm T., Hytönen V. P., Slotte J. P., and Kulomaa M. S. , 2003, "Enhancing the thermal stability of avidin. Introduction of disulfide bridges between subunit interfaces.," *The Journal of Biological Chemistry,* 278(4), pp. 2479-2483.

[9] Howarth M., Chinnapen D. J., Gerrow K., Dorrestein P. C., Grandy M. R., Kelleher N. L., El-Husseini A., and Ting A. Y. , 2006, "A monovalent streptavidin with a single femtomolar biotin binding site.," *Nature Methods*, 3(4), pp. 267-273.

[10] Chivers C. E., Crozat E., Chu C., Moy V. T., Sherratt D. J., and Howarth M. , 2010, "A streptavidin variant with slower biotin dissociation and increased mechanostability," *Nature methodes*, 7(5), pp. 391-393.

[11] Wu S., and Wong S. , 2005, "Engineering soluble monomeric streptavidin with reversible biotin binding capability.," *The Journal of Biological Chemistry,* 280(24), pp. 23225-23231.

[12] Bruch R. C., and White H. B. , 1982, "Compositional and structural heterogeneity of avidin glycopeptides.," *Biochemistry,* 21(21), pp. 5334-41.

[13] Delange R. J. , 1970, "Egg White Avidin," *The Journal of Biological Chemistry,* 245(5), pp. 907-916.

[14] Green N. M., and Toms E. J. , 1970, "Purification and crystallization of avidin.," *Biochemical Journal,* 118(1), pp. 67-70.

[15] Green M. N. , 1975, "Avidin," *Adv. Protein Chemistry*, 29, pp. 85-133.

[16] Green M. N. , 1963, "Avidin," *Journal of Biochemistry*, 89, pp. 599-609.

[17] Marttila A. T., Laitinen O. H., Airenne K. J., Kulik T., Bayer E. A., Wilchek M., and Kulomaa M. S. , 2000, "Recombinant NeutraLite Avidin : a non-glycosylated , acidic mutant of chicken avidin that exhibits high affinity for biotin and low non-specific binding properties," *FEBS Letters*, 467, pp. 31-36.

[18] Schechter B, Silberman R, Arnon R W. M. , 1990, "Tissue distribution of avidin and streptavidin injected to mice," *Europien Journal of Biochemistry*, 189, pp. 327-331.

[19] Ashwell G., and Harford J. , 1982, "Carbohydrate-specific receptors of the liver.," *Annual Review of Biochemistry,* 51, pp. 531-54.

[20] Breitz H. B., Weiden P. L., Beaumier P. L., Axworthy D. B., Seiler C., Su F. M., Graves S., Bryan K., and Reno J. M. , 2000, "Clinical optimization of pretargeted radioimmunotherapy with antibody-streptavidin conjugate and 90Y-DOTA-biotin.," *Journal of Nuclear Medicine*, 41(1), pp. 131-40.

[21] Subramanian N., and Adiga P. R. , 1997, "Mapping the common antigenic determinants in avidin and streptavidin.," *Biochemistry and Molecular Biology International,* 43(2), pp. 375-382.

[22] Meyer D. L., Schultz J., Lin Y., Henry A., Sanderson J., Jackson J. M., Goshorn S., Rees A. R., and Graves S. S. , 2001, "Reduced antibody response to streptavidin through site-directed mutagenesis," *Protein Science,* 10, pp. 491-503.

[23] Szekeres Z., Herbáth M., Angyal A., Szittner Z., Virág V., Balogh P., Erdei A., and Prechl J. , 2010, "Modulation of immune response by combined targeting of complement receptors and low-affinity Fcgamma receptors.," *Immunology Letters,* 130(1-2), pp. 66-73.

[24] Laitinen O. H., Nordlund H. R., Hytönen V. P., Uotila S. T., Marttila A. T., Savolainen J., Airenne K. J., Livnah O., Bayer E. A., Wilchek M., and Kulomaa M. S. , 2003, "Rational design of an active avidin monomer.," *The Journal of Biological Chemistry,* 278(6), pp. 4010-4.

[25] Salmaso S., Semenzato A., Bersania S., Chinol M., Paganelli G., and Caliceti P. , 2005, "Preparation and characterization of active site protected poly (ethylene glycol)–avidin bioconjugates," *Biochemica and Biophysica Acta,* 1726, pp. 57 - 66.

[26] Skea D. L., and Barber B. H. , 1993, "Studies of the adjuvant-independent antibody response to immunotargeting. Target structure dependence, isotype distribution, and induction of long term memory.," *The Journal of Immunology,* 151(7), pp. 3557-3568.

[27] Carayanniotis G., and Barber B. H. , 1987, "Adjuvant-free IgG responses induced with antigen coupled to antibodies against class II MHC," *Nature,* 327, pp. 59-61.

[28] Carayanniotis G., Vizi E., Parker J. M., Hodges R. S., and Barber B. H. , 1988, "Delivery of synthetic peptides by anti-class II MHC monoclonal antibodies induces specific adjuvant-free IgG responses in vitro," *Molecular Immunology,* 25(9), pp. 907-9011.

[29] Skea D. L., Douglas A. R., Skehel J. J., and Barber B. H. , 1993, "The immunotargeting approach to adjuvant-independent immunization with influenza haemagglutinin.," *Vaccine*, 11(10), pp. 994-1002.

[30] Tiong G. K., Gill H. S., Lofthouse S., and Puri N. K. , 1993, "Comparison of conventional adjuvants and 'adjuvant-free' monoclonal antibody targeting for stimulating antibody responses against a conjugate of luteinizing hormone releasing hormone and avidin.," *Vaccine,* 11(4), pp. 425-430.

[31] Carayanniotis G., Skea D. L., Luscher M. A., and Barber B. H. , 1991, "Adjuvant-independent immunization by immunotargeting antigens to MHC and non-MHC determinants in vivo," *Molecular Immunology*, 28(3), pp. 261-267.

[32] Frleta D., Demian D., and Wade W. F. , 2001, "Class II-targeted antigen is superior to CD40-targeted antigen at stimulating humoral responses in vivo.," *International Immunopharmacology*, 1(2), pp. 265-75.

[33] Kuroda K., Liu H., Kim S., Guo M., Navarro V., and Bande N. H. , 2010, "Saporin toxin-conjugated monoclonal antibody targeting prostate-specific membrane antigen has potent anticancer activity," *The Prostate,* 70(12), pp. 1286-1294.

[34] Muzykantov V. R., Christofidou-Solomidou M., Balyasnikova I., Harshaw D. W., Schultz L., Fisher A. B., and Albelda S. M. , 1999, "Streptavidin facilitates internalization and pulmonary targeting of an anti-endothelial cell antibody (platelet-endothelial cell adhesion molecule 1): a strategy for vascular immunotargeting of drugs.," *Proc. Natl. Acad. Sci. USA,* 96(5), pp. 2379-2384.

[35] Wiewrodt R., Thomas A. P., Cipelletti L., Christofidou-Solomidou M., Weitz D. A., Feinstein S. I., Schaffer D., Albelda S. M., Koval M., and Muzykantov V. R. , 2002, "Size-dependent intracellular immunotargeting of therapeutic cargoes into endothelial cells.," *Blood*, 99(3), pp. 912-922.

[36] Shuvaev V. V., Dziubla T., Wiewrodt R., and Muzykantov V. R. , 2004, "Streptavidin-biotin crosslinking of therapeutic enzymes with carrier antibodies: nanoconjugates for protection against endothelial oxidative stress.," *Methods In Molecular Biology*, 283, pp. 3-19.

[37] Rusakiewicz S., Aubert G., Clark R. E., Madrigal A. J., Dodi A. I., and Travers P. J. , 2009, "Soluble HLA/peptide monomers cross-linked with co-stimulatory antibodies onto a streptavidin core molecule efficiently stimulate antigen-specific T cell responses.," *Cancer immunology immunotherapy CII*, 58(9), pp. 1459-1470.

[38] Kipriyanov S. M., Breitling F., Little M., and Dübel S. , 1995, "Single-chain antibody streptavidin fusions: tetrameric bifunctional scFv-complexes with biotin binding activity and enhanced affinity to antigen.," *Human Antibodies And Hybridomas*, 6(3), pp. 93-101.

[39] Dübel S., Breitling F., Kontermann R., Schmidt T., Skerra a., and Little M. , 1995, "Bifunctional and multimeric complexes of streptavidin fused to single chain antibodies (scFv).," *Journal of immunological methods*, 178(2), pp. 201-9.

[40] Adamova E., Walsh M., Gosselin D., Hale K., Preissler M., Graziano R., and Gosselin E. , 2005, "Enhanced antigen-specific antibody and cytokine responses when targeting

antigen to human Fcgamma receptor type I using anti-human Fcgamma receptor Type I-streptavidin fusion protein in an adjuvant-free system," *Immunol. Invest,* 34(4), pp. 417-29.

[41] Angyal A., Szekeres Z., Balogh P., Neer Z., Szarka E., Virag V., Medgyesi D., Prechl J., and Sarmay G. , 2009, "CD16/32-specific biotinylated 2.4G2 single-chain Fv complexed with avidin-FITC enhances FITC-specific humoral immune response in vivo in a CD16-dependent manner," *International Immunology*, (8), pp. 1-10.

[42] Wang WW, Das D S. M. , 2009, "A versatile bifunctional dendritic cell targeting vaccine vector," *Vaccine*, 6(1), pp. 158-172.

[43] Savage P., Cowburn P., Clayton A., Man S., Mcmichael A., Lemoine N., Epenetos A., and Ogg G. , 2002, "Induction of viral and tumour specific CTL responses using antibody targeted HLA class I peptide complexes," *British Journal of Cancer,* 86, pp. 1336 - 1342.

[44] Savage P., Dyson J., Milrain M., Mathews D., King B., Chan H. T., Barber L., Epenetos A., Ogg G., McMichael A., Glennie M. J., and French R. R. , 2007, "Immunotherapy with antibody-targeted HLA class I complexes: results of in vivo tumour cell killing and therapeutic vaccination.," *Tumour biology the journal of the International Society for Oncodevelopmental Biology and Medicine*, 28(4), pp. 205-211.

[45] Robert B., Guillaume P., Luescher I., Romero P., and Mach J. P. , 2000, "Antibody-conjugated MHC class I tetramers can target tumor cells for specific lysis by T lymphocytes.," *European Journal of Immunology,* 30(11), pp. 3165-3170.

[46] Schultz J., Lin Y., Sanderson J., Zuo Y., Stone D., Mallett R., Wilbert S., and Axworthy D. , 2000, "A tetravalent single-chain antibody-streptavidin fusion protein for pretargeted lymphoma therapy.," *Cancer research,* 60(23), pp. 6663-9.

[47] Zhang M., Zhang Z., Garmestani K., Schultz J., Axworthy D. B., Goldman C. K., Brechbiel M. W., Carrasquillo J. A., and Waldmann T. A. , 2003, "Pretarget radiotherapy with an anti-CD25 antibody-streptavidin fusion protein was effective in therapy of leukemia/lymphoma xenografts.," *Proceedings of the National Academy of Sciences of the United States of America,* 100(4), pp. 1891-1895.

[48] Lin Y., Pagel J. M., Axworthy D., Pantelias A., Hedin N., and Press O. W. , 2006, "A genetically engineered anti-CD45 single-chain antibody-streptavidin fusion protein for pretargeted radioimmunotherapy of hematologic malignancies.," *Cancer Research,* 66(7), pp. 3884-3892.

[49] Sano T., Pandori M. W., Chen X., Smith C. L., and Cantor C. R. , 1995, "Recombinant core streptavidins. A minimum-sized core streptavidin has enhanced structural stability and higher accessibility to biotinylated macromolecules.," *The Journal of biological chemistry,* 270(47), pp. 28204-9.

[50] Hainsworth J. E., Harrison P., and Mather S. J. , 2006, "Novel preparation and characterization of a trastuzumab-streptavidin conjugate for pre-targeted radionuclide therapy.," *Nuclear Medicine Communications,* 27(5), pp. 461-471.

[51] Yao Z., Zhang M., Axworthy D. B., Wong K. J., Garmestani K., Park L., Park C. W., Mallett R. W., Theodore L. J., Yau E. K., Waldmann T. A., Brechbiel M. W., Paik C. H., Pastan I., and Carrasquillo J. A. , 2002, "Radioimmunotherapy of A431 xenografted mice with pretargeted B3 antibody-streptavidin and (90)Y-labeled 1,4,7,10-

tetraazacyclododecane-N,N',N",N'"-tetraacetic acid (DOTA)-biotin.," *Cancer Research*, 62(20), pp. 5755-5760.

[52] Ferreira C. S., Matthews C. S., and Missailidis S. , 2006, "DNA aptamers that bind to MUC1 tumour marker: design and characterization of MUC1-binding single-stranded DNA aptamers.," *Tumour biology the journal of the International Society for Oncodevelopmental Biology and Medicine,* 27(6), pp. 289-301.

[53] Bruno J. G. , 2009, "Aptamer-biotin-streptavidin-C1q complexes can trigger the classical complement pathway to kill cancer cells.," In vitro Cellular and Developmental Biology. *Animal*, pp. 107-113.

[54] Walsh M. C., Banas J. A., Mudzinski S. P., Preissler M. T., Graziano R. F., and Gosselin E. J. , 2003, "A two-component modular approach for enhancing T-cell activation utilizing a unique anti-FcgammaRI-streptavidin construct and microspheres coated with biotinylated-antigen.," *Biomolecular Engineering*, 20(1), pp. 21-33.

[55] Klegerman M. E., Wassler M., Huang S., Zou Y., Kim H., Shelat H. S., Holland C. K., Geng Y., and McPherson D. D. , 2010, "Liposomal modular complexes for simultaneous targeted delivery of bioactive gases and therapeutics.," *Journal of controlled release : official journal of the Controlled Release Society,* 142(3), pp. 326-31.

[56] Schilbach K., Kerst G., Walter S., Eyrich M., Wernet D., Handgretinger R., Xie W., Rammensee H., Müller I., Bühring H., and Niethammer D. , 2005, "Cytotoxic minor histocompatibility antigen HA-1-specific CD8+ effector memory T cells: artificial APCs pave the way for clinical application by potent primary in vitro induction.," *Blood*, 106(1), pp. 144-149.

[57] Oosten L. E., Blokland E., Van Halteren A. G., Curtsinger J., Mescher M. F., Falkenburg J. H., Mutis T., and Goulmy E. , 2004, "Artificial antigen-presenting constructs efficiently stimulate minor histocompatibility antigen-specific cytotoxic T lymphocytes.," *Blood,* 104(1), pp. 224-226.

[58] Nash a. D., Lofthouse S. a., Barcham G. J., Jacobs H. J., Ashman K., Meeusen E. N., Brandon M. R., and Andrews a. E. , 1993, "Recombinant cytokines as immunological adjuvants.," *Immunology and cell biology*, 71 (Pt 5)(May), pp. 367-79.

[59] Andrews A., Bowles V., Brandon M., Lofthouse S., and Nash A. , 1994, "Vaccine," Production and in vivo use of recombinant ovine IL-1 beta as an immunological adjuvant, 12(1), pp. 14-22.

[60] Lofthouse S. a., Andrews a. E., Elhay M. J., Bowles V. M., Meeusen E. N., and Nash a. D. , 1996, "Cytokines as adjuvants for ruminant vaccines.," *International journal for parasitology,* 26(8-9), pp. 835-42.

[61] Heath a. W., and Playfair J. H. , 1990, "Conjugation of interferon-gamma to antigen enhances its adjuvanticity.," *Immunology,* 71(3), pp. 454-6.

[62] Klinman D. M., Barnhart K. M., and Conover J. , 1999, "CpG motifs as immune adjuvants.," *Vaccine*, 17(1), pp. 19-25.

[63] Klinman D. M., Currie D., Gursel I., and Verthelyi D. , 2004, "Use of CpG oligodeoxynucleotides as immune adjuvants.," *Immunological reviews*, 199, pp. 201-16.

[64] Haas K. M., Toapanta F. R., Oliver J. a., Poe J. C., Weis J. H., Karp D. R., Bower J. F., Ross T. M., and Tedder T. F. , 2004, "Cutting edge: C3d functions as a molecular

adjuvant in the absence of CD21/35 expression.," *Journal of immunology* (Baltimore, Md. : 1950), 172(10), pp. 5833-7.

[65] Lee Y., Haas K. M., Gor D. O., Ding X., Karp D. R., Greenspan N. S., Poe J. C., and Tedder T. F. , 2005, "Complement component C3d-antigen complexes can either augment or inhibit B lymphocyte activation and humoral immunity in mice depending on the degree of CD21/CD19 complex engagement.," *Journal of immunology* (Baltimore, Md. : 1950), 175(12), pp. 8011-23.

[66] Henson S. E., Smith D., Boackle S. a., Holers V. M., and Karp D. R. , 2001, "Generation of recombinant human C3dg tetramers for the analysis of CD21 binding and function.," *Journal of immunological methods*, 258(1-2), pp. 97-109.

[67] Elpek K. G., Yolcu E. S., Franke D. D., Lacelle C., Schabowsky R., and Shirwan H. , 2007, "Ex vivo expansion of CD4+CD25+FoxP3+ T regulatory cells Based on Synergy between IL-2 and 4-1BB Signaling 1," *The Journal of Immunology*, 179, pp. 7295-7304.

[68] Sharma R. K., Elpek K. G., Yolcu E. S., Schabowsky R., Zhao H., Bandura-Morgan L., and Shirwan H. , 2009, "Costimulation as a platform for the development of vaccines: a peptide-based vaccine containing a novel form of 4-1BB ligand eradicates established tumors," *Cancer res.*, 69(10), pp. 4319-26.

[69] Sharma R. K., Schabowsky R., Srivastava A. K., Elpek K. G., Madireddi S., Zhao H., Zhong Z., Miller R. W., Macleod K. J., Yolcu E. S., and Shirwan H. , 2010, "4-1BB Ligand as an Effective Multifunctional Immunomodulator and Antigen Delivery Vehicle for the Development of Therapeutic Cancer Vaccines.," *Cancer Research,* 70(4), pp. 1-10.

[70] Kilinc M. O., Mukundan L., Yolcu E. S., Singh N. P., Suttles J., and Shirwan H. , 2006, "Generation of a multimeric form of CD40L with potent immunostimulatory activity using streptavidin as a chaperon," *Experimental and Molecular Pathology*, 80, pp. 252 - 261.

[71] Moro M., Gasparri A. M., Pagano S., Bellone M., Tornaghi P., Veglia F., Corti A., Casorati G., and Dellabona P. , 1999, "Induction of Therapeutic T-Cell Immunity by Tumor Targeting with Soluble Recombinant B7-Immunoglobulin Costimulatory Molecules 1," *Cancer*, (15), pp. 2650 -2656.

[72] Hussey S. L., and Peterson B. R. , 2002, "Efficient delivery of streptavidin to mammalian cells: clathrin-mediated endocytosis regulated by a synthetic ligand.," *Journal of the American Chemical Society*, 124(22), pp. 6265-73.

[73] Wikman M., Friedman M., Pinitkiatisakul S., Andersson C., Hemphill A., Lövgren-Bengtsson K., Lundén A., and Ståhl S. , 2005, "General strategies for efficient adjuvant incorporation of recombinant subunit immunogens.," *Vaccine,* 23(17-18), pp. 2331-5.

[74] Wikman M, Friedman M, Pinitkiatisakul S, Hemphill A, Lövgren-Bengtsson K, Lundén A, and Ståhl S, 2005, "Applying biotin-streptavidin binding for iscom (immunostimulating complex) association of recombinant immunogens.," *Biotechnol. Appl. Biochem.*, 41(Pt 2), pp. 163-74.

[75] Pinitkiatisakul S., Mattsson J. G., Wikman M., Friedman M., Bengtsson K. L., Ståhl S., and Lundén A. , 2005, "Immunisation of mice against neosporosis with recombinant NcSRS2 iscoms.," *Veterinary parasitology*, 129(1-2), pp. 25-34.

[76] Pinitkiatisakul S., Friedman M., Wikman M., Mattsson J. G., Lövgren-Bengtsson K., Ståhl S., and Lundén A. , 2007, "Immunogenicity and protective effect against murine

cerebral neosporosis of recombinant NcSRS2 in different iscom formulations.," *Vaccine*, 25(18), pp. 3658-68.

[77] Lofthouse S. a., Kajihara M., Nagahara S., Nash a., Barcham G. J., Sedgmen B., Brandon M. R., and Sano a. , 2002, "Injectable silicone implants as vaccine delivery vehicles.," *Vaccine*, 20(13-14), pp. 1725-32.

[78] Phillips W., Klipper R., and Goins B. , 2000, "Novel method of greatly enhanced delivery of liposomes to lymph nodes.," *J. Pharmacol. Exp. Ther.*, 295(1), pp. 309-13.

[79] Pogue G., Lindbo J., Garger S., and Fitzmaurice W. , 2002, "Making an ally from an enemy: plant virology and the new agriculture.," *Annu. Rev. Phytopathol.*, 40, pp. 45-74.

[80] Smith M. L., Lindbo J. A., Dillard-telm S., Brosio P. M., Lasnik A. B., Mccormick A. A., Nguyen L. V., and Palmer K. E. , 2006, "Modified Tobacco mosaic virus particles as scaffolds for display of protein antigens for vaccine applications," *Virology,* 348, pp. 475 - 488.

[81] Corinti S., Chiarantini L., Dominici S., Laguardia M., Magnani M., and Girolomoni G. , 2002, "Erythrocytes deliver Tat to interferon-gamma-treated human dendritic cells for efficient initiation of specific type 1 immune responses in vitro.," *J. Leukoc. Biol.*, 71(4), pp. 652-8.

[82] Novak E. J., Liu a. W., Nepom G. T., and Kwok W. W. , 1999, "MHC class II tetramers identify peptide-specific human CD4(+) T cells proliferating in response to influenza A antigen.," *The Journal of clinical investigation*, 104(12), pp. R63-7.

[83] Maus M. V., Riley J. L., Kwok W. W., Nepom G. T., and June C. H. , 2003, "HLA tetramer-based artificial antigen-presenting cells for stimulation of CD4 Ƨ T cells," *Clinical Immunology*, 106, pp. 16 -22.

[84] Broz P., Benito S., Saw C., Burger P., Heider H., Pfisterer M., Marsch S., Meier W., and Hunziker P. , 2005, "Cell targeting by a generic receptor-targeted polymer nanocontainer platform.," *J. Control Release.*, 102(2), pp. 475 - 88.

[85] Parrott M., Adams K., Mercier G., Mok H., Campos S., and Barry M. , 2003, "Metabolically biotinylated adenovirus for cell targeting, ligand screening, and vector purification.," *Mol. Ther.*, 8(4), pp. 688-700.

[86] Maguire C., Sapinoro R., Girgis N., Rodriguez-Colon S., Ramirez S., Williams J., and Dewhurst S. , 2006, "Recombinant adenovirus type 5 vectors that target DC-SIGN, ChemR23 and alpha(v)beta3 integrin efficiently transduce human dendritic cells and enhance presentation of vectored antigens.," *Vaccine*, 24(5), pp. 671-82.

[87] Schabowsky R., Sharma R. K., Madireddi S., Srivastava A., Yolcu E. S., and Shirwan H. , 2009, "ProtEx technology for the generation of novel therapeutic cancer vaccines.," *Experimental and molecular pathology*, 86(3), pp. 198-207.

[88] Singh N. P., Yolcu E. S., Taylor D. D., Gercel-Taylor C., Metzinger D. S., Dreisbach S. K., and Shirwan H. , 2003, "A novel approach to cancer immunotherapy: tumor cells decorated with CD80 generate effective antitumor immunity.," *Cancer research,* 63(14), pp. 4067-73.

[89] Yolcu E. S., Askenasy N., Singh N. P., Cherradi S. L., and Shirwan H. , 2002, "Cell membrane modification for rapid display of proteins as a novel means of immunomodulation: FasL-decorated cells prevent islet graft rejection.," *Immunity,* 17(6), pp. 795-808.

[90] Askenasy N., Yolcu E., Wang Z., and Shirwan H. , 2003, "Display of Fas Ligand Protein on Cardiac Vasculature as a Novel Means of Regulating Allograft Rejection.," *Circulation*, 107(11), pp. 1525-31.

[91] Askenasy N., Yolcu E. S., Yaniv I., and Shirwan H. , 2005, "Review in translational hematology Induction of tolerance using Fas ligand : a double-edged immunomodulator," *Blood,* 105(4), pp. 1396-1404.

[92] Franke D., Yolcu E., Alard P., Kosiewicz M., and Shirwan H. , 2007, "A novel multimeric form of FasL modulates the ability of diabetogenic T cells to mediate type 1 diabetes in an adaptive transfer model.," *Molecular Immunology*, 44(11), pp. 2884-92.

[93] Yolcu E., Gu X., Lacelle C., Zhao H., Bandura-Morgan L., Askenasy N., and Shirwan H. , 2008, "Induction of tolerance to cardiac allografts using donor splenocytes engineered to display on their surface an exogenous fas ligand protein.," *J. Immunol.*, 181(2), pp. 931-9.

[94] Askenasy N., Yolcu E., Shirwan H., Wang Z., and Farkas D. , 2003, "Cardiac allograft acceptance after localized bone marrow transplantation by isolated limb perfusion in nonmyeloablated recipients.," *Stem Cells*, 21(3), pp. 200-207.

[95] Singh N., Miller R., Yolcu E., Killinc M., Oechsli M., Huseby R., Taylor D., Perry M., Larocca R., and Shirwan H. , 2006, "Primary tumor cells resected from cancer patients and decorated with a novel form of CD80 protein serve as effective antigen-presenting cells for the induction of autologous T cell immune responses ex vivo.," *Hum. Gene Ther.,* 17(3), pp. 334-46.

[96] Sharma R., Yolcu E., Elpek K., and Shirwan H. , 2010, "Tumor cells engineered to codisplay on their surface 4-1BBL and LIGHT costimulatory proteins as a novel vaccine approach for cancer immunotherapy.," *Cancer Gene Ther.* 17(10), pp. 730-741.

[97] Oosten L. E., Blokland E., Van Halteren A. G., Curtsinger J., Mescher M. F., Falkenburg J. H., Mutis T., and Goulmy E. , 2004, "Artificial antigen-presenting constructs efficiently stimulate minor histocompatibility antigen-specific cytotoxic T lymphocytes.," *Blood,* 104(1), pp. 224-226.

In: Biochemistry Research Updates
Editor: Simon J. Baginski

ISBN 978-1-61209-700-8

Chapter 3

ELECTROCHEMICAL AND OPTICAL BIOSENSORS BASED ON STREPT(AVIDIN)-BIOTIN AFFINITY

XinRan Cheng and Kagan Kerman
University of Toronto, Canada

ABSTRACT

Streptavidin is a ~53kDa tetrameric protein purified from the bacterium *Streptomyces avidinii.* It is well known to contain four identical subunits, each containing a single binding site that binds very tightly to the small molecule biotin, also known as vitamin H. This high affinity (K_{assoc}= $10^{14}M^{-1}$) has been applied widely in biotechnology and biosensors. This review will cover the latest trends in the application of streptavidin-biotin system in DNA sensors and immunosensors using electrochemical and optical techniques. Streptavidin is sometimes used in connection with magnetic beads in DNA sensing. By immobilizing the biotinylated DNA probes on streptavidin-conjugated magnetic beads, the target DNA can be easily isolated from a complex matrix using an external magnet. Electrochemical and fluorescence-based immunosensors utilize the biotinylated antibodies with streptavidin-conjugated enzymes (horseradish peroxidase, alkaline phosphatase, etc.) or streptavidin-conjugated fluorescent dyes as the source of analytical signals. Surface plasmon resonance (SPR) and Localized SPR, which provide label-free analysis of dynamic surface events, are valuable optical tools for studying interactions between biomolecules. In immunosensors, SPR is typically applied to detect the formation of a sandwich-type immune complex comprised of a capturing antibody immobilized on a sensor surface, the target biomolecule, and a detecting antibody. The well-studied streptavidin-biotin system serves as an excellent model for the development of LSPR biosensors using metal nanoparticles. This review will also describe the recent fundamental spectroscopic studies that reveal key relationships governing the LSPR spectral location and its sensitivity to the local environment, including changes on the substrate due to binding events. The immobilization of streptavidin on gold nanoparticles is detected with the changes in the peak wavelength and intensity of the LSPR absorption band. Further attachment of the biotinylated ligands on streptavidin-conjugated gold nanoparticles can be detected at high sensitivity and selectivity. The results from these biosensor studies guide the design of new sensing experiments, illustrated through applications of nanomaterials, in which researchers use both the streptavidin-biotin affinity and the advanced transduction techniques to detect molecules of chemical and biological relevance.

1. INTRODUCTION

The earliest research on avidin stems from nutritional studies in attempts to understand the malnutrition of rats that fed on large quantities of egg whites. It was later found that avidin was forming complexes with vitamin H and thus, depleting it and interfered with the animal's nutrition. [1] In 1940, the vitamin was chemically identified as biotin. Biotin is therefore, the lacking substance that resulted in dermatitis observed in the rats. In 1975, avidin was isolated from chicken egg white by Green. [2] They are also found in tissues of birds.

Avidin is a tetrameric glycoprotein with four identical subunits weighing 67KDa. The avidin monomer contains 128 amino acid residues and it is soluble in aqueous solutions. It is also stable over wide temperature and pH ranges. Avidin has one disulfide bridge per subunit and they could only be reduced after the tetramer is dissociated into subunits. [3]

Biotin is a vitamin that occurs naturally in every living cell. Only the bicyclic ring is required to be intact for the interaction with avidin. It has a molecular mass of 244.31 Da and its molecular composition is $C_{10}H_{16}N_2O_3S$. [4] Unlike soluble avidin, biotin only has a solubility of 220μg/mL in water at 25^0C. Biotin also has very insignificant chromophoric properties as seen from its absorbance of 0.111 at 250nm for every 1mg/mL. The carboxyl group on the valeric acid side chain is not involved in avidin binding and can be used for conjugation with proteins by biotinylation reagents. Since biotin is a small molecule, its conjugation to macromolecules does not affect conformation, size, or functionality. [5] Its zero absorbance of light from 260-280nm implies that the concentration of biotinylated proteins and DNA could be monitored directly at these wavelengths without interference. [4]

Streptavidin are produced from *Streptomyces avidinii* and was isolated by Chaiet and Wolf in 1964 [6] in an antibiotic screen. Streptavidin and avidin are closely related tetrameric proteins as thirty three percent of the core streptavidin are similar to the residues found in avidin.[4] Avidin and streptavidin both have the distinct characteristic of binding four biotin units with extremely high affinities (Kd $\sim10^{-13}$ and $\sim10^{-15}$) respectively despite the differences in amino acid composition. This is because crucial amino acid sequences and structures required for biotin binding are highly conserved. This binding corresponds to about 18mg of biotin bound per gram of streptavidin, which was equivalent to 1 mole of biotin bound per 14Kg of streptavidin. [6]The native form of streptavidin was isolated and shown to have a molecular weight of 75KDa. [7] A subunit of streptavidin comprises 159 amino acids, which is not much more than the subunit of avidin.

There are many differences between streptavidin and avidin. Unlike avidin, streptavidin's primary structure of its subunit does not have a disulfide bond. Avidin is very soluble in aqueous solutions but streptavidin is not. In turn, streptavidin is more resistant to dissociation by guanidine hydrochloride and denaturation by acids as compared to avidin. [2] The isoelectric points of avidin and streptavidin are pI ~10 and 5 respectively. Because avidin is a glycoprotein, it has higher nonspecific binding properties compared to streptavidin and thus, streptavidin is preferred in most applications using biotinylated molecules instead. In cases of steric hindrances, avidin may be preferred as studies show that binding sites on streptavidin is deeper and less accessible. [8] This implies that long-chain (LC) biotinylation reagents would be preferred in a streptavidin-biotin system.

1.1. Biotin Derivatives

The valeric acid side chain of biotin can be easily derivatized to add different reactive groups in order to attach biotin to other groups without affecting strept(avidin) binding to the biotin. Using these reactive groups, biotin can be conveniently attached to most proteins, DNA and other molecules. Biotinylation usually targets functional groups typically like sulfhydryls, carboxyls, oxidized carbohydrates groups of nucleic acid proteins and primary amines. For biomolecules, the most frequently used reagents are N-hydroxysuccinimide (NHS) esters of biotin [9] that react with primary amines under mild conditions forming amides. This may be because primary amines are accessible and available for biotinylation for most proteins. NHS esters can be divided into water-soluble and water-insoluble classes. The former could be used in an aqueous mixture while the latter are usually dissolved in an organic solvent like dimethyl sulfoxide (DMSO) or dimethylformamide (DMF) before they are aliquoted into aqueous mixtures. Because NHS esters hydrolyze easily, they are not recommended for long term storage.

Sulfo-NHS-LC-Biotin (Figure 1.) is commonly used for labeling monoclonal and polyclonal antibodies. However, if the antibody's antigen binding site is lysine rich, amine reactive reagents like NHS may not be appropriate for biotinylation of the antibody. Biotin-Hydrazide that reacts with aldehyde becomes useful in this case.[10] By oxidizing carbohydrates on antibodies or other glycoproteins with periodate, aldehydes could be generated and used for biotinylation by forming a hydrazone linkage. Also, since these carbohydrates are mostly found in the Fc region of antibodies, the biotinylation becomes restricted to a region away from the binding site of the antibodies. [4]

For proteins, peptides, DNA that do not contain reactive functional groups like carbohydrates, aldehydes, amines, etc. may be labeled with photo-reactive biotinylation reagents. For example, the biotinylation of alkaline phosphatase and tubulin were done with retention of activity using photobiotin[11,12].

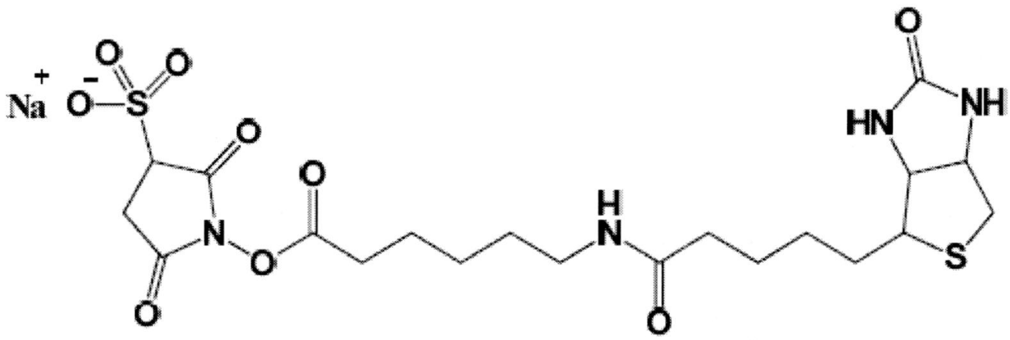

Figure 1. Structure of Sulfo-NHS-LC-Biotin. Sulfo-NHS group leaves after reaction with a primary amine group that conjugates the biotin via an amide bond.

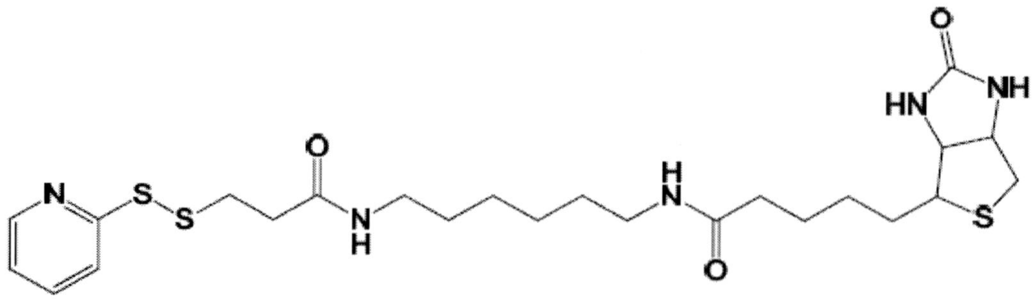

Figure 2. Structure of biotin-HPDP. It conjugates biotin to sulfhydryl groups on proteins by making the pyridine a leaving group.

An aryl azide group is chemically inert until it gets exposed to ultraviolet light, which then forms a temporary reactive aryl nitrene group. [13,14]This group reacts and inserts spontaneously with electron dense sites like double bonds or active hydrogen bonds. Consequently, the resultant conjugate would ideally have a biotin 'handle' accessible to avidin binding. Therefore, photactivatable biotin is useful in DNA or RNA biotinylation since these molecules do not contain readily reactive functional groups but are rich in electron dense region. [11] Also, the resulting biotinylated probes produced are stable under standard hybridization conditions and give reproducible results for a minimum of five months.[4] However acidic conditions and reducing agents should be avoided because they may inactivate the aryl azide group. Another popular bioconjugate technique used to attach biotin to DNA is to use biotin-d 2′-deoxyuridine, 5′-triphosphate (dUTP) derivatives formed by modification of the C-5 position of uridine. The most common derivative is biotin-11-dUTP with eleven atoms as its spacer arm. A DNA polymerase is then used to add the labeled monomer to an existing strand in a nonradioactive manner.[15,16,17] It was also suggested by Folsom et al. that the sensitivity of nucleic acid targets detection can be improved by using RNA instead of DNA probes. [18]

N-[6-(Biotinamido)hexyl]-3'-(2'-pyridyldithio)propionamide (biotin-HPDP) (Figure 2.) [19], 3-W-maleimidopropionyl)biocytin (MPB), 1-biotinamido-4-[4'- (maleimidomethyl) cyclohexane-carboxamido]butane (biotin-BMCC) and N-iodoacetyl-N-biotinylhexylene-diamine (iodoacetyl-LC-biotin) are biotin derivatives that react with sulfhydryl groups. Peptides or proteins containing cysteine on their surface can be biotinylated easily using these reagents. However, excess mercaptoalkanes and other thiol compounds must be absent prior to biotinylation. [4] Targeting sulfhydryls for modification usually result in derivatization away from active centers or binding sites, promoting binding with strept(avidin) with minimal steric hinderance. A common reagent for such conjugation is biotin-HPDP. The valeric acid side chain of biotin is attached to a 1,6-diaminohexane spacer group that in turn attaches to the PDP that has a sufyhydryl-reactive group.[20] Both MPB and biotin-BMCC have maleimide ends and react with sulfhydryl groups to form stable thioester bonds. BMCC has a spacer arm of length 32.6Å which is longer than that of MPB's 27.6 Å. Both spacer arms are however, sufficiently long enough to bind with strept(avidin) probes. Iodoacetyl-LC-biotin is also a sulfhydryl-reactive reagent that is attached to biotin's valeric acid side chain. The sulfyhydryl reactivity is attributed to its iodoacetyl group. A strong thioester bond is formed after it couple to sulfhydryl containing molecules. Iodoacetyl-LC-biotin has been used to localize the SH 1 thiol of myosin by applying the avidin–biotin complex using electron microscopy.[21] Modifications to lengthen the spacer arm of iodoacetyl-LC-biotin can assure better binding with strept(avidin) probes. However, iodoacetyl-LC-biotin containing reagents should be protected by light because photolysis could possibly liberate the iodine and thus, reduces activity of the reagent. [20]

1.2. Strept(Avidin) Conjugates/Modifications

Labelled strept(avidin) are becoming increasingly popular because they are generally less expensive than labeled antibodies, commercially available and simple to prepare in most cases. Enzymes, fluorescent markers and gold nanoparticles are commonly used for detection. Strept(avidin) can be conveniently conjugated to these particles without compromising their biotin binding ability. [4]

Metalloproteins such as ferritin conjugated to avidin produces a very sensitive reagent for use in electron microscopy due to its high electron density.[22] Glutaraldehyde is a common compound that is conjugated with strept(avidin). It is typically used to conjugate to enzymes or proteins due to its versatility. Also, enzyme or ferritin-labelled avidin conjugates can be produced using it. A two-step protocol performed by Suffin et al. and Avrameas [23, 24] ensures unidirectionality of the crosslinking.

Gold colloids are very common in the biosensing field. [25, 26] Therefore, gold conjugates of strept(avidin) can be very useful. For example, succinylated avidin was labeled with gold nanoparticles for the study of internalization of biotin labeled toxin penicillin. [27] Also, protein phosphorylation detection has been described by Kerman and coworkers by using adenosine triphosphate with biotin conjugated to the γ-phosphate and streptavidin coated gold nanoparticles to bind to the phosphorylated protein. The reduction signal of the gold nanoparticles was monitored using differential pulse voltammetry. [28]

Radioisotope-labeled ligands for the avidin-biotin assay produce some of the lowest detection limits and could be applied to the study of pharmacological effects.[29, 30] However, there are many issues with radioisotope labeling in handling and disposal. Therefore, fluorescent and enzyme-labelled substances were conjugated to streptavidin to circumvent the problems faced by radioactive labeling.

Fluorescent labels like fluorescein and its derivatives are also common labels that could be used to study cell membranes by electron microscopy. [31] Fluorophores absorb energy when exposed to incident light of a particular wavelength. The electrons get excited to a higher energy state but then lose energy as emitted light as they relax back to their previous energy level. This emitted light is applied in fluorescence microscope to view tissues stained with fluorochromes. [4] In a past innovation, a fluorescent non-natural amino acid 2-anthrylalanine, (containing anthraniloyl group) was incorporated into the primary amino acid sequence of streptavidin during in vitro translation using a four base codon/anticodon. Most of the streptavidin mutants retained the biotin binding affinity. [32, 33]

Extravidin, NeutrAvidin and NeutraLite are deglycosylated version of avidin. Despite removal of the carbohydrate, the biotin binding affinity is retained while non-specific charge-charge binding property of avidin is minimized [34] due to its near neutral pI of 6.3. Lectin binding is also reduced to minimal level after the carbohydrates are removed.

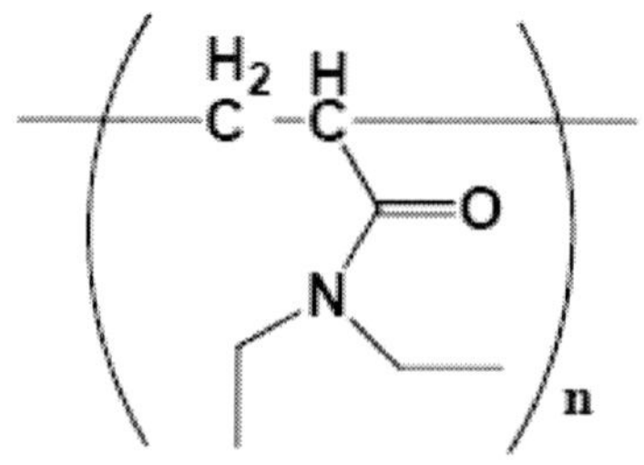

Figure 3. Structure of poly(N, N-diethylacrylamide) used in thermosensitive polymers.

Immobilized monomeric avidin was developed in order to purify biotinylated proteins. It is unique among other biotin binding proteins because it binds reversibly to biotin and allows gentle elution. In this manner, the recovery of biotinylated molecules is made convenient.

NitrAvidin was created by nitrating the Tyr^{33} residue *in vitro*. This resulted in the reduced pK_a of the phenol group of streptavidin. This modification produced a pH-dependent system whereby biotin is strongly bound at acidic pH (4-5) while released when pH is raised. NeutrAvidin was designed to display high specific binding to biotin. It has been found that covalent binding of NeutrAvidin onto carboxyl-derivatized polymers produces a higher coating density than immobilization via a biotinylated interlayer.[35] Also, by linking a copolymer of acrylic acid and N-isopropylacrylanide (NIPA Am) to an added cysteine residue on a modified streptavidin, another pH-sensitive streptavidin could be produced.[36]

Other modifications can result in temperature sensitive streptavidin. This is done by adding thermosensitive polymers like poly(N, N-diethylacrylamide) (Figure 3.) also known as PDEAAm to control biotin binding. Below the lower critical solution temperature of PDEAAm, the polymer is in its hydrated and extended state and acts as a shield that blocks the binding of large biotinylated proteins while above this temperature, it collapses, exposing the binding site, thereby allowing biotin binding. [37]

1.3. Biotin/ Strept(Avidin) Coated Surfaces

Both biotin and strept(avidin) may be bound directly to surfaces like oxidized silicon, gold surfaces, filter membranes and even magnetic beads via a covalent linkage. For example, a typical biosensing surface for the immobilization of biotin and strept(avidin) is the silicon dioxide surface. By using mercaptopropyltrimethoxysilane's (MPTMS) sulfhydryl group, m-maleimidobenzoyl-N-hydroxysuccinimide ester (MBS) could be attached. Then, the nucleophilic attack by the amine group of strept(avidin) at the carbonyl group of the MBS can result in a covalent binding of strept(avidin) on the silicon dioxide surface. [38] Similarly, biotin could also be conveniently attached to the silicon dioxide's functionalized MPTMS surface by using iodoacetyl-LC-biotin or sulfo-NHS-biotin [39] that will react with the thiol group. Other chemical or photochemical linkage of the biotinylated molecules could also be formed depending on the functional groups available on the biosensing surface of attachment. [40]

Self-assembled monolayers (SAMs) are formed when surfactant molecules spontaneously forms a homogeneous, highly-ordered film in the substrate surface. Specific interactions have been shown between a SAM of biotinylated thiols and streptavidin, suggesting the possibility of combining a well-ordered SAM structure with the avidin-biotin system. [41, 42] SAMs have therefore been used extensively to create biotin and strept(avidin) surfaces. For example, a simple method for constructing a biotinylated SAM used a mixture of biotinylated alkanethiol.[43] A classical example of self assembled monolayer (SAM) is the reaction of alkanethiols with a gold surface. [44] Therefore, SAMs offer ease of functionalization and their orderly arrays also make them ideal models in many fields. Sulfur gold interactions are semi covalent and thus, thiols have a strong affinity for gold surfaces. SAMs for optical waveguides modifications can also be prepared through biotinylated alkanethiolates on gold and silver coatings used in Surface Plasmon Resonance (SPR). When a layer of biotin is immobilized, a secondary layer of streptavidin could be formed above it. This secondary layer appears to be more robust and its surface interference is minimized as compared to directly bound strept(avidin)on the sensing surface. [45, 46] If the strept(avidin) is in direct contact with a hydrophobic surface, denaturation of the protein may be promoted. With this first buffing layer of biotin, the strept(avidin) will then be protected. [47, 48] Furthermore, the biotin layer can help orient the strept(avidin) by binding to two sites of the protein, leaving two binding site exposed on the opposite side of the protein accessible for biotinylated molecules. Another indirect immobilization strategy to bind an oriental streptavidin monolayer is to use a cysteine derivative described by Reznik et al. [49]

Magnetic particles are widely used in various biomedical and environmental applications. A major advantage of magnetic particles is the ease of separation of target molecules from complex matrices using a strong magnet. Magnetic particles are increasingly popular in DNA extraction [50, 51] and antigen detection [52, 53]. Among these various magnetic particle conjugates developed to date, streptavidin-coated magnetic beads have shown great potential in biosensing due to the stability of the biotin-streptavidin system and the resulting ability to separate target molecules efficiently [54,55,56,57]. The synthesis of streptavidin immobilized magnetic particles is tedious and time consuming but conveniently, they are now commercially available. For example, Kerman and Rambihar made used of streptavidin-coated superparamagnetic iron oxide beads to detect the lectin-glycoprotein interactions using fluorescein isothiocyanate (FITC) as the label. A sandwich assay was made by binding biotinylated wheat germ agglutinin onto the streptavidin coated beads before detecting the interactions of target glycoprotein. Then a detecting biotinylated lectin was added to bind on top of the attached glycoprotein before introduction of Extravidin-conjugated FITC dye onto the complex. The magnetic beads allow efficient and convenient washing of the bound glycoprotein from unbound ones to reduce noise signals by using a stronger magnet to retain the beads during washing. [58]

1.4. Biosensors

A biosensor is a self-contained integrated device that is capable of providing specific quantitative or semi-quantitative analytical information using a biological recognition element which is in direct spatial contact with a transduction element.[59] (Figure 4.) Biosensors are generally devices that detect, record, convert, process and transmit information with regards to a physiological process. They are being applied in various fields including medicine, biotechnology, quality control and environmental monitoring.

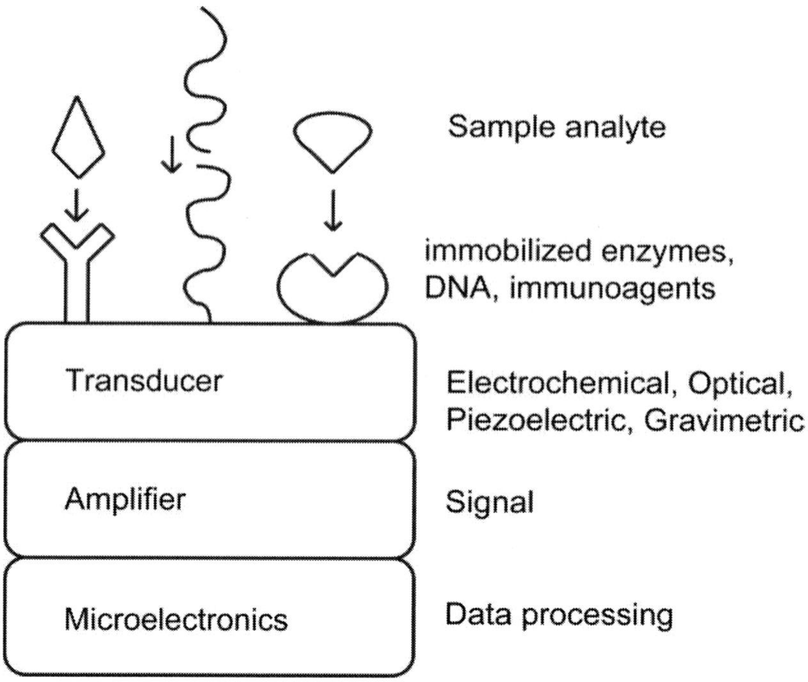

Figure 4. General components of a typical biosensor comprising of the biological recognition element, transduction element, amplifier and a microelectronic processor.

Biorecognition elements are biological molecules having either catalytic nature or binding property. Proteins are one of the most frequently used classes of molecules that possess biorecognition capability. They are commonly immobilized by direct binding via a covalent bond formed between functional groups of the protein and electrophilic groups on the sensor surface. [60]

Some important properties of biosensors are their sensitivity, ease of calibration, selectivity and dynamic response.[61] The interest in developing small devices for biomedical use is growing at an alarming rate. The potential applications of these devices are limitless when it comes to point-of-care and critical care testing. Ongoing research with affinity biosensors based on electrochemical, optical, or piezoelectric (e.g. quartz crystal balance) transducers will definitely improve bioanalysis in a clinical setting. [62]

A couple of examples of current biosensors are pregnancy test kit that detects human chorionic gonadotrophin (hCG) protein in urine and glucose monitoring device [63] for diabetic patients. Typical sensing techniques for biosensors includes fluorescence, DNA microarray, surface plasmon resonance, impedence spectroscopy, scanning probe microscopy, quartz crystal microbalance, surface enhanced Raman Spectroscopy and other electrochemical methods.

2. STREPT(AVIDIN)/BIOTIN BASED ELECTROCHEMICAL BIOSENSORS

The advantage of using electrochemical techniques for detection is that only a small sample size is required. Also, the detection system could be easily miniaturized without compromising its sensitivity. Furthermore, electrochemical systems tend to be cost effective and portable, which is much favored by many point of care testing diagnosis.

An interesting electrochemical detection of avidin is performed by Sugawara et al. using biotin labelledwith an electroactive compound, daunomycin. Avidin was indirectly detected based on the decrease in electrochemical signal as the labeled biotin binds to avidin and subsequently lose their electroactivity as the daunomycin gets coated by avidin. The advantage of this method is that a free labeled biotin would not have to be separated from a bound one. [64]

Another creative electrochemical detection of dissolved biotinylated-HRP, biotin and avidin was performed by Vreeke et al. Direct detection of the particles were able to be done by employing an avidin-containing, electron-conducting hydrogel film on the surface of an electrode rotating in a hydrogen peroxide solution and then measuring the electrocatalytic reduction of the peroxide. The sensitivity and detection limits were sufficient for widely performed assays. [65]

2.1. DNA Sensors

Electrochemistry of nucleic acids has led to heightening interest in the development of biosensors. After the discovery of anodic peak specific for guanine, [66] voltammetry of nucleic acids became an important milestone for electrochemical monitoring of native double-stranded (ds) and denatured single-stranded (ssDNA). Electrical DNA biosensors have become an attractive technique for detecting and identifying oligonucleotides and other target analytes because of their low cost, ease of miniaturization and direct readout of electrical signals. [67] Therefore, Biosensors employing DNA have been developed extensively in the past decade. [68,69] Electrochemical biosensors for detecting DNA make use of the affinity between a ssDNA and its complementary strands. This allows for both direct and indirect measurement methods. Direct approaches exploit the intrinsic electroactivity of nucleobases while indirect methods mostly detect electroactive labels that are conjugated or interchelated to the DNA. [67] The analysis of specific DNA sequences of hybridization has an impact on clinical diagnosis and treatment, genetic disease research and environmental monitoring.

Direct electrochemical techniques like linear sweep voltammetry (LSV) and cyclic voltammetry (CV) however, are destructive and cause irreversible damage to the DNA immobilized on the substrate electrode due to the properties of guanine and adenine. [70] Therefore, the indirect method of using redox indicator-based electrochemistry becomes an important alternative. Different redox indicators such as methylene blue [71, 72], and metal complexes [68] have been reported to differentiate native and denatured form of DNA.

Zhou et al. did a study on different oligonucleotide immobilization methods (direct chemical bonding, avidin–biotin interaction or electrostatic adsorption on polyelectrolyte films) and found that the DNA sensing films designed by binding biotinylated DNA probe to an avidin coated surface provide fast sensor response and high hybridization efficiencies. Quartz crystal microbalance (QCM) is known to provide extremely sensitive mass change detection because of its decrease in resonant frequency upon the increase of a given mass on the quartz crystal electrode at a sub-nanogram level [73]. Because of this, QCM is a direct and ultra-sensitive method to assess the immobilization of biomolecules and to monitor affinity reactions such as antigen–antibody or DNA hybridisation reaction in real time

[74].An 18-mer DNA probe that is complementary to the genetic β-thalassaemia mutations site was immobilized on the QCM electrodes via covalent bonding or electrostatic adsorption on polyelectrolyte films to form mono-/multi-layered sensing films by self-assembly process. Subsequently, the results of hybridization on the QCM sensor constructed by the different immobilization methods were compared. The results show that the optimum concentration range of target DNA is 50-70μg/mL. Furthermore, the advantages of this technique over conventional fluorescence-labeling methods are that pre and post-treatments are unnecessary to modify DNA probes, in-situ monitoring of hybridization is possible, fast results and relatively cheap instrument required. It is promising to say that the QCM DNA sensor with its direct detection of polymerase chain reaction (PCR) amplified products provide a rapid, low cost and convenient diagnostic method for β-thalassaemia and other genetic diseases. [75]

Avidin-biotin system is well known for its stability and strong affinity for each other. However, also using microgravimetric QCM, Dupont-Filliard et al. came up with a sensor that makes biotin a renewable grafting unit. The steps in creating this biosensor are as follows: 1) electropolymerization of monomers bearing biotin entities to form a biotinylated polypyrrole film. 2) Avidin is then added to bind with biotin, leaving three sites available for interacting with other biotinylated molecules like oligonucleotide-biotin in this case. 3) The complementary oligonucleotide-biotin then binds to the oligonucleotide of the previous layer. 4) Eventually, the sensor was dipped into a solution that has streptavidin conjugated with R-phycoerythrin (a fluorophore marker) whereby the binding could be detected by QCM transduction as well as couple with fluorescence detection. The intriguing part of this research is that the support could be reused by denaturing the matrix with a denaturant. The biotin-avidin interactions were broken but the polypyrrole-biotin matrix remains intact. And a similar fluorescent signal could be obtained after regenerating the previous layers. An electrochemical DNA biosensor system that is reliable, sensitive and reusable like this is definitely a promising innovation in clinical diagnosis. [76]

Recently, Arora and coworkers designed an ultrasensitive DNA hybridization biosensor based on polyaniline (PANI) electrochemically deposited onto Pt disc electrode. The system made use of the well known biotin–avidin as indirect coupling agent to immobilize single-stranded 5'-biotin end-labeled polydeoxycytidine (BdC) probes and 5'-biotin end-labeled 35 base-long oligonucleotide probe (BdE) to detect its complementary target. PANI films were first activated by polymerized chronopotentiometrically in a three-electrode electrochemical cell before adding monomer solution containing 0.1M aniline subjected to a constant current. The avidin was then added with activation solution, 1-Ethyl-3-(3-dimethylaminopropyl) carbodiimide (EDC) and NHS to bind to the activated PANI surface with the active carboxylated group of avidin. The BdC and BdE were then added to form BdC-avidin-PANI and BdE-avidin-PANI bioelectrodes. Both direct electrochemical oxidation of guanine and redox indicator methylene blue (MB) were used and characterized by the electrochemical technique differential pulse voltammetry (DPV) and impedance spectroscopy. It was found that hybridization detection using MB resulted in about 100 times lower detection limit when compared with direct electrochemical oxidation of guanine. Also the reusability of the biosensor using methylene blue can go up as high as seven times. [77] Similarly, the genoelectodes were shown to detect the presence of complementary sequences in *E.Coli* genomic DNA and its cell lysate without the use of PCR and external reagent in a similar electrode setup by Malhotra and coworkers. The fast, ultrasensitive and direct detection of the bacteria cells with DPV has strong implications in the fields of point of care testing in clinical diagnosis and food/water monitoring. [81]

SAMs with oligo(ethylene glycol) (OEG)-terminated thiols are known to be highly protein-resistant and have the property of repelling non-specific adsorption with high efficiency. Interestingly, it is fully compatible with electrochemical biosensing as the monolayer allows electron transfer to the electrode surface with ease. [78] Microbes like *Escherichia coli (E.Coli)*, *Campylobacter*, *Salmonella*, and *Listeria monocytogenes* are responsible for most the food-borne infection outbreaks. [79, 80] Contemporary methods for microbes detection (microbiological/ biochemical/ immunological detection kits, southern hybridization, PCR, etc.) are slow and labor intensive unlike the proliferating electrochemical biosensing techniques. Making use of the remarkable properties of the SAM, Zhang et al. designed a sandwich electrochemical biosensor for sequence specific detection of *E. Coii* DNA targets to a limit of 1pM in 'dirty' conditions (Human serum). Thiolated capture probes for one end of the *E. Coli* targets were immobilized on the gold substrate via two hour incubation at room temperature before treating with OEG to obtain mixed SAMs. The 250 base pair PCR product of the *E. Coli* DNA was then made to hybridize partially with the shorter capture probe. Then a biotinylated reporter probe compliment to the other end of the *E. Coli* DNA was incubated to form the sandwich structure at the electrode surface. Thereafter, an HRP-avidin conjugate was introduced and its strong affinity for biotin was utilized to bind to the reporter probe. With the help of an electroactive cosubstrate TMB, HRP enzyme catalyzes the reduction of hydrogen peroxide which is then detected by CV and amperometric measurements. Because no biotin was present before reporter probe binding, the electrochemical signal can be ascribed directed to the presence of *E. Coli* DNA on the electrode. Such non fouling surface is essential in the specific detection of bacterial pathogens. [78] A general schematic that makes use of such hybridization systems is shown in Figure 5.

Recently, with the advancement and popularity of electrochemical DNA biosensors, Liu et al. has come up with an enzyme based E-DNA sensor for detection of DNA targets up to the femtomolar level. This is done by replacing the classical signal amplification redox molecules with redox enzymes and inorganic nanoparticles labels. [82] A more comprehensive review of E-DNA ssensor could be found in Ricci and Paxco's paper. [83] In Liu's paper, amine modified biotin was reacted with activated (with EDC and NHS) 11-mercaptoundecanoic acid on gold surface. After which, a layer of streptavidin was added to the surface to bind with the biotin before adding another layer of 3'-biotinylated stem-loop DNA probes that has 5'-digoxigenin (DIG) affinity label to the immobilized streptavidin. Eventually, the probes were used to detect a 250 base pair target from the *E. Coli* genomic DNA. After the hybridization of the complementary target strand, the probe forms a rigid duplex that lifted the DIG, which was initially blocked from bulky anti-DIG-horseradish peroxidase (HRP) conjugate due to steric hindrance near the surface of the electrode, and become accessible to the anti-DIG-HRP. 3,3',5,5'-Tetramethylbenzidine (TMB) was then used as an electron shuttle to couple and initiate the catalytic reduction of hydrogen peroxide by the HRP even in order to trigger an electrochemical signal that could be detected by CV and amperometric measurements using a multichannel potentiostat. The detection limit for this work is 10 femtomolar which is extremely impressive. False positive results are also minimized due to its signal-on nature. [82]

Exploiting techniques like CV, DPV, scanning electron microscopy (SEM) and energy dispersive X-ray specrtroscopy, Sadik and coworkers were able to provide evidence for the success of a novel electrochemical detection scheme for DNA-organic molecules binding. Indirect electrochemical detection methods do not show a significant measurable response in the presence of DNA mismatches. Silver ions could form complexes with DNA bases and they were used to instill electrical functionality along the DNA molecule. In their work, a monolayer of silver deposited on a gold electrode could be oxidized to produce silver ions and electrons together with a reversible redox signal. [84] Streptavidin was adsorbed to the underpotential deposited silver monolayer according to work of Osa et al.[85] before addition of biotinylated dsDNA. With a layer of dsDNA present on the electrode, the oxidized silver gets dispersed and upon reduction, returns to the electrode surface whereby the reduction current is measured. With the binding of its interchelator, a structural change in the DNA structure results which is then detected by a decrease in reduction current measured because the flow of electrons is hindered by the low molecular weight organic molecules on the electrode. [84] Theoretically, if the sensor could detect structural changes of DNA, the stem-loop system could be employed in this sensor whereby the target DNA could be detected with greater ease.

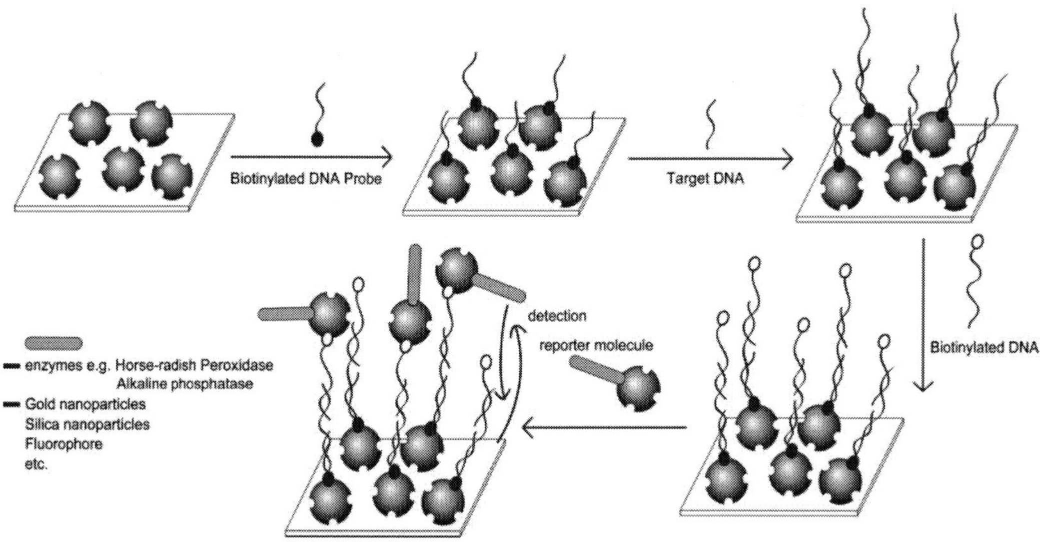

Figure 5. General schematic of a DNA sensor that utilizes streptavidin and biotin affinity to detect the hybridization of complementary DNA oligonucleotides electrochemically.

We have talked about detecting DNA targets by immobilizing a probe on an electrode surface. Chaumpluk et al. came up with a creative way to quantitatively detect target DNA (*Cry*IA(b) from maize) in food samples using LSV without having to immobilize capture probes. The trace targets in a food sample were first amplified using 5' thiolated forward primer and 5' biotinylated reverse primer. This way, the amplified sequence would have a thiol end and a biotinylated end. S1 nuclease was then used to reduce non-target signals by cleaving other ssDNA in the sample. After which, a ferrocene conjugated with streptavidin was used to bind with the biotinylated amplified DNA. Magnetic trapping was performed to accumulate signals before they form a SAM on the gold electrode using their thiol ends. Eventually Hoechst 33258 was used to aggregate DNA on the electrode surface to enhance the signal directly measured by LSV. They were able to show a linear dependence of the LSV anodic peak on the concentration of the target DNA on a screen printed gold electrode surface. This system does not require tedious immobilization of probes or DNAs onto the surface of the electrode and is promising in point of care testing. [86]

The development of DNA sensors are crucial in the detection of clinically, environmentally or security-relevant nucleic acid targets. Attractive features of electrochemical DNA sensors include high sensitivity, portability, fast response as well as low power/cost/sample requirements. Most importantly, their simple interfacial designs at the electrodes to detect low levels of DNA stand out amongst other methods.

2.2. Immunosensors

Generally, a sensor consists of a sensing element and a transducer. For immunosensors, the sensing element is usually immobilized antibodies or antigens. There are many methods developed for immobilizations of the immunoagents involved on different substrates.[87, 88, 89] The transducer plays the role of converting the binding event into a measurable signal. Transduction has been widely performed by electrochemical techniques [90, 91] Electrochemical detection of immunointeraction can be performed with or without labeling. The classical electrochemical immunosensing method is to use amperometric immunosensing, where enzyme-labeled proteins produce an electroactive product from binding of substrate. [91] Direct detection could be performed by CV, impedence spectroscopy, differential pulse voltammetry (DPV), square wave voltammetry SWV), etc. These techniques detect a change in the capacitance or resistance of the substrate electrode surface induced by binding events.

Hianik et al. [92, 93] studied the effect of addition of 2,4-dichlorophenoxyacetic acid (2,4-D) to different antibody-modified membranes. Metal supported Bilayer lipid membranes (BLM) have been formed on polymer coated stainless steel from 2% phospholipid solution in a mixture of *n*-decane and butanol and by immersion of a gold electrode modified with an octadecanethiol SAM into a 2% solution of phospholipids in *n*-hexane. Avidin was conjugated with swine anti-human IgE using glutaraldehyde. This antibody was then added to the membrane exploiting the biotinylated phospholipid. Detection of 2,4-D binding has been performed by measuring the impedance at $E^0 = 40$ mV and $f = 1$ kHz. With BLMs on gold the detection limit was 0.5 μM, but with good reproducibility. No contribution of nonspecific adsorption to the signal has been found.

Recently, Barton and coworkers fabricated a label-free immunosensor for the stroke marker protein: Neuron Specific Enolase (NSE). As we know, diagnosing stroke immediately after its initial symptoms is cardinal. The delay in obtaining accurate diagnosis may waste much time that are limited in availability in which the brain has to respond to reperfusion. This increases the risk of hemorrhage significantly. Therefore, successful treatment requires sensitive and fast diagnosis. NSE may elevate in concentration in serum due to cerebral damage coupled with increased permeability of the blood brain barrier. NSE had been reported to correlate with the extent of neurological damage. Thus, its detection is significant in the field of stroke diagnosis. In their work, poly(1,2-diaminobenzene) was electrodeposited on a conductive carbon electrode (on a commercial screen printed electrode) to form an insulating layer. The insulated layer was then sonochemically ablated to form a micropore array before growing a conductive polyaniline protrotrusion layer modified with a biotinylating reagent. Binding of Neutravidin to the biotinylated layer allows binding of specific biotinylated monoclonal anti-NSE antibody. Then, NSE samples of different concentrations were introduced to the sensor surface and the binding event was monitored by alternating current impedance measurements. The electron-transfer resistance of $[Fe(CN)6]^{3-/4-}$ in the buffer solution was measured by the impedance spectroscopy. A larger resistance implies a greater NSE concentration bound to the electrode surface. A linear plot was calibrated for NSE concentration from 0-10pg/mL. The low limits of detection for NSE demonstrated the promising application of this immunosensor assembly in a clinical setting.[94]

Instead of modifying the biotinylated layer on polyaniline, a polypyrrole biotin layer was directly fabricated on the electrode surface when Ouerghi and coworkers detect human IgG. Pyrrole biotin monomers were oxidized at 0.85V in CH_3CN and tetrabutylammonium perchlorate to form the biotinylated pyrrole layer.[95] Specific binding of avidin was then performed before incubation in a biotinylated anti-human IgG antibody. Varying concentrations of human IgG was then introduced to bind to the assay. Impedance spectroscopy was also used to monitor the concentration of human IgG binding to the electrode. The immunosensor had a linear dynamic range of 10-80 ng/mL and low limit of detection of 10 pg/mL. [96] The sensing layers could even be regenerated successfully under acidic conditions.

Cholera is significant in countries in Asia, Africa and Latin America. The endemic scourge prompted the detection of anti-cholera toxin antibody. [97] The development of cholera vaccines requires biosensing tools for the detection of the corresponding antibody that has immunization effects. Ionescu and coworkers set out to detect the anti-cholera toxin with an innovative photoelectrochemical method. Electronic detection in photoelectrochemical procedures is simpler in instrumentation and more convenient to miniaturize as compared to conventional optical detection methods. In their work, a novel ruthenium(II) tris-(bioyridyl) polymer film was designed to detect immunoreactions via the evolution of its photoelectrochemical properties. Electro-oxidation of a biotinylated tris-(bioyridyl) ruthenium(II) complex bearing four pyrrole groups generated a photosensitive biotinylated polypyrrole film. The photoexcitation of the modified electrode potentiostated at 0.5V against saturated calomel electrode, results in a cathodic photocurrent in the presence of pentaaminechloro cobalt(III) chloride which is an oxidative quencher. Biotinylated cholera toxin was then strongly bound to this polypyrrole film using avidin as a bridging molecule based on biotin-avidin affinity. The photocurrent was recorded with an X/t recorder. The intensity of photocurrent (I_{ph}) increases with increasing Ru film thickness, indicating that the irradiation process and the subsequent quenching by the oxidative quencher take place at the film surface and within the whole film structure. The binding of avidin and subsequent biotinylated toxin on the biotinylated film resulted in a decrease in photocurrent due to the increase in steric hindrances against the diffusion of quencher molecules to the underlying Ru film. As expected, the binding of anti-cholera toxin antibody also resulted in a marked decrease of 28% in I_{ph}, indicating its specific binding. Using such detection methods, the limit of detection was found to be 0.5 μg /mL. [98]

Detecting the same anti-cholera toxin but using a permeable biotinylabted polypyrrole film, Ionescu and coworkers were able to obtain a much lower limit of detection than Haddour. Interestingly, biotinylated-pyrrole and pyrrole-lactobioamide monomers were copolymerized to form a film that improves the hydrophilic character of the organic film, and hence improves its permeability of electroactive species. Avidin was then incubated to bind to the biotinylated film before adding biotinylated pentameric cholera toxin B subunit protein as the recognition layer. Rabbit anti-cholera toxin antibody was then bound to the cholera toxin as the primary antibody. After that, HRP-labeled IgG anti-cholera antibody binds to the primary antibody as a secondary antibody layer. Eventually the HRP enzyme catalyzes the oxidation of hydroquinone (redox reactive species) in the presence of hydrogen peroxide and is detected using CV. Therefore the permeability of the anchoring layer is crucial for this assay to work. The amperiometruc detection of cholera toxin was also carried out and a linear variation was observed over the range 1-200μg and the limit of detection was 50ng/mL. This result appears to be more promising than the previous photoelectrochemical method. [97]

Tumor necrosis factor-α (TNF- α) is a biomarker for neonatal listeriosis, severe meningococcemia, HIV infection, systemic erythema, etc. It is at a typically low level in human serum samples. Therefore, it is crucial for the development of a sensitive biosensor in order to detect its presence. The current methods of detection like immune-PCR assay, ELISA, fluorescence, etc. are tedious, time consuming and may require sophisticated instrumentation. As mentioned previously, electrochemical detection enables, fast, low-cost, sensitive and simple detection of biomolecules. Indeed, Wang and coworkers were able to design a novel electrochemical immunosensor for TNF- α on poly-(guanine)-functionalized silica nanoparticles. (Figure 6.) When guanine is used as the electroactive reporter in this work, the guanine oxidation is mediated by $Ru(bpy)^{3+/2+}$, which catalyzes the oxidation process in order to enhance the oxidation signal. Silica nanoparticles were first functionalized with polyguanine and an avidin by EDC/ NHS approach. A screen printed carbon electrode was coated with avidin in the same manner before adding biotinylated anti-TNF-α antibody. After immobilization of the antibody on the electrode surface, varying concentrations of TNF-α was incubated.

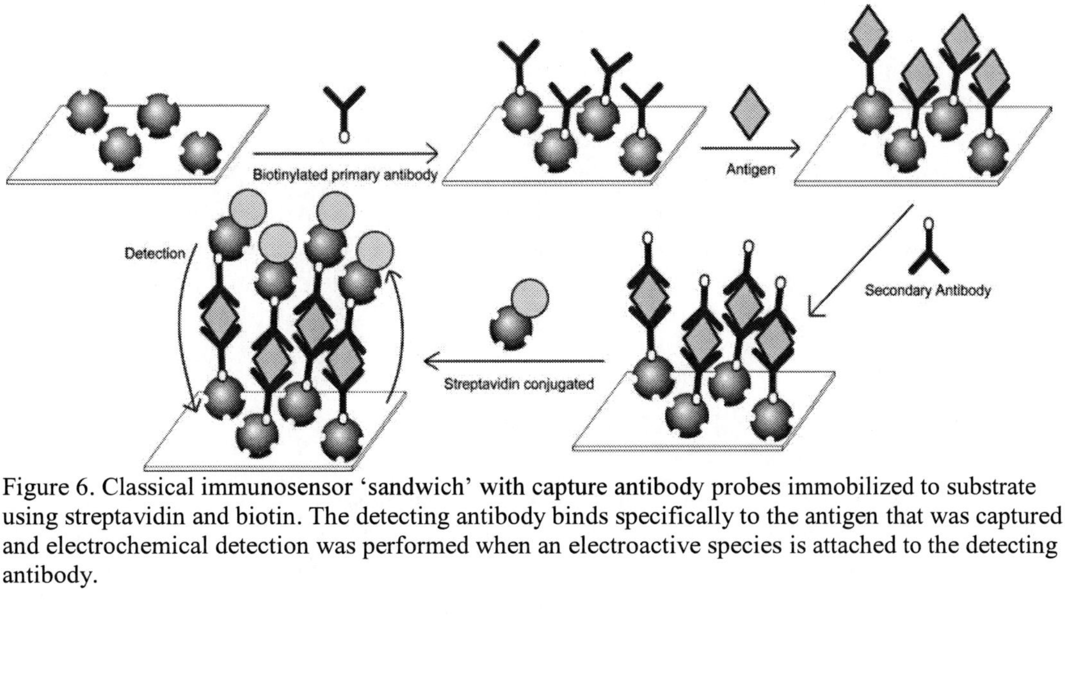

Figure 6. Classical immunosensor 'sandwich' with capture antibody probes immobilized to substrate using streptavidin and biotin. The detecting antibody binds specifically to the antigen that was captured and electrochemical detection was performed when an electroactive species is attached to the detecting antibody.

Another layer of biotinylated anti-TNF-α antibody was incubated to bind to the avidin of the silica nanoparticles. After the complete formation of the sandwich, $Ru(bpy)^{2+}$was added and catalytic current resulted from guanine oxidation is measured by SWV, CV and chronocoulometric voltammetry. $Ru(bpy)^{2+}$ is oxidized to $Ru(bpy)^{3+}$ at the electrode surface while reduced back to $Ru(bpy)^{2+}$ by guanine on the silica particles. The oxidation of guanine is irreversible thus the oxidation measured is directly related to the concentration of TNF-α. The detection limit is found to be extremely sensitive of 2 pM. This label-free method therefore, has a very high potential to be feasibly applied to the diagnosis of other biomarkers [99]

In short, the applications of strept(avidin) in electrochemical immunosensors provide the obvious advantage of sensitivity, simplicity and speed. Strept(avidin) and biotin systems form very good linker groups as well as labels for detection. There are uncountable ways of applying this high affinity system with creativity to create powerful biosensors that be used to alleviate health issues in the world.

3. Strept(Avidin)/Biotin Based Optical Biosensors

There are many different optical biosensors used by scientists today. These include examples like fluorescent microscopy applied in quantum dots and L(SPR) measured from gold/ silver surfaces. Because aggregation of colloidal silver and gold particles causes a change in color of the sol, [100] strept(avidin) could be detected using colloidal gold and silver. Silver and gold colloidal particles could be biotinylated by self assembly of a biotin disulfide molecule and tetrameric strept(avidin) binding to the biotin on different metal colloids would then result in the crosslinking of the particles, inducing a color change that could be measured using optical absorption spectroscopy. [101] It was found however, that silver sol showed a larger color change upon reaction with avidin.

Optical sensors can make use of superparamagnetic beads by measuring changes in optical transmittance of a solution. Unlike tedious and time consuming methods like fluorescent imaging in enzyme linked immunosorbent assays (ELISA) [102], recently a rapid and quantitative biorecognition method was developed by Park et al. using these beads. By applying a rotating external field to an aqueous solution with biotinylated magnetic beads, the beads will rotate in phase with the field. The binding of strept(avidin) on the beads will result in a longer chain of magnetic beads to rotate. These longer chains were detected by increases in the amplitude of oscillating optical transmittance signal of the magnetic colloid solution. The detection limit is as low as 6.7ng/mL. This sensitive and rapid system has high potential in the field of optical biosensing. [103]

3.1. SPR and LSPR in Avidin-Biotin Detection

However, the main focus of this review will be on the application of strept(avidin) and biotin in optical biosensing by SPR and Localized SPR (LSPR). The first experimentally documented observation of surface plasmon resonance (SPR) as an optical phenomenon was by Prof Robert Wood in the early 1902. [104] The energy of photons is directly absorbed by the free electron constellation of a thin noble metal layer (gold or silver) and converted into an evanescent field, which is propagated through the dielectric medium close to metal. The phenomenon occurs only for a specific wavelength at the SPR of incidence. At the SPR angle, (a specific resonance angle) the intensity of the reflected light abruptly drops. [105] The refractive index above the metal film that is on the non-illuminated side of the prism affects the SPR angle. Surface plasmon resonance (SPR) is a phenomenon that is measured by the change in the refractive index at a surface as a function of the mass of the material bound to the surface. At the substrate surface, photons whose energy and momentum resonate with the collective oscillations of the delocalized surface electrons are termed surface plasmons. Surface plasmon resonance occurs during total internal reflection when the optical evanescent wave oscillates with the electron cloud in the metal film at the sensing surface. (Figure 7.) When biomolecules are bound on the detection surface of the sensor, the attachment process will change the refractive index of the metal film and consequently change the SPR angle. When an interaction event occurs on the metal film surface, the change in the SPR angle is continuously adjusted and detected by the SPR instrument and can be plotted on a sensorgram in real time. [106] The first ligand modification of SPR for biosensing was established in 1983 by Liedberg and Lundström. The thin film surface for SPR was formed by evaporating silver onto a glass slide followed adsorption of immunoglobulin G (IgG) to the surface. A goniometer and photodiode apparatus were used to determine the SPR angle. A shift in the angle was observed upon injection of the surface with anti-IgG. Liedberg was then able to achieve sensitivity levels of 0.2 μg/ml. [107] The commercialized BIAcore instrument allows for the real-time SPR measurement of the kinetics of noncovalent binding between a biorecognition element and its target analyte. [108, 109, 110]

Noble metal nanoparticles also display a strong UV-visible absorption band that is not present in the spectrum of their bulk metal. This absorption band results when the incident photon frequency is resonant with the collective oscillation of the conduction electrons and is known as LSPR. (Figure 8.) LSPR excitation results in wavelength-specific absorption with extremely large molar extinction coefficients, resonant Rayleigh scattering with high efficiency equivalent to that of 10^6 fluorophores, and the amplified local electromagnetic fields near the surface of the nanoparticles. These properties are responsible for the intense signals observed in LSPR measurements. However, the absorbance shows an exponential decay with decreasing photon energy, known as Mie scattering, onto which an LSPR band is superimposed. The large shift in wavelength induced by a monolayer of small molecules on silver nanoprisms promoted the detection of biological molecules. [111]

This wavelength of the surface plasmon absorption bands is highly dependent on the type of material and the size and shape of the nanostructures as well as their environment. Other than nanoprisms, LSPR occurs in other metallic nanostructures such as nanotriangles, nanoislands and nanoparticles. Not surprisingly, strept(avidin)-biotin is a model system that was chosen to be tested for LSPR nanosensors after their success in SPR measurements[112, 113].

There was apparent red shift in the LSPR measurement after each modification of the silver nanoparticles surface (incubation of biotin and subsequently streptavidin). It was also found that the limit of detection of streptavidin could go as low as 1pM (solution concentration). [111]

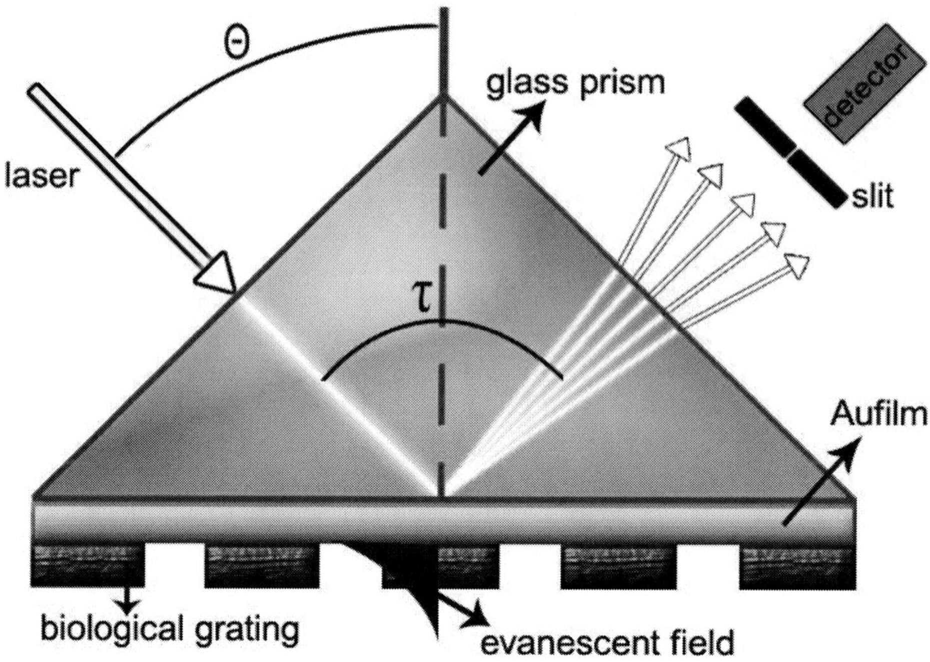

Figure 7. Schematics of an prism-based SPR system. The sensor consists of a prism coupler and a thin metal film that is SPR-active.

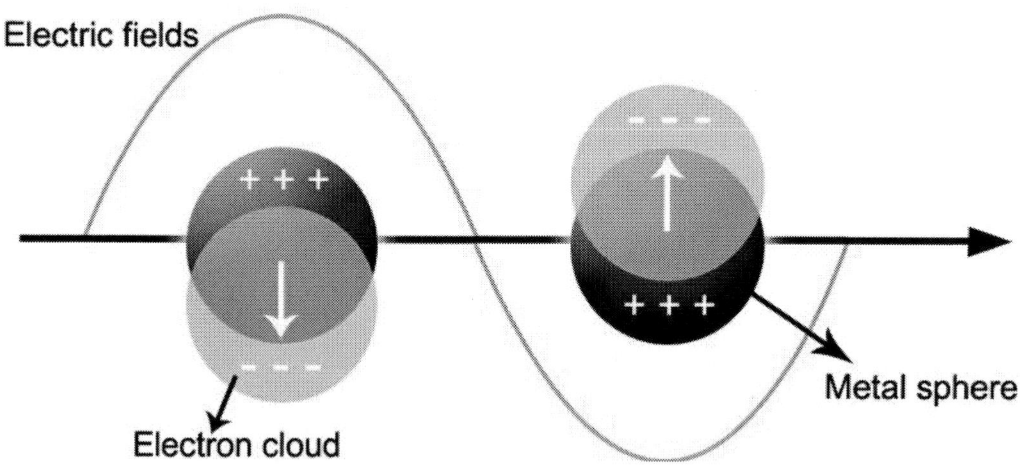

Figure 8. Schematic diagram of propagating plasmons after excitation by incident light.

Another experiment that applied the avidin-biotin model was performed by Heip H. and coworkers, who designed an interference LSPR (iLSPR) biosensor, based on photonic structure of gold nanoparticles and silicon dioxide/silicon (SiO_2/Si) multilayers. They integrated the iLSPR substrate into a microfluidic based biosensor for label free, real time monitoring of biotin-avidin interaction. It was found that the 11-amino-1-undecanethiol hydrochloride binding to gold nanoparticles, sulfosuccinimidyl 1-D biotin and avidin binding could be observed clearly in real time. Also, increments in avidin concentration also produced increasing change in reflectance. [114]

3.2. DNA Sensors

The unique complementary recognition features of nucleic acids can provide a key to understanding their specific binding properties with ligands, metal ions, and proteins. This information can also provide much insight into the workings of anticancer drugs and the mechanisms by which DNA damage by toxins occurs. Thus, the sequence-specific detection of DNA hybridization has attracted considerable interest various fields including clinical diagnostics, environmental/food/water monitoring, and even anti-bioterrorism. [115, 116] Consequently, various optical, piezoelectric and electrochemical techniques have been utilized for the development of DNA sensors. In olden times, widely employed fluorescent indicators like ethidium bromide that interchelate between dsDNA are simple and convenient. However, these intercalators may suffer from nonspecific affinity for ssDNA giving a high background signal. [117] Therefore, the value of using the sandwich system employing avidin-biotin system by introducing a DNA probe for the target sequence detected by optical sensors like SPR and LSPR becomes prominent.

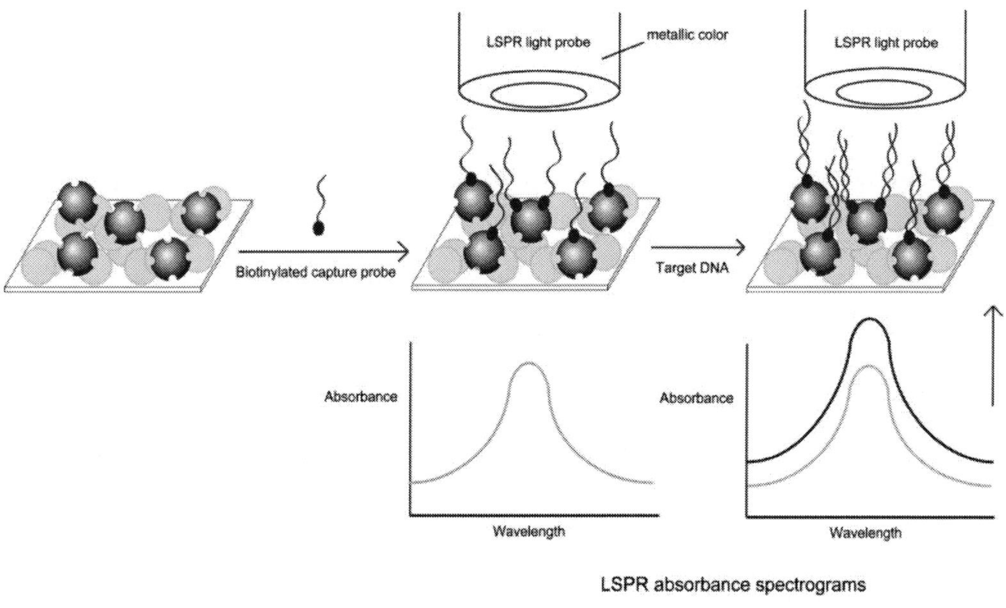

Figure 9. Typical detection of DNA hybridization using a label-free and non-destructive LSPR spectroscopy.

Fluorescence from QDs may be sensitive and convenient as a biosensor, but it still requires some sort labeling in order to bind with the QDs. However, LSPR enables a label-free detection of biomolecular molecules. Recently, Endo and coworkers was able to use LSPR to detect single base-pair mismatch in DNA-DNA and DNA-PNA hybridization. A SAM was first formed on the substrate with DDA before activation by EDC. NHS was then used to immobilize streptavidin. After that, biotinylated DNA/PNA probes were bound to the substrate via avidin-biotin linkage. Using this LSPR label-free biochip, the optical characteristics were monitored for the detection of specific DNA-DNA and DNA-PNA hybridization reactions. The DNA detection limit was 1pM and the detection was able to be performed in a short time. As a result LSPR label-free is a convenient technique that promises low cost, small sample volume, sensitive and fast detection. [118]

Tamiya and coworkers had also performed DNA sensing using LSPR measurement on gold capped nanoparticles. In Tamiya's experiment, peptide nucleic acids (PNA) were synthesized and immobilized to recognize target single stranded DNA(ssDNA) that is related to the tumor necrosis factor (TNF-α) which is the basis of many important diseases like ischemia and reperfusion injury. PNA are neutral, achiral nucleic acid whose backbone is composed of repeating N-(2-aminoethyl)-glycine units linked by peptide bonds. They bind to DNA and RNA with extra high affinity because they do not have charged phosphate groups present in DNA and RNA that repel each other in the hybridized form. PNAs are also resistant to enzymes such as proteases and nucleases. Therefore, PNAs are definitely good candidates for DNA binding. The experimental setup employed streptavidin and biotinylated probes to detect target DNA. A 4, 4'-Dithiodibutyric acid (DDA) was self assembled on the gold substrate before it gets activated by EDC and NHS for the covalent binding of streptavidin. After that, biotinylated ssDNA probes were incubated to bind to the streptavidin. Target DNA was then added to see an increase in the LSPR measurement. (Figure 9.) In this experiment, to simulate real life situation, both the normal TNF promoter's ssDNA and the defective ssDNA with single nucleotide polymorphism at position 308 (mutation affects the capacity to secrete TNF in response to stimuli) was used as target DNA. Gold nanoparticles layer was deposited by thermal evaporator onto a nanoparticles layer of silica bound on gold deposited glass substrate before immobilizing the ssPNA. It was found that immobilized PNA (ssPNA) biosensor detection limit was a remarkable concentration of 0.677pM. The mismatch DNA was also detected when an excess amount of 1.0μM was added to the immobilized probe, showing a slight red shift. Comparable experiments employing complimentary ssDNA as probes immobilized on the substrate surface had a higher detection limit of 0.803pM and the standard deviation was larger than ssPNA probes due to ssDNA's lower affinity for each other compared to PNA-DNA. Detecting single base pair mismatch implies extremely high sensitivity of the probe system and this could be used further in monitoring interactions of biomolecules (e.g. proteins, whole cells, receptors, etc.) with high detection capacity. [26] The ease of fabrication of this optical biosensor combined with the label free property of this technique also give huge promise to offer cost effective and convenient use in a miniaturized package.

A key characteristic of living organisms is their capability to grow, which requires the replication of DNA at the cellular level. The DNA polymerase catalyzes the lengthening of the growing end of a DNA chain by forming a phosphodiester bond between the deoxyribonucleoside triphosphate and the primer. Tsoi and coworkers described the interactions between DNA polymerases and DNA using the SPR biosensor in real time using DNA as the biosensing probe. The Klenow part of the *E. Coli* polymerase I, bacteriophage T4 polymerase and bacteriophage T7 polymerase were the chosen polymerases to work on matched or mismatched template DNA-primer duplexes. Biotinylated phospholipid monolayer was first formed on the gold electrode by depositing phospholipid dipalmitoyl-L-α-phosphatidyl ethanolamine (DPPE) and N-biotinylated DPPE in various ratio. The surface was then equilibrated with an HSM buffer (10mM Hepes, 150mMNaCl, and 1mMMgCl2, pH 7.5) before incubation with an avidin layer. A biotinylated DNA half duplex that contains the primers was then added to the matrix. The SPR angle was shown to increase with each binding event in real time. After testing their DNA probes with the three different polymerases, they were able to observe single base discrimination involved in protein-DNA interactions based on the SPR angle change. The results have depicted the capability of using the SPR biosensor to study molecular recognition events and the potential to detect DNA-DNA interactions too. [119]

Caruso and coworkers employed QCM to detect DNA hybridization on single and multi layered films of biotin-DNA. The gold QCM was exposed to dithiodipropionic acid before activation by EDC and NHS. Avidin was then incubated to be firmly immobilized on the surface. The biotinylated DNA was then injected, while the QCM surface gets measured by frequency changes using the QCM instrument. The experiment found that the detection of 30-mer oligonucleotide was possible. [120] The same group later tested the DNA binding to surface using SPR. Even though they did not use avidin biotin system in this case, they were able to detect a 21-mer oligonucleotide. They found that a modified gold surface using an ionizable thiol compound enabled the spontaneous attachment of DNA to the sensing surface through electrostatic attractions. SPR was used to confirm the oligomer's presence on the gold surface. [121] Piscevic et al. were able to detect a 10-mer oligonucleotide hybridize to its complement sequence immobilized by forming a thiolated DNA-SAM on a solid support using SPR. Parallel multi site detection of the DNA binding was possible with SPR as well. [122] Therefore combining the information above, it is theoretically possible to make a DNA sensor to detect DNA targets using SPR measurements. And by using the avidin-biotin system described by Caruso, a more robust sensing layer of DNA probes may be fabricated as compared to the SAMs used by Piscevic.

This theory is supported by Olofsson and coworkers' work. Much research has been done using UV-vis spectroscopy applied in the detection of DNA hybridization upon aggregation of colloidal gold nanoparticles that gives a large change in the collective electromagnetic response of the nanoparticles suspension. Inspired by such efforts, Oloffson designed a surface that enables aggregation-free colorimetric detection of biorecognition events. To get high sensitivity, dense biotin phospholipids were used to form a lipid bilayer supported on the walls of a quartz cuvette. In this way, a dense layer of avidin conjugated colloidal gold could be immobilized on the cuvette surface and this would reduce non specific binding to the surrounding substrate. Since most biotin binding sites of the avidin on the gold colloid are still available, biotinylated DNA probes could be bound to the surface. Using Mie theory and de Feiter's formalism, the changes in optical extinction was related to the coupled mass measured and the result was used to predict the LSPR shifts observed in binding events. A 20-mer DNA probe was used to hybridize with a label-free fully complementary DNA. This hybridization could be observed from LSPR spectrums to see a slightly larger (+0.7%) and red shift (0.3nm) peak after hybridization. The predictions were found to match the observed LSPR under certain parameters. This experimental observations support the capability of using the LSPR system as a DNA biosensor and point toward the possibility to perform detection array formats of antibody-antigen recognition. [123]

3.3. Immunosensors

Among different transduction techniques, optical methods like SPR can be rather popular because of its label-free and non-destructive property and high sensitivity. However, a disadvantage of SPR systems is that they detect an aspecific parameter i.e. the refractive index. Therefore selectivity must be introduced, for example, applying immunochemical experiments. [124] Optical immunosensors based on SPR were developed for the measurement of antigens, which could bind to antibodies immobilized on biosensor surfaces. LSPR have also become widely used for the development of miniaturized high-throughput immunosensors. [111]

One of the most typical immunosensor that made use of biotin avidin affinity is illustrated by Morgan and Taylor in 1991. (Figure 10)They showed that the interaction between immoblilized antibody and corresponding antigen could be easily monitored using surface plasmon resonance spectroscopy to the nanomolar range. A biotinylated gold surface was used to bind a monolayer of streptavidin. Biotinylated monoclonal antibodies to sex hormone binding globulin (SHBG) were then bound to the streptavidin before incubation of SHBG of varying concentrations. A linear dependence on the concentration of SHBG was observed for the shift in resonance angle of SPR and the detection nanomolar range was possible. The avidin-biotin system here not only acts a linker system, but it also helped in eliminating non-specific binding. [125]

A typical optical immunosensor that made used of quantum dots was reported by Kerman *et al.* Colloidal semiconductor quantum dots (QDs) have dimensions in the ~1-10nm range. The quantum confinement of hole carriers and electrons in a space smaller than the Bohr exciton radius results in these nanoscale spherical particles with size dependent tunable luminescent emission. [126] When QDs' inorganic core is coated with a layer of wider band gap semiconducting material, the QD photoluminescence quantum yield will be enhanced. Also, the use of appropriate surface coating ligands will cause QDs to disperse in a variety of solvents which enables a surface ready for bioconjugation. [127] Photoluminescence of QD is utilized in biochemical applications as labels in bioanalysis and diagnostics.

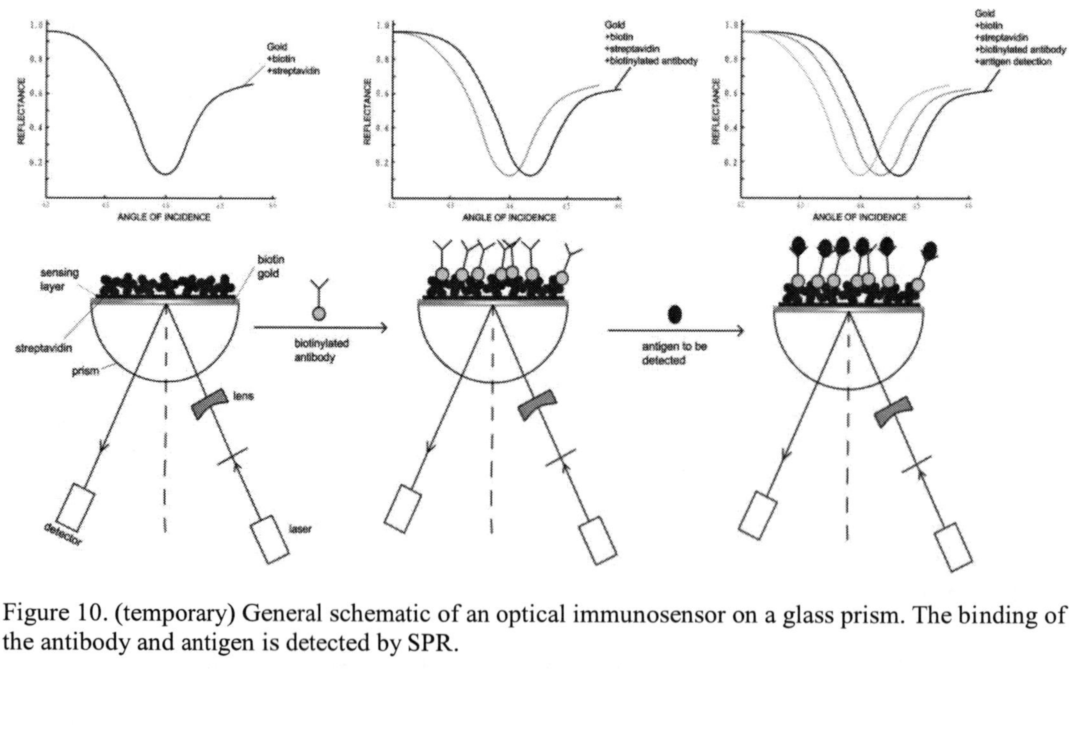

Figure 10. (temporary) General schematic of an optical immunosensor on a glass prism. The binding of the antibody and antigen is detected by SPR.

In Kerman's paper, possible application of QD-labeled biomolecules in a sandwich-type immunoassay for TPSA detection in connection with fluorescence imaging was explored. The sensing surface comprises of a covalently bound protein-A layer which then binds to the Fc region of total prostate-specific antigen-monoclonal antibodies (TPSA-mAb). TPSA which is the target protein was then introduced to the substrate to bind to the capturing antibody. After that, biotinylated detecting antibody binds to the TPSA specifically and then in turn, binds to Quantum dot 525 Streptavidin conjugates via the strong biotin-avidin affinity. Optical sensing of the surface fluorescence was performed in undiluted human serum and the detection limit was 0.25ng/mL. This assay has the potential to circumvent the tedious labeling of secondary antibodies in ELISA that may cause structural change in epitope and suppression in its specific recognition ability. [126]

The sensitivity in detecting binding events using LSPR is expected to be lower when large molecules like proteins are used as receptors on the metallic sensing surface because the penetration depth of the LSP field in the metal structure is limited by few tens of nanometers. In order to circumvent this problem, the sensitivity of the LSPR has to be drastically improved. In a recent study by Lee and coworkers, 4-chloro-1-napththol (4-CN) was used for precipitation on the gold nanoislands (NI) surface to amplify the change in LSPR signal to detect interferon-gamma (IFN–γ). [128] IFN–γ is a important cytokine secreted by immune cells in response to pathogens. The levels of IFN–γ provide crucial diagnostic information on the ability of the body to activate an immune response. For example, high levels of IFN–γ in HIV patients produced by their T-helper cells and cytotoxic T-lymphocytes could imply slow progression of the disease. [129]Glass chips were coated with gold films using electron beam evaporator, forming a 60 Å thickness of gold NI. The chips were then incubated in 11-mercaptoundecanoic acid before activation with NHS and EDC. The chips were then incubated inanti-IFN–γ in order to form a layer of IFN–γ layer followed by blocking with BSA and ethanolamine. Various concentrations of IFN–γ were added to the 'sandwich' in order to be captured by the antibodies. A mixture of biotinylated IFN–γ detection antibody was then introduced to the matrix to bind to the antigen before adding a HRP conjugated avidin solution that binds to the detection antibody. The reaction signal is amplified by adding hydrogen peroxide together with 4-CN to the chip surface in order to induce an immunoprecipitation reaction. HRP mediates the conversion of 4-CN to benzo-4-chlorocyclohexadienone (an insoluble precipitate) in the presence of the peroxide. The LSPR results show that precipitation drastically improved the sensitivity of the LSPR system. Before precipitation, there was no significant change in the centroid position of the LSP band. The large increase in local dielectric constant near the gold Nis results from the high polarizability of benzene and chloride group of the precipitate, thus dramatically enhancing the LSPR signal. The avidin-biotin system played a crucial role in the linking of HRP to the detection antibody in this study. (130)

SPR on the other hand has a much higher penetration depth than LSPR. However, the amplification of the signal for SPR is quite different from LSPR because it measures the refractive index on the sensor surface. SPR could detect large molecules on planar surfaces but its disadvantage is to detect low concentrations of smaller molecules. Thus in order to improve the sensitivity of SPR, Wink and coworkers, decided to use liposome immunosorbent assay (LISA) for application in microtiter plates.[124] LISA utilizes liposomes that entraps biomarker molecules like carboxyfluorescein to detect antigen binding as the biomarker is released and quantified upon treatment with detergent. [131] The size of liposome vesicles are large (up to about 1μm), thus a considerable shift in resonance angle in SPR is expected. In Wink's experiment, capture monoclonal IFN–γ antibody was adsorbed physically onto a thin polystyrene that covers a gold surface. After adding IFN–γ samples to bind to the capture antibodies, a biotinylated detection antibody was added to detect the IFN–γ that was bound. Avidin was then used as a bridging molecule to link the immune-sandwich with biotinylated liposomes. The SPR system was then used to monitor changes on the surface in real time, showing increase in SPR angle after every incubation step, with the most drastic one after introduction of liposomes. It was shown that when liposomes are used, the detection limit was improved by about 4×10^4 times to 100pg/mL as compared to no liposomal amplification. Because it is easy to biotinylate phospholipids, a liposome could be easily biotinylated by using biotinylated phospholipids. The biotinylation process of antibodies is also well established. Therefore, such convenience motivated the exploitation of avidin-biotin affinity in the binding mechanism of the liposomes to biotinylated antibody. [124]

Recently, an optical biosensor for IFN–γ incorporated fluorescence resonance energy transfer (FRET) and SPR in its development. Tuleuova and coworkers designed a sensing strategy that does not require multiple reagents and washing steps for real time detection of antigen binding to immunosensors. The design was based on an optimal aptamer beacon for FRET based detection of IFN–γ using SPR studies for optimization. The aptamer contains a fluorophores while a acceptor complimentary strand hybridizes with the aptamer to form a DNA duplex. Binding of IFN–γ disrupts the DNA duplex and results in fluorescence. 3' biotinylated aptamer with or without spacer and 5' biotinylated with or without spacer were used as variants to bind to a streptavidin coated surface. SPR experiments show that the 3'biotinylated aptamer without spacer had the lowest K_d and was used to construct the DNA duplex-based aptamer beacon. Eventually IFN–γ was introduced to bind to the aptamer, which then displaces the quencher strand. Fluorescence was detected at increasing intensities with the increase in the IFN–γ concentration. The disruption in the aptamer/anti-sense duplex by IFN–γ was proven by SPR detection of similar signal intensity after IFN–γ addition to both the channels containing aptamer only probes and aptamer duplex probes. Fluorescence microscopy was then used to observe a linear range from 5nM to 100nM of the antigen with a limit of detection of 5nM IFN–γ. This one-step immunoassay is fast and has high potential to be used in the dynamic monitoring of cytokine release. Ease of biotinylation of the aptamer is important in the immobilization of the probe to the streptavidin coated surface. [129]

Using a SAM that had a non fouling background by mixing oligo(ethylene glycol) alkanethiol and biotinylated alkanethiol, Ladd and coworkers bound a layer of streptavidin atop the SAM. Biotinylated single-stranded oligonucleotides were then bound on top of the streptavidin layer. A complimentary oligonucleotide conjugated with monoclonal anti-human chorionic gonadotropin (hCG) antibody was then added to the setup. By exploiting the specificity of DNA hybridization, the antibody gets immobilized onto the matrix too. The binding even was detected by SPR and the hCG limit of detection was 0.5ng/mL which was lower than the lowest detection limit of 5ng/mL established using biotinylated antibodies. This work has shown the potential of a DNA-antibody conjugate to convert a DNA array into a protein array that is site-specific by immobilization of DNA. [132]

The optical systems discussed in this section are mostly based on the relatively new emerging SPR and LSPR techniques, that applied the avidin-biotin system to detect DNA and protein based molecules. The two techniques are promising in the sense that high sensitivity could be achieved despite the absence of labeling reagents.

CONCLUSION

The case studies explored in this review emphasize on some specific electrochemical and optical detection techniques and they are far from complete. As we could already see, the useful information we could receive from electrochemical and optical biosensors range from the ultra sensitive detection of DNA using QCM to interferon-γ detection using SPR. Other molecules that were detected include *E. Coli* DNA, β-thalassaemia mutations, tumor necrosis factors, etc. which are all important in the field of clinical diagnosis. The strept(avidin)-biotin system plays a cardinal role in all these detections because they provide a bridging effect that helps form the immuno-sandwiches and the linkages of the reporter to hybridized DNA. Few, if not any, of the detection approaches described here are unique to their biomolecules. Instead, it is the special characteristics of strept(avidin) and biotin that gave researchers different opportunities to apply them in biosensors. The strong affinity between strept(avidin) and biotin ensure specificity in the binding interactions. Also, the most important advantage of the biotin-strept(avidin) system is that as long as one of them is immobilized, we could be sure that the other binding product would not be washed away. Most of the detection techniques and assays are based on the ease of biotinylation or strept(avidin) conjugation of alkanethiols, phospholipids, oligonucleotides, HRP enzyme, quantum dots, magnetic beads, gold nanoparticles and antibodies. With the development and advancement of new bioconjugation techniques, it is safe to say that more novel and creative detection assays could be fabricated in the near future.

ACKNOWLEDGMENT

We are very grateful to Yun Zhong for the designs of the graphic art.

REFERENCES

[1] Boas, M.A. The effect of desiccation upon the nutritive properties of egg-white. *Biochem. J.,* 1927,21, 712-724.

[2] Green, N. M. Avidin. In Advances in Protein Chemistry, New York: Academic Press; 1975.

[3] Green, N.M.; Avidin. 4. Stability at extremes of pH and dissociation into sub-units by guanidine hydrochloride. *Biochem. J.,* 1963, 89, 609-620.

[4] Savage MD, Mattson G, Desai S, Neilander G, Morgensen S, Conklin EJ. Avidin-biotin chemistry: A handbook. Rockford, Illinois : Pierce; 1992 .

[5] Rusmini, F; Zhong Z; Feijen J. Protein Immobilization Strategies for Protein Biochips,

Biomacromolecules, 2007, 8, 1775-1789.

[6] Chalet, L; Wolf, F J. The properties of streptavidin, a biotin-binding protein produced by *Streptomycetes. Arch Biochem. Biophys,* 1964, 106, 1-5.

[7] Bayer, E.; Ben-Hur, H.; Gitlin, G.; Wilchek, M. An improved method for the single-step purification of streptavidin. *J. Biochem. Biophys. Meth.,* 1986,13, 103-112.

[8] Gitlin, G.; Khait, I.; Bayer, E.A.; Wilchek, M.; Muszkat, K.A. Studies on the biotin-binding sites of avidin and streptavidin. A chemically induced dynamic nuclear polarization investigation of the status of tyrosine residues. *Biochem. J.,* 1989, 259, 493-498.

[9] Becker, J.M.; Wilchek, M.; Katchalski, E. Irreversible inhibition of biotin transport in yeast by biotinyl-p-nitrophenyl ester. *Proc. Natl. Acad. Sci. USA,* 1971, 68(10), 2604-2607.

[10] Orth, R. N.; Clarck, T. G.; Craighead, H. G. Avidin-Biotin Micropatterning Methods for Biosensor Applications. *Biomed. Microdevices.* 2003, *5*, 29-34.

[11] Forster, A.C. ; McInnes, J.L.; Skingle, D.C.; Symons, R.H. Non-radioactive hybridization probes prepared by the chemical labeling of DNA and RNA with a novel reagent, photobiotin . *Nucleic Acid Res.* 1985, 13, 745 – 761.

[12] Lacey, B. ; and Grant , W.N. Photobiotin as a sensitive probe for protein labeling . *Anal. Biochem.*1987, 163 , 151 – 158 .

[13] Diamandis, E.P.; Christopoulos, T.K. The biotin-(strept)avidin system: principles and applications in biotechnology. *Clin. Chem.,* 1991,37(5), 625-636.

[14] Dontha, N.; Nowall, W.B.; Kuhr, W.G. Generation of Biotin/Avidin/Enzyme Nanostructures with Maskless Photolithography. *Anal. Chem.* 1997, 69, 2619-2625

[15] Kumar, A. ; Tchen , P.; Roullet, F.; and Cohen, J. Nonradioactive labeling of synthetic oligonucleotide probes with terminal deoxynucleotidyl transferase . *Anal. Biochem.* 1988, 169, 376 – 382.

[16] Saiki, R.K.; Gelfand, D.H.; Stoffel, S.; Scharf, S.J.; Higuchi, R.; Horn, G.T.; Mullis, K.B.; Erlich, H.A. Primer-directed enzymatic amplification of DNA with a thermostable DNA polymerase. *Science,* 1988, 239, 487-491.

[17] Lo, Y.-M.D.; Mehal, W.Z.; Fleming, K.A. Rapid production of vector-free biotinylated probes using the polymerase chain reaction. Nucl. *Acids Res.,* 1988,16(17), 719.

[18] Folsom, V.; Hunkeler, M.J.; Hates, A.; Harding, J.D. Detection of DNA targets with biotinylated and fluoresceinated RNA probes. *Anal. Biochem.,* 1989,182, 309-314.

[19] Pierrat, S.; Zins, I.; Breivogel, A.; Sonnichsen, C. Self-assembly of Small Gold Colloids with Functionalized Gold Nanorods. *Nana Lett.* 2007, 7(2), 259-263.

[20] Hermanson, G.T. *Bioconjugate Techniques.* Second Edition. San Diego: Academic Press; 1996.

[21] Sutoh, K.; Yamamoto, K.; Wakabayashi, T. Electron microscopic visualization of the SH1 thiol of myosin by the use of an avidin–biotin system. *J. Mol. Biol.* 1984, 178, 323 – 339.

[22] Papermaster, D.S.; Schneider, B.G.; Zorn, M.A.; Kraehenbuhl, J.P. Immunocytochemic allocalization of opsin in outer segments and golgi zones of frog photoreceptor cells. An electron microscope analysis of cross-linked albumin-embedded retinas. *J. Cell Biol.,* 1978, 77, 196-210.

[23] Suffn, S.C.; Muck, K.B.; Young, J.C.; Lewin, K.; Porter, D.D. Improvement of the glucose oxidase immunoenzyme technic. Use of a tetrazolium whose formazan is stable

without heavy metal chelation. *Am. J. Clin. Pathol.,* 1979, *71, 492-496.*

[24] Avrameas, S. Coupling of enzymes to proteins with glutaraldehyde. Use of conjugates for the detection of antigens and antibodies. *Immunochem.,* 1969,6, 43-52.

[25] Ye, J.; Bonroy, K.; Nelis, D.; Frederix, F.; D'Haen, J.; Maes, G.; Borghs G. Enhanced localized surface plasmon resonance sensing on three-dimensional gold nanoparticles assemblies. *Colloids Surf., A,* 2008. 321(1-3), 313-317.

[26] Endo, T.; Kerman, K.; Nagatani, N.; Takamura, Y.; Tamiya, E. Label-Free Detection of Peptide Nucleic Acid-DNA Hybridization Using Localized Surface Plasmon Resonance Based Optical Biosensor. *Anal. Chem.* 2005, 77, 6976-6984.

[27] Morris, R.E.; Saelinger, C.B. Visualization of intracellular trafficking: use of biotinylated ligands in conjunction with avidin-gold colloids. *J. Histochem. Cytochem.,* 1984, 32(l), 124-128.

[28] Kerman, K.; Chikae, M.; Yamamura, S.; Tamiya, E. Gold nanoparticle-based electrochemical detection of protein phosphorylation. *Anal. Chim. Acta.* 2007, 588, 26-33.

[29] Cole, S.R.; Ashman, L.K.; Ey, P.L; Biotinylation:an alternative to radioiodination for the identification of cell surface antigens in immunoprecipitates. *Mol.Immunol.* 1987, 24, 699-705.

[30] Livaniou, E.; Evangelatos, G. P.; Ithakissios, D. S. Biotin radioligand assay with an $^{123-}$I-labeled biotin derivative, avidin, and avidin double-antibody reagents. *Clin. Chem.* 1987, 33, 1983-1988.

[31] Heitzmann, H.; Richards, F.M.; Use of the avidin-biotin complex for specific staining of biological membranes in electron microscopy. *Proc. Natl. Acad. Sci. USA,* 1974, *71 (9),* 3537-3541.

[32] Taki, M.; Hohsaka, T.; Murakami, H.; Taira, K.; Sisido, M. A non-natural amino acid for efficient incorporation into proteins as a sensitive fluorescent probe. FEBS Lett. 2001, 507, 35-38

[33] Murakami, H.; Hohsaka, T.; Ashizuka, Y.; Hashimoto, K.; Sisido, M. Site-directed incorporation of fluorescent nonnatural amino acids into streptavidin for highly sensitive detection of biotin. *Biomacromolecules,* 2000, 1, 118-25.

[34] Marttila, A. T.; Airenne, K. J.; Laitinen, O. H.; Bayer, E. A.; Wilchek, M.; Kulomaa, M. S. Engineering of chicken avidin: a progressive series of reduced charge mutants. *FEBS Lett.* 1998, 441, 313-317.

[35] Vermette, P.; Gengenbach, T.; Divisekera, U.; Kambouris, P.A.; Griesser H.J.; Meagher, L. Immobilization and surface characterization of NeutrAvidin biotin-binding protein on different hydrogel interlayers. *J. Colloid Interface Sci.* 2003, 259, 13-26.

[36] Hytonen, V.P., Laitinen, O.H., Airenne, T.T., Kidron, H., Meltola, N.J., Porkka, E.J., Horha, J., Paldanius, T., Maatta, J.A., Nordlund, H.R., et al. Efficient production of active chicken avidin using a bacterial signal peptide in *Escherichia coli. Biochem. J.,* 2004, 384, 385–390

[37] Ding, Z.; Fong, R. B.; Long, C. J.; Stayton, P. S.; and Hoffman, A. S. Size-dependent control of the binding of biotinylated proteins to streptavidin using a polymer shield. *Nature* 2001, 411, 59-62.

[38] Kossek, S.; Padeste, C.; Tiefenauer, L. Immobilization of streptavidin for immunosensors on nanostructured surfaces. *J. Mol. Recognit.* 1996, 9, 485-487.

[39] Breitenstein, M.; Holzel, R.; Bier, F.; Immobilization of Different biomolecules by

Atomic Force Microscopy. *J. Nanobiotechnol.* 2010, DOI: 10.1186/1477-3155-8-10.
[40] Hermanson, G.T.; Mallia, A.K.; Smith, P.K. Immobilized Affinity Ligand Techniques. San Diego: Academic Press; 1992.
[41] Kuramitz, H.; Sugawara, K.; Tanaka, S. Electrochemical Sensing of Avidin-Biotin Interaction using Redox Markers. *Electroanalysis* 2000, 12(16), 1299-1303.
[42] J. Spinke, M. Liley, F. Schmitt, H. Guder, L. Angermaier and W. Knoll. Molecular recognition at self-assembled monolayers: Optimization of surface functionalization. *J. Chem. Phys,*. 1993, 99, 7012.
[43] Piscevic, D.; Knoll, W.; Tarlov, M.J. Surface plasmon microscopy of biotin-streptavidin binding reactions on UV-photopatterned alkanethiol self-assembled monolayers. *Supramol. Sci.,* 1995, 2(2), 99-106.
[44] Nuzzo, R.G..; Allara D.L. Adsorption of bifunctional organic disulfides on gold surfaces. *J. Am. Chem. Soc.,* 1983, 105(13), 4481-4483.
[45] Wu, F. B.; He, Y. F.; Han S. Q. Matrix interference in serum total thyroxin (T4) time-resolved fluorescence immunoassay (TRFIA) and its elimination with the use of streptavidin-biotin separation technique. *Clin. Chim. Acta* 2001, 308, 117-126.
[46] Biebricher, A.; Paul, A.; Tinnefeld, P.; Gölzhäuser, A.; Sauer, M. Controlled three-dimensional immobilization of biomolecules on chemically patterned surfaces. *J. Biotechnol.,* 2004, 112, 97-107.
[47] Peluso, P.; Wilson, D.S.; Do, D.; Tran, H.; Venkatasubbaiah, M.; Quincy, D.; Heidecker, B.; Poindexter, K.; Tolani, N.; Phelan, M.; Witte, K.; Jung, L.S.; Wagner. P.; Nock, S. Optimizing antibody immobilization strategies for the construction of protein microarrays. *Anal. Biochem.*, 2003, 312, 113-124
[48] Su, X.D.; Wu, Y.J.; Robelek, R.; Knoll, W. Surface plasmon resonance spectroscopy and quartz crystal microbalance study of streptavidin film structure effects on biotinylated DNA assembly and target DNA hybridization. *Langmuir.* 2005, 21(1), 348-353.
[49] Reznik, G.O.; Vajda, S.; Cantor, C.R.; Sano, T. A Streptavidin Mutant Useful for Directed Immobilization on Solid Surfaces. *Bioconjugate Chem.* 2001, 12, 1000-1004.
[50] Davies, M. J.; Taylor, J. I.; Sachsinger, N.; Bruce, I. J. Isolation of plasmid DNA using magnetite as a solid-phase adsorbent. *Anal. Biochem.* 1998, 262, 92–94.
[51] Matsunaga, T.; Okamura, Y.; Tanaka, T. Biotechnological application of nano-scale engineered bacterial magnetic particles. *J. Mater. Chem.* 2004, 14, 2099–2105.
[52] Fan, A.; Lau, C.; Lu, J. Magnetic bead-based chemiluminescent metal immunoassay with a colloidal gold label. *Anal. Chem.* 2005, 77, 3238–3242.
[53] Matsunaga, T.; Maeda, Y.; Yoshino, T.; Takeyama, H.; Takahashi, M. Ginya, H.; Aasahina, J.; Tajima, H. Fully automated immunoassay for detection of prostate-specific antigen using nano-magnetic beads and micropolystyrene bead composites, 'Beads on Beads.' *Anal. Chim. Acta* 2007, 597, 331– 339.
[54] de Boer, E.; Rodriguez, P.; Bonte, E. Krijgsveld, J.; Katsantoni, E.; Heck, A.; Grosveld, F.; Strouboulis, J. Efficient biotinylation and single-step purification of tagged transcription factors in mammalian cells and transgenic mice. *Proc. Natl. Acad. Sci. USA* 2003, 100, 7480–7485.
[55] Liao, S.; Seeman, N. C. Translation of DNA signals into polymer assembly instructions. *Science* 2004, 306, 2072–2074.
[56] Maruyama, K.; Takeyama, H.; Nemoto, E.; Tanaka, T.; Yoda, K.; Matsunaga, T. Single

nucleotide polymorphism detection in aldehyde dehydrogenase 2 (ALDH2) gene using bacterial magnetic particles based on dissociation curve analysis. *Biotechnol. Bioeng.* 2004, 87, 687–694.

[57] Nakagawa, T.; Hashimoto, R.; Maruyama, K.; Tanaka, T.; Takeyama, H.; Matsunaga, T. Capture and release of DNA using aminosilane-modified bacterial magnetic particles for automated detection system of single nucleotide polymorphisms. *Biotechnol. Bioeng.* 2006, 94, 862–868.

[58] C, Rambihar.; K, Kerman.; Magnetic bead-based fluorometric detection of lectin-glycoprotein interactions, *Talanta*, 2010, 81, 1676-1680.

[59] IUPAC, A definition of biosensors. Electrochemical biosensors: proposed definitions and classification. *Biosens. Bioelectron.*, 1996, 11, 4

[60] Malachovska, V.; Pilla, P.; Cusano, A.; Netti, P.A.; Giordano, M.; The Development of Self-Assembled Monolayer Coatings as a Platform for Fiber-Optic Based Biosensors. 2010 June 28. Available from: www.aidic.it/nanomaterials/webpapers/42Malachovska.doc *Nanomaterials.*

[61] Canh, T.M. Biosensors. London: Chapman and Hall and Masson; 1993.

[62] Wang, J. Electroanalysis and Biosensors.*Anal. Chem.*, 1999, 71(12), 328-332.

[63] Palleschi, G.; Faridnia, M.H.; Lubrano, G.J.; Guilbault, G.G. Ideal hydrogen peroxide-based glucose sensor. Appl. Biochem. Biotechnol., 1991, 31(1), 21-35.

[64] Sugawara, K.; Tanaka, S.; Nakamura, H. Electrochemical Assay of Avidin and Biotin Using a Biotin Derivative Labeled with an Electroactive Compound. *Anal. Chem.,* 1995, 67, 299-302.

[65] Vreeke, M.; Rocca, P.; Heller, A. Direct Electrical Detection of Dissolved Biotinylated Horseradish Peroxide, Biotin, and Avidin. *Anal. Chem.,* 1995, 67, 303-306.

[66] Palecek, E.; Jelen, F.; Trnkova, L. Cyclic Voltammetry of DNA at a mercury electrode: an anodic peak specific for guanine. *Gen. Physiol. Biophys*. 1986, 5 (3), 315–329.

[67] Privett, B.J.; Shin, J.H.; Schoenfisch, M.H. Electrochemical Sensors. *Anal. Chem.,* 2008, 80, 4499-4517.

[68] Drummond, T. G.; Hill, M. G.; Barton, J. K. Electrochemical DNA sensors. *Nat. Biotechnol.* 2003, *21*, 1192-1199.

[69] Kerman, K.; Kobayashi, M.; Tamiya, E. Recent trends in electrochemical DNA biosensor technology. *Meas. Sci. Technol.* 2004, *15*, R1-R11.

[70] Wang Z, Liu D, Dong S (2000) Study on adsorption and oxidation of calf thymusDNA at glassy carbon electrode. *Electroanalysis* 12:1419

[71] Erdem, A.; Kerman, K.; Meric, B.; Ozsoz, M. Methylene blue as a novel electrochemical hybridization indicator. *Electroanalysis* 2001, *13*, 219-223.

[72] Kerman, K.; Ozkan, D.; Kara, P.; Meric, B.; Gooding J. J.; Ozsoz, M. Voltammetric determination of DNA hybridization using methylene blue and self-assembled alkanethiol monolayer on gold electrodes. *Anal. Chim. Acta* 2002, *462*, 39-47.

[73] Lucklum, R.; Henning, B.; Hauptmann, P.; Schierbaum, K.D.; Vaihinger, S.; Gopel, W. Quartz microbalance sensors for gas detection *Sens. Actuat. A.* 1991, 25, 705-710.

[74] Tombelli, S.; Mascini, M. Piezoelectric Quartz Crystal Biosensor: Recent Immobilisation Schemes. *Anal. Lett.,* 2000, 33(11), 2129-2151.

[75] Zhou X.C.; Huang, L.Q.; Li, F.Y. Microgravimetric DNA sensor based on quartz crystal microbalance: comparison of oligonucleotide immobilization methods and the application in genetic diagnosis. *Biosens. Bioelectron.* 2001, 16, 85-95.

[76] Dupont-Fillard, A.; Roget, A.; Livache, R.T.; Billon, M. Reversible oligonucleotide immobilisation based on biotinylated polypyrole film. *Anal. Chim. Acta.,* 2001, 449, 45-50.

[77] Arora, K.; Prabhakar, N.; Chand, S.; Malhotra, B.D. Ultrasensitive DNA hybridization biosensor based on poly aniline. *Biosens. Bioelectron.* 2007, 23, 613-620.

[78] Zhang, J.; Lao, R.; Song, S.; Yan, Z.; Fan, C. Design of an Oligonucleotide-Incorporated Nonfouling Surface and Its Application in Electrochemical DNA Sensors for Highly Sensitive and Sequence-Specific Detection of Target DNA. *Anal. Chem.*2008, 80, 9029-9033.

[79] Brooks, B. W.; Robertson, R. H.; Wallace, C. L. L.; Pfahler, W. Monoclonal antibodies specific for *Campylobacter fetus* lipopolysaccharides. *Vet. Microbiol.* 2002, *87*, 37-49.

[80] Alocilja, E. C.; Radke, S. M. Market analysis of Biosensors for food safety. *Biosens. Bioelectron.* 2003, *18*, 841-846.

[81] Arora, K.; Prabhakar, N.; Chand, S.; Malhotra, B.D. *Escherichia coli* Genosensor based on polyanline. *Anal. Chem.* 2007, 79, 6152-6158.

[82] Liu, G.; Wan, Y.; Gau, V.; Zhang, J.; Wang, L.; Song, S.; Fan, C. An Enzyme-Based E-DNA Senor for Sequence-Specific Detection of Femtomolar DNA Targets.

[83] Ricci, F.; Plaxco, K.W. E-DNA sensors for convenient, label-free electrochemica detection of hybridization. *Microchim. Acta,* 2008, 63, 149-155.

[84] K'Owino, I.O.; Agarwal, R.; Sadik, O.A. Novel Electrochemical Detection Scheme for Binding Interactions Using Monodispersed Reactivity of Silver Ions. *Langmuir.* 2003, 19, 4344-4350.

[85] Hoshi, T.; Anzai, J.; Osa, T. Controlled deposition of glucose oxidase on platinum electrode based on an avidin/biotin system for the regulation of output current of glucose sensors. *Anal. Chem.* 1995, 67, 770-774.

[86] Chaumpluk, P.; Kerman, K.; Takamura, Y.; Tamiya, E. Accumulation of amplified target DNAs using thiol/biotin labellin, SI nuclease, and ferrocene-sreptavidin-magnetic system and a direct detection of specific DNA signals with screen printed gold electrode. *Sci. Technol. Adv. Mater.* 2007, 8, 323-330.

[87] Rao, S. V.; Anderson, K. W.; Bachas, L. G. Oriented immobilization of proteins, *Mikrochim.Acta* 1998, *128*, 127-143.

[88] Scouten, W. H.; Luong, J. H. T.; Brown, R. S. Enzyme or protein immobilization techniques for applications in biosensor design, *Trends biotechnol.* 1995, *13*, 178-185.

[89] Taylor, R. F.; Marenchic, I. G.; Spencer, R. H. Antibody- and receptor-based biosensors for detection and process control, *Anal. Chim. Acta* 1991, *249*, 67-70.

[90] Ghindilis, A. L.; Atanasov, P.; Wilkins, M.; Wilkins, E. Immunosensors: Electrochemical sensing and other engineering approaches, *Biosens. Bioelectron.* 1998, *13*, 113-131.

[91] Skládal, P. Advances in electrochemical immunosensors, *Electroanalysis*, 1997, *9*, 737-745.

[92] Hianik, T.; Passechnik, V. I.; Sokolikova, L.; Snejdarkova, M.; Sivak, B.; Fajkus, M.; Ivanov, S. A.; Franek, M. Affinity biosensors based on solid supported lipid membranes. Their structure, physical properties and dynamics, *Bioelectrochem. Bioenerg.* 1998, *47*, 47-55.

[93] Hianik, T.; Snejdarkova, M.; Sokolikova, L.; Meszar, E.; Krivanek, R.; Tvarozek, V.; Novotny, I.; Wang, J. Immunosensors based on supported lipid membranes, protein

films and liposomes modified by antibodies., *Sens. Actuators* 1999, *B57*, 201-212.

[94] Barton, A.C.; Davis, F.; Higson, S.P. Labeless immunosensor assay for the stroke marker protein neuron specific enolase based upon an alternating current impedance protocol. *Anal Chem.* 2008, 80, 9411-9416.

[95] Cosnier, S.; Galland, B.; Gondran, C.; Lepellec A. Electrogeneration of biotinylated functionalized polypyrroles for the simple immobilization of enzymes, *Electroanalysis* 1998, 10, ,808– 813.

[96] Ouerghi, O.; Touhami, A.; Jaffrezic-Renault, N.; Martelet, C.; Ben Ouada, H.; Cosnier, S. Impedimetric Immunosensor Using Avidin-Biotin for Antibody Immobilization. *Bioelectrochemistry* 2002, 56, 131-133.

[97] Ionescu, R.E.; Gondran, C.; Gheber, L.A.; Cosnier, S.; Marks, R.S.; Construction of amperometric immunosensors based on the electrogeneration of a permeable biotinylated polypyrrole film. *Anal Chem.* 2004, 76(22), 6808-13.

[98] Haddour, N.; Chauvin, J.; Gondran, C.; Cosnier, S.; Photoelectrochemical immunosensor for label-free detection and quantification of anti-cholera toxin antibody. *J. Am. Chem. Soc.* 2006, 128(30), 9693-8.

[99] Wang, J.; Liu, G.; Engelhard, M.H.; Lin, Y. Sensitive Immunoassay of a Biomarker Tumor Necrosis Factor-α Based on Poly(guanine)-Functionalized Silica Nanoparticle Label. *Anal. Chem.* 2006, 78, 6974-6979.

[100] Mulvaney, P. Surface Plasmon Spectroscopy of Nanosized Metal Particles. *Langmuir* 1996, *12*, 788-800.

[101] Sastry, M.; Lala, N.; Patil, V.; Chavan, S.P.; Chittiboyina, A.G. Optical Absrption Study of the Biotin-Avidin Interactions on Colloidal Silver and Gold Particles. *Langmuir.,* 1998, 14, 4138-4142.

[102] Murakami, A.; Nakaura, M.; Nakatsuji, Y.; nagahara, S.; Tran- Cong, Q.; Makino, K. Fluorescent-labeled oligonucleotide probes: detection of hybrid formation in solution by fluorescence polarization spectroscopy. *Nucleic Acids Res.* 1991, *19*, 4097–4102.

[103] Park, S.Y.; Handa, H.; Sandhu, A. Magneto-Optical Biosensing Platform Based on Light Scattering from Self-Assembled Chains of Functionalized Rotating Magnetic Beads. *Nano Lett.* 2010, 10, 446-451.

[104] Habauzit, D.; Chopineau, J.; Roig, B. SPR-based biosensors: a tool for biodetection of hormonal compounds. *Anal. Bioanal. Chem.* 2007, 387, 1215-1223.

[105] Kretschmann E, Raether H Radiative decay of nonradiative surface plasmons excited by light. *Z. Naturforsch.* 1968, 23, 2135–2136

[106] Tsoi, P.Y.; Yang, J.; Sun, Y.T.; Sui S.F.; Yang, M.S. Surface plasmon resonance study of DNA polymerases binding to template/primer DNA duplexes immobilized on supported lipid monolayers, *Langmuir* 2000, 16, 6590–6596.

[107] Liedberg, B.; Nylander C.; and Lundstrom, I. Biosensing with surface plasmon resonance – how it all started. *Biosens. Bioelectron.* 1995, 10, i-ix.

[108] Malmqvist, M. Surface plasmon resonance for detection and measurement of antibody-antigen affinity and kinetics. *Curr. Opin. Immunol.* 1993, 5, 282-286.

[109] O'Shannessy, D. J. Determination of kinetic rate and equilibrium binding constants for macromolecular interactions: a critique of the surface plasmon resonance literature. *Curr. Opin. Biotechnol.* 1993, 5, 65-71.

[110] Schuck, P. Reliable determination of binding affinity and kinetics using surface plasmon resonance biosensors. *Curr. Opin. Biotechnol.* 1997, 8, 498-502.

[111] H. M. Hiep *et al.* A localized surface plasmon resonance based immunosensor for the detection of casein in milk. *Sci. Technol. Adv. Mater.* 2007, 8, 331.

[112] Jung, L.S.; Campbell, C.T.; Sticking Probabilities in Adsorption of ... from Liquid Ethanol Solution onto Gold *J. Phys. Chem. B.* 2000, 104, 11168–11178.

[113] Pérez-Luna, V.H.; O'Brien, M.J.; Opperman, K.A.; Hampton, P.D.; López, G.P.; Klumb, L.A.; Stayton, P. S. Molecular recognition between genetically engineered streptavidin and surface-bound biotin. *J. Am. Chem. Soc.* 1999, 121, 6469–6478.

[114] Hiep, H.M.; Yoshikawa, H.; Saito, M.; Tamiya, E. An Interference Localized Surface Plasmon Resonance Biosensor Based on the Photonic Structure of Au Nanoparticles and SiO2/Si Multilayers. *ACS Nano*, 2009, 3(2), 446-452.

[115] Debouck, C.; Goodfellow, P.N. DNA microarrays in drug discovery and development. *Nat. Genet.* 1999, 21. 48-50.

[116] Heller, M. J.; DNA microarray technology: devices, systems and applications *Annu. Rev. Biomed. Eng.* 2002, 4, 129–153.

[117] Millan, K.M.; Mikkelsen, S.R. Sequence-selective biosensor for DNA based on electroactive hybridization indicators. *Anal. Chem.* 1993, 65, 2317-2323.

[118] Endo, T.; Kerman, K.; Nagatani, N.; Tamiya, E. Excitation of localized surface plasmon resonance using a core-shell structured nanoparticle layer substrate and its application for label-free detection of biomolecular interactions. *J. Phys.: Condens. Matter.* 2007, 19, 215201-1-215201/10.

[119] Tsoi, P.Y.; Yang, J.; Sun, Y.T.; Sui, S.F.; Yang, M.S. Surface plasmon resonance study of DNA polymerases binding to template/primer DNA duplexes immobilized on supported lipid monolayers, *Langmuir* 2000, 16, 6590–6596.

[120] Frank C.; Rodda, E.; Furlong, N. QCM Study of DNA Immobilization and Hybridization for Nucleic Acid Sensor Development. *Anal. Chem.* 1997, 69, 2043-2049.

[121] Frank C.; Rodda, E.; Furlong, N.; Haring, V. DNA binding and hybridization on gold and derivatized surfaces. *Sens. Actuators, B.* 1997, 41, 189-197.

[122] Piscevic, D.; Lawall, R.; Veith, M.; Liley, M.; Okahata, Y.; Knoll, W. Oligonucleotide hybridization observed by surface plasmon optical techniques. *Appl. Surf. Sci.*, 1995, 90, 425–436.

[123] Olofsson, L.; Rindzevicius, T.; Pfeiffer, I. Kall, M. Hook, F. Surface-based gold-nanoparticle sensor for specific and quantitative DNA hybridization detection, *Langmuir.* 2003, 19, 10414-10419.

[124] Wink, T.; Van Zuilen, S. J.; Bult, A.; Van Bennekom W P. Liposome-mediated enhancement of the sensitivity in immunoassays of proteins and peptides in surface plasmon resonance spectrometry. *Anal. Chem.* 1998, *70*, 827-832.

[125] Morgan, H.; and Taylor, D. M. A surface plasmon resonance immunosensor based on the streptavidin-biotin complex. *Biosens. Bioelectron.* 1992, 7, 405–410.

[126] Kerman, K.; Endo, T.; Tsukamoto, M.; Chikae, M.; Takamura Y.; Tamiya, E. Quantum dot-based immunosensor for the detection of prostate-specific antigen using fluorescence microscopy. *Talanta.* 2007, 71, 1494–1499.

[127] Lingerfelt, B. M.; Mattoussi, H.; Goldman, E. R.; Mauro, J. M.; Anderson, G. P. Preparation of quantum dot-biotin conjugates and their use in immunochromatography assays. *Anal. Chem.* 2003, *75*, 4043-4049.

[128] Patolsky, F.; Zayats, M.; Katz, E.; Willner, I. Precipitation of an insoluble product on

enzyme monolayer electrodes for biosensor applications: characterization by faradaic impedance spectroscopy, cyclic voltammetry, and microgravimetric quartz crystal microbalance analyses. *Anal. Chem.* 1999, 71, 3171-3180.

[129] Tuleuova, N. *et al.* Development of an aptamer beacon for detection of interferon-gamma. *Anal. Chem.* 2010, 82, 1851-1857.

[130] Lee T.H. *et al.* Signal Amplification by Enzymatic Reaction in an Immunosensor Based on Localized Surface Plasmon Resonance (LSPR). *Sensors.* 2010, 10, 2045-2053.

[131] Rongen, H. A. H.; Van der Horst, H. M.; Hugenholtz, G. W. K.; Bult, A.; Van Bennekom, W. P. Development of a liposome immunosorbent assay for human interferon-γ *Anal. Chim. Acta.* 1994, 287, 191-199.

[132] Ladd, J.; Boozer, C.; Yu, Q.; Chen, S.; Homola, J.; Jiang, S.; DNA-directed protein immobilization on mixed self-assembled monolayers via a streptavidin bridge. *Langmuir.* 2004, 20(19), 8090-8095.

In: Biochemistry Research Updates
Editor: Simon J. Baginski
ISBN 978-1-61209-700-8

Chapter 4

CARDIAC (PATHO)PHYSIOLOGICAL ACTIONS OF THE CLASSICAL MU-, DELTA-, AND KAPPA-OPIOID RECEPTOR SYSTEM

Craig S. Bolte[1,2], Garrett J. Gross[3] and Jo El J. Schultz[1]
[1]Department of Pharmacology and Cell Biophysics, University of Cincinnati, US
[2]Department of Neonatology, Division of Pulmonary Biology, Cincinnati Childrens' Hospital Research Foundation, US
[3]Department of Pharmacology and Toxicology, Medical College of Wisconsin, US

I. INTRODUCTION

A. Cardiac Opioid Peptides and Classical Opioid Receptor System

The endogenous opioidergic system is known to express both receptors and peptides within the cardiovascular system, including within the heart itself (Barron, 2000). Precursors for all three families of endogenous opioid peptides (EOPs), prodynorphin (PDYN), proenkephalin (PPE) and proopiomelanocortin (POMC) are generated in the heart and alternative processing results in a myriad of mature peptides (Barron, 1999). Biosynthesis of the opioid peptide precursors, prodynorphin and proenkephalin, occurs predominantly in ventricular cardiomyocytes, not in the atria (Howells et al., 1986; Springhorn and Claycomb, 1989, 1992; Younes et al., 2000); whereas, processing of proopiomelanocortin to the cardiac β-endorphin peptide occurs in atrial myocytes (Millington et al., 1999). PPE has been shown to be more highly expressed in the adult heart (McLaughlin and Wu, 1998) and higher in the ventricles than in the atria (Barron, 2000; Weil et al., 1998), with left-sided preference (Weil et al., 1998; Weil et al., 2006). Similarly, POMC has been reported to only be detectable in the adult heart (McLaughlin and Wu, 1998) and to be preferentially localized to the ventricles compared to the atria (Millington et al., 1999). However, PDYN is reported to only be detected in the atria (McLaughlin and Wu, 1998). Similar to the biosynthesis of cardiac peptides from their prohormone precursors, cardiac opioid peptides are spatially distributed;

levels of enkephalins are higher in ventricular tissue (Barron et al., 1995; Springhorn and Claycomb, 1992) and β-endorphin peptide is predominantly found in atrial tissue (Millington et al., 1999). In atrial tissue, enkephalins are packaged in vesicles, but are found unpackaged in ventricles (Barron, 1999, 2000; Barron et al., 1995; Millington et al., 1999).

All three classical opioid receptors (OR), mu (MOR), delta (DOR) and kappa (KOR), are also expressed in the heart and like the EOPs, OR expression is temporally and spatially regulated. In the rodent heart, MOR and KOR are expressed in the embryonic and perinatal heart but MOR expression wanes rapidly postnatally as the DOR can first be detected. Thus, in the juvenile and adult heart the DOR and KOR are expressed and the MOR cannot be detected (Pugsley, 2002; Schultz and Gross, 2001). Binding studies further demonstrated that ORs exhibit significantly higher expression in the atria than the ventricles and on the left side compared to the right (Krumins et al., 1985).

B. Classical Opioid Receptor Signaling

Opioid receptors belong to the family of G-protein coupled receptors (Barron, 2000). The G_i and G_o types of G-proteins are thought to couple to most opioid receptors and mediate receptor signaling (Law and Loh, 1999; Standifer and Pasternak, 1997). Known functions of opioid receptor signaling include: 1) inhibition of both N- and L-type calcium channels, 2) inhibition of cyclic AMP (cAMP) formation and accumulation, 3) increased protein kinase C (PKC) activity, 4) decreased activity of the sodium-hydrogen exchanger, 5) stimulation of potassium channels (particularly the inward rectifying K^+ channel), and 6) increased phospholipase C (PLC) turnover leading to inositol triphosphate (IP_3) formation and release of Ca^{2+} from internal stores (Law and Loh, 1999). More recently, signaling through opioid receptors has been shown to stimulate mitogen-activated protein kinase (MAPK) pathways, especially ERK1/2. While our understanding of opioid receptor signaling in neurons is relatively complete, our knowledge of opioid receptor signaling in the heart is limited.

C. Cardiovascular Actions of the Opioidergic System

The endogenous opioidergic system has been shown to participate in cardiovascular physiology and pathophysiology (Bodnar and Hadjimarkou, 2003; Holaday, 1983; Pugsley, 2002). The cardiovascular actions that opioids exhibit are mediated both by opioid receptors located centrally in specific regions and nuclei of the brain that regulate the control of cardiovascular function and peripherally, by tissue-specific opioid receptors (Holaday, 1983; Pugsley, 2002). Opioids have been implicated in several significant cardiovascular disease states including in the cardio-depressant effects that accompany circulatory shock (Holaday, 1983), hypertension (Dumont and Lemaire, 1988; Feuerstein and Siren, 1987), ischemia-induced arrhythmias (Bian et al., 2000; Lee and Wong, 1986; Liu et al., 1988; Pugsley et al., 1992a; Sheng et al., 1994; Wong and Huang, 1988; Yu et al., 1999a), and protecting the heart against myocardial infarction (Schultz et al., 1998; Schultz and Gross, 2001; Schultz et al., 1996; Schultz et al., 1995; Schultz et al., 1997). In addition to the endogenous opioidergic system, opiates have long been known to decrease heart rate as well as blood pressure (Barron, 2000).

As might be implicated by their wide distribution, opioid receptors are involved in a variety of systemic processes including reproduction, respiration, feeding and nociception (Bodnar and Hadjimarkou, 2003); however, this chapter will focus on the activity of the classical opioid receptor system in cardiac diseases of ischemia-reperfusion injury and heart failure.

II. Role of the Classical Opioid Receptor System in Ischemia-Reperfusion Injury and Cardioprotection

A. Opioid-Induced Cardioprotection

The first evidence that opioids produced potent cardioprotective effects via activation of opioid receptors was published by Schultz and colleagues (Schultz et al., 1996; Schultz et al., 1995; Schultz et al., 1997) in intact rat hearts subjected to 30 minutes of index ischemia and 2 hours of reperfusion. Ischemic preconditioning (IPC) was used as a basis of comparison and was produced by three 5 minutes of left coronary occlusion interspersed with 5 minutes of reperfusion prior to the prolonged index ischemic period. IPC produced an approximately a 70% reduction in infarct size. Similarly, three 5 minute infusions of the opioid agonist, morphine, at 100 μg/kg per infusion period followed by a 5 minute washout period before the index ischemic period also reduced infarct size in this rat model approximately 65%, nearly the same as IPC. Interestingly, both of these effects of IPC and morphine were blocked by the ATP-sensitive potassium (K_{ATP}) channel blocker glibenclamide (Schultz et al., 1996). A confirmative study was then published by the Van Winkle group in rabbits (Chien and Van Winkle, 1996) which led to a surge of interest on studying the cardioprotective role of opioids, primarily as they relate to their mechanism in IPC and the receptors involved and distal signaling pathways. To obtain a good background on the major findings with morphine and other endogenous opioids in IPC and as a protectant in other organ systems as well, please refer to three excellent reviews already published (Gross, 2003; Peart et al., 2005; Schultz and Gross, 2001). For the remainder of this section of the chapter, focus will be on more recent advances and future perspectives in cardioprotection via activating delta and kappa opioid receptors with an emphasis on new aspects of opioid signaling, opioid-induced postconditioning (POC) and chronic protective effects during and following long-term treatment with opioids.

B. Classical Opioid Receptor Signaling in Cardioprotection and Ischemia-Reperfusion Injury

1. Delta Opioid Receptor Signaling

It has become apparent after many previous studies that adenosine, bradykinin and opioids are the three major triggers of the heart's response to an IPC stimulus (Peart et al., 2005) and that these agonists are coupled to G protein-coupled receptors (GPCRs). Although many subsequent steps in the transduction pathway are similar for all three agonists, there are some notable exceptions which may make opioids unique. Following receptor binding, the delta opioid receptor produces the src-dependent transactivation of the epidermal growth factor receptor (EGFR) followed by stimulation of the mitochondrial (mito) K_{ATP} channel and the release of a small amount of reactive oxygen species (ROS) (Forster et al., 2007). The ROS serve a signaling function and are thought to activate an isoform of PKC, most likely PKC-ε, which may phosphorylate PI3K, Akt, ERK/1/2 and the target of rapamycin (mTOR) to produce cardioprotection (Gross et al., 2004). Inhibition of all these kinases blocks the cardioprotection produced by delta opioid activation.

More recently, it was demonstrated that morphine, a nonselective opioid agonist and BW373U86, a delta selective opioid agonist inhibited glycogen synthase kinase 3-β phosphorylation (GSK-3β) at ser^9 when these opioids were given either before ischemia or just prior to reperfusion. No changes were observed in the phosphorylation of GSKα (Ser^{21} or Tyr^{279}) or GSKβ (Tyr^{216}). Interestingly, several direct GSK inhibitors, SB216763 (SB21) or SB415286 (SB41), mimicked opioid-induced cardioprotection; however, their effects were not blocked by wortmannin, LY 294002 or rapamycin which suggests that they act downstream of the PI3K pathway (Gross et al., 2004).

As opposed to many signaling molecules that are activated when phosphorylated, GSK3β phosphorylation inactivates the enzyme and serves to keep the mitochondrial permeability transition pore (MPTP) in a closed position and results in cardioprotection (Juhaszova et al., 2004; Saotome et al., 2009). Adenosine has many of the same downstream effects of bradykinin and opioids which converge at the MPTP site but it does not have ROS as an initial trigger or stimulate EGFR transactivation.

As far as mechanisms responsible for opioid-induced cardioprotection are concerned, both GSK3β (Gross et al., 2004) and the K_{ATP} (Gross et al., 2007; Liang and Gross, 1999; Schultz et al., 1998; Schultz et al., 1996) channel have been shown to mediate opioid-induced cardioprotection, but it has not been clear as to whether the K_{ATP} channel is upstream or downstream from GSK3β and if it is a trigger or end effector or both. In a recent study, Gross and group (Gross et al., 2007) administered morphine and BW373U86, a nonselective delta selective opioid agonist, respectively, 5 minutes prior to reperfusion in anesthetized rats. Both opioids significantly reduced infarct size and this effect was abrogated by either the sarcolemmal (sarc) selective K_{ATP} channel blocker HMR-1098 and by the mitochondrial (mito) K_{ATP} antagonist, 5-hydroxydecanoic acid (5-HD). Two K_{ATP} channel openers, P-1075, a sarc K_{ATP} selective agonist and the mito selective K_{ATP} opener, BMS-191095, were each administered 5 minutes prior to reperfusion after 25 minutes of ischemia. Both P-1075 and BMS-191095 significantly reduced infarct size, an effect that was blocked by the PI3K inhibitor, wortmannin (Gross et al., 2007). Cardioprotection produced by the direct GSK3β inhibitor, SB216763 given 5 minutes before reperfusion was completely blocked by the combination of HMR-1098 or 5-HD, respectively. These data suggest that both the sarc K_{ATP} channel and the mito K_{ATP} channel are involved in acute opioid-induced cardioprotection and that the GSK signaling pathway regulates cardioprotection via K_{ATP} channel opening.

There is also convincing evidence that the JAK/STAT pathway is essential for opioid-induced cardioprotection and that JAK 2 mediates the activation of STAT3, Akt and GSK3β in intact rat hearts and H9c2 cardiac cells (Gross et al., 2006). In these studies, rats were treated with vehicle, morphine, the δ-opioid agonist fentanyl isothiocynate (FIT), or the GSK3β inhibitor SB-216763 10 min before ischemia. Five minutes before opioid or SB21 treatment, some rats received the putative JAK2 inhibitor, AG-490, and the putative JAK3 inhibitor ZM-449829. H9C2 cardiomyoblast cells were also used to investigate FIT-induced signaling via molecular analysis. Morphine induced the phosphorylation of JAK2, but not JAK1, in the area at risk. Morphine, FIT, and SB21 also reduced infarct size compared with vehicle when administered before ischemia (Gross et al., 2006). AG-490 abrogated opioid-induced cardioprotection; whereas, ZM-449829 had no effect. Cardioprotection was afforded by SB21 even in the presence of AG-490. Morphine phosphorylated STAT3, Akt, and GSK-3β, and phosphorylation was abrogated by AG-490. FIT stimulation of H9C2 cells also caused a time-dependent phosphorylation of STAT3, Akt, and GSK-3β, and this effect was abrogated by AG-490. STAT3 phosphorylation was also dependent on phosphatidylinositol 3-kinase (PI3K) activation in both tissue and H9C2 cells. These data suggest that opioid-induced cardioprotection occurs via the JAK2 regulation of PI3K pathway-dependent STAT3, Akt, and GSK-3β, with GSK-3β contributing a central role in acute opioid-induced protection.

2. Kappa Opioid Receptor Signaling

There is less evidence to support a role for the κ-opioid receptor as opposed to the δ-opioid receptor in mediating the beneficial effects of acute ischemic preconditioning (IPC) to reduce infarct size. In fact, Aitchison and investigators (Aitchison et al., 2000) found that a κ-opioid agonist, bremazocine, actually increased infarct size in isolated rat hearts, an effect which was blocked by nor-BNI. On the other hand, Wang and group (Wang et al., 2001) preconditioned the isolated rat heart with two 5 minute periods of ischemia interspersed with 5 minutes of reperfusion prior to a 30 minute period of ischemia and determined infarct size. Ischemic PC produced a marked reduction in infarct size, an effect which was attenuated with either the selective κ-opioid antagonist, nor-BNI or by the selective δ-opioid antagonist, naltrindole. U50,448H, a selective κ-opioid receptor agonist or DADLE, a selective δ-opioid agonist, both produced marked reductions in infarct size. Based on these findings, these authors concluded that both δ- and κ-opioid receptors are involved in IPC in the isolated rat heart. Based upon the conflicting results of Aitchison et al. (Aitchison et al., 2000) and Wang et al. (Wang et al., 2001), further studies are needed to clarify the relative importance of these two receptors in acute cardioprotection and IPC in different animal species. One possibility for the differences in the results of these two studies may be related to the use of different κ agonists, U50,448H and bremazocine. Although U50,448H is a selective κ agonist, it is also known to inhibit L-type calcium channels in ventricular myocytes independent of its opioid receptor binding properties (Zhang et al., 1996). This function might be partially responsible for at least a portion of its cardioprotective effect, an action bremazocine has not been shown to possess.

More recent studies from our laboratory lend credibility to support a beneficial effect of kappa receptor activation on myocardial stunning and infarct size reduction in mouse and rat hearts (Peart et al., 2004; Peart et al., 2008; Peart and Gross, 2004b). To this end, we employed various concentrations of selective κ- and δ-opioid receptor agonists in an isolated murine heart model undergoing 20 minute global ischemia followed by 45 minute reperfusion (Peart and Gross, 2004b). Left ventricular developed pressure (LVDP) returned to 50.7 ± 2.1% of baseline values in untreated hearts. Infusion of the non-selective opioid agonist, morphine, led to a marked improvement in post-ischemic contractile recovery where LVDP returned to 65.5 ± 2.4% of baseline function. The δ-opioid selective agonist BW373U86 hydrochloride elicited maximal protection of 63.9 ± 3.4% recovery of LVDP. This effect was blocked by the δ-opioid selective antagonist, BNTX. Furthermore, administration of the κ-opioid selective agonist U50,488H also produced a marked improvement in contractile recovery leading to a 72.5 ± 5.3% recovery of LVDP. This degree of protection was also abolished by the κ-opioid receptor antagonist, nor-BNI. These data suggest that exogenous activation of δ- and κ-opioid receptors afford protection against myocardial stunning in the isolated murine heart, an effect attenuated by selective κ-and δ- receptor antagonists.

In a subsequent study (Peart et al., 2004), we examined three different κ-opioid receptor agonists and their effects upon infarct size and arrhythmia development. Male Sprague-Dawley rats were subjected to 30 minutes of occlusion followed by 90 minutes of reperfusion. Opioid receptor agonists were administered 10 minutes before the onset of ischemia, while the opioid antagonists were given 20 minutes before occlusion. Untreated rats exhibited an infarct size (IS as a percent of the area-at-risk, AAR) of 52.4 ± 2.7%. Pretreatment with the δ-opioid receptor agonist, BW373U86, limited infarct development to 37.2 ± 1.8%, which was reversed by the selective δ-opioid receptor antagonist, BNTX. All three κ-opioid receptor agonists studied, U50,488H, ICI 204,448, and BRL 52537 significantly reduced infarct size to levels comparable to that of BW373U86. The infarct-sparing effects of U50,488H and ICI 204,448 were abolished by the selective κ-opioid receptor antagonist, nor-BNI; however, nor-BNI failed to inhibit the cardioprotective effects of BRL 52537. Furthermore, U50,488H and BRL 52537, but not ICI 204,448, significantly reduced the incidence of arrhythmias. These effects were not blocked by nor-BNI. These data demonstrate that κ-opioid receptor activation provides a similar degree of infarct size reduction as δ-opioid receptor activation. κ-opioid receptor agonists also reduced arrhythmogenesis; however, these responses appear to be independent of κ-opioid receptor activation.

Finally, the temporal properties of κ-opioid receptor-mediated cardioprotection are less well characterized than δ-opioid receptor responses (Peart et al., 2008). Our study was aimed at delineating the time course of κ-opioid receptor-mediated protection in two experimental models: an *in vivo* rat model of regional myocardial infarction and an *ex vivo* Langendorff perfused murine heart model. In the rat model, the selective κ-opioid receptor agonist U50,488H administered either 10 minutes prior to ischemia or 5 minutes prior to reperfusion, significantly reduced infarct size compared to untreated rats. Administration of U50,488H 10 seconds after the onset of reperfusion failed to elicit protection. Cardioprotection with U50,448H administered immediately prior to reperfusion was abolished by a κ-opioid receptor antagonist given 10 minutes prior to reperfusion. In the *ex vivo* murine model, untreated hearts exhibited 28 ± 2% IS/AAR. Infusion of U50,488H significantly limited infarct size in mouse hearts when applied at the onset of reperfusion, yet failed to afford protection when infused prior to ischemia. Additionally, in both models studied, treatment with either wortmannin or 5-hydroxydecanoate abrogated the protective effects of U50,488H applied just prior to reperfusion. In summary, κ-opioid receptors afford cardioprotection primarily when activated prior to and not after reperfusion.

The signaling pathways, which mediate κ-opioid receptor-induced cardioprotection have been less well studied as compared to the pathways involved in δ-opioid receptor-induced protection. Ventura and colleagues (Ventura et al., 1992) initially showed that U50,448H increased both inositol 1,4,5-triphosphate (IP_3) and cytosolic-free calcium in rat ventricular myocytes and that stimulation of phospholipase C was also involved. Subsequently, Sheng and group (Sheng et al., 1997) observed that pertussis toxin abolished these effects of U50,448H, which suggests the involvement of a $G_{i/o}$ protein in transducing these effects of κ-receptor stimulation. Zhang and Wong (Zhang and Wong, 1998) also showed that U50,448H suppressed the accumulation of cyclic AMP via the phosphoinositol/calcium pathway in rat ventricular myocytes. Wenzlaff and investigators (Wenzlaff et al., 1998) demonstrated that κ-receptor stimulation with either U50,448H or the endogenous peptide dynorphin A reduced cell shortening in electrically driven rat ventricular myocytes and that these effects were blocked by pretreatment with pertussis toxin, naloxone and nor-BNI. DADLE, a μ-and δ-opioid agonist had no demonstrable effect on contractility in these myocytes. These investigators suggested that the endogenous release of dynorphins from the heart during hypoxia or ischemia may protect the heart similar to that already described for another endogenous mediator, adenosine. Two papers by Pyle and colleagues (Pyle et al., 2001; Pyle et al., 2000) support this hypothesis presented by Wenzlaff et al. (Wenzlaff et al., 1998). Pyle and group (Pyle et al., 2001; Pyle et al., 2000) showed in isolated rat hearts that U50,448H decreased the actin-myosin cycling rate in myocytes, which would be expected to lead to a conservation of ATP and a cardioprotective effect during ischemia. These investigators (Pyle et al., 2001) also showed that this effect was partially mediated via a PKC-activated pathway, which led to a reduction of actomyosin-induced ATP consumption. The importance of these specific pathways remains to be determined in acute κ-opioid receptor-induced cardioprotection.

C. Opioids and Postconditioning

Since opioids have been shown to reduce infarct size when administered just prior to reperfusion (Gross et al., 2004) it seems reasonable to hypothesize that opioids may be a component of the recently discovered phenomenon of ischemic postconditioning (POC). Ischemic postconditioning (Kin et al., 2004; Zatta et al., 2008) shares many of the factors that regulate the signaling cascade for ischemic preconditioning (IPC) except that the multiple occlusion reperfusion cycles occur immediately after the index ischemic period instead of prior to ischemia. Similar to IPC, ROS, K_{ATP} channels, PKC-ε, GSK-3β and the reperfusion injury salvage kinase (RISK) are all involved as mediators of IPC and POC; however, POC may be more clinically relevant since it can be used at the onset of reperfusion when most of the damage occurs in the ischemic/reperfused heart.

Although adenosine appears to be an important mediator of the POC phenotype (Kin et al., 2005), it is likely that other mediators may be involved in the protection produced by POC. In this regard, Zatta and colleagues (Zatta et al., 2008) recently investigated a role for opioid receptors in POC in anesthetized rats subjected to three 10 second cycles of reperfusion/occlusion at the onset of reperfusion after 30 minutes of ischemia. Naloxone (NL), its peripherally restricted analog naloxone methiodide, δ-opioid receptor antagonist naltrindole (NTI), κ-opioid receptor antagonist nor-binaltorphimine (nor-BNI), and μ-opioid receptor antagonist H-D-Phe-Cys-Tyr-D-Trp-Arg-Thr-Pen-Thr-NH^2 (CTAP) were administered intravenously 5 minutes before reperfusion. POC reduced infarct size; yet, none of the antagonists alone altered infarct size. All antagonists abrogated POC protection at higher doses. However, blockade of infarct sparing by POC was lost, when the doses of NL, NTI, nor-BNI, and CTAP were lowered. Representative small, intermediate, and large enkephalins in the ischemic region were quantified (fmol/mg protein). I/R reduced proenkephalin (58 ± 9 vs. 33 ± 4; $p<0.05$) and sum total of measured enkephalins, including proenkephalin, peptide B, methionine-enkephalin, and methionine-enkephalin-arginine-phenylalanine (139 ± 17 vs. 104 ± 7; $p<0.05$) compared with shams. POC increased total enkephalins (89 ± 8 vs. 135 ± 5; $p<0.05$) primarily by increasing proenkephalin (33 ± 4 vs. 96 ± 7; $p<0.05$). Therefore, the infarct-sparing effect of POC appeared to involve μ-opioid receptors and δ-opioid receptors, and preservation of enkephalin synthesis in the ischemic region. However, since there is strong evidence that μ-opioid receptors do not exist in adult rat myocardium, it appears unlikely that μ-opioid receptors are mediating the effect of POC in myocytes, although other cell types (e.g., neurons, endothelial cells) in the heart may be involved.

D. Opioids and Second Window of Cardioprotection

Baxter and co-workers (Baxter et al., 1994) originally demonstrated a second window of cardioprotection that occurred 24 to 48 hr following an ischemic preconditioning (IPC) stimulus or following administration of an adenosine type 1 receptor agonist in rabbits. This same group (Imagawa et al., 1997) also demonstrated that delayed cardioprotection produced by IPC could be blocked by either a PKC or tyrosine kinase inhibitor, suggesting that some important similarities exist between the early phase and late phases of IPC. It has been hypothesized that this delayed protection may result from translocation of PKC from the cytosol to the nucleus (Baxter et al., 1995) and the subsequent transcription and synthesis of cardioprotective molecules, such as inducible nitric oxide synthase and cyclooxygenase-2 (Dawn and Bolli, 2002). Although no previous work had been performed showing a delayed cardioprotective effect of opioids, the work of Ventura and colleagues (Ventura et al., 1998; Ventura et al., 1995) demonstrated that κ-opioid receptor stimulation resulted in a translocation of PKC to the nucleus and produced increased opioid peptide gene transcription. Furthermore, Gutstein and colleagues (Gutstein et al., 1997) have studied the involvement of opioids on the mitogen-activated protein kinase (MAPK) pathway, and have shown that opioids may induce the activation of extracellular signal-regulated kinase (ERK) and p38 MAPK. Based upon these intriguing findings, our laboratory examined the possibility that opioids might produce delayed cardioprotection in our rat model of ischemia/reperfusion injury. Fryer and group (Fryer et al., 1999) found that TAN-67, a selective δ-opioid agonist, had no cardioprotective effect to reduce infarct size 12 hr after administration; however, it produced a marked cardioprotective effect at 24 to 48 hr following drug administration, which disappeared at 72 hr. These cardioprotective effects were blocked by pretreatment with BNTX, a selective δ_1 antagonist. In addition, it was also possible to block these protective effects of TAN-67 by glibenclamide, a nonselective K_{ATP} channel antagonist, and by 5-HD, a mitochondrial selective K_{ATP} channel blocker, when the antagonists were administered 30 or 5 minutes prior to the long ischemic period following 48 hr of TAN-67 pretreatment, respectively (Fryer et al., 1999). These results suggest that δ_1-opioid receptor activation 24-48 hr prior to an ischemic insult results in a delayed cardioprotective effect that appears to be mediated by the mitochondrial K_{ATP} channel. Results obtained by Jiang and investigators (Jiang et al., 2004) and Patel and co-workers (Patel et al., 2004) suggested that inducible nitric oxide synthase (iNOS) and cyclooxygenase-2 were both involved in opioid-induced delayed protection. Jiang (Jiang et al., 2004) used an iNOS inhibitor, S-methylthiourea and iNOS knockout mice to arrive at their conclusion in the presence and absence of morphine treatment, whereas, Patel and group (Patel et al., 2004) produced delayed protection by administering two selective δ-agonists, BW373U86 and SNC-121 24 hours prior to the index ischemic period in the presence or absence of two iNOS inhibitors, S-methylthiourea and aminoguanidine or the COX-2 inhibitor, NS-398. Bolli and colleagues have confirmed these findings using the δ-agonist, BW373U86 , in the conscious rabbit heart (Kodani et al., 2002). Work of Wu and colleagues (Wu et al., 1999) revealed that κ-opioid receptor-induced cardioprotection occurs in two phases: the first window occurred ~ 1 hr after receptor activation and the second window developed 16 to 20 hr following drug administration in isolated rat ventricular myocytes. This κ-opioid receptor-mediated delayed cardioprotection was attenuated by inhibiting PKC. More studies are required to uncover the signaling pathways involved in κ-receptor mediated delayed cardioprotection.

E. Chronic Opioid-Induced Cardioprotection

It has recently been observed by our laboratory that chronic morphine exposure results in a profound ischemia-tolerant phenotype in mouse hearts (Peart and Gross, 2004c). In this study from our laboratory, mice were subcutaneously implanted with 75 mg morphine slow-release pellets for up to 5 days. A preconditioning-like effect occurred as soon as 3 days after morphine implantation; however, analgesic tolerance (hot plate test) did not occur until 4 days. Moreover, this cardioprotective phenotype persisted for at least 48 hours after complete opioid withdrawal. Interestingly, this chronic effect of morphine occurred in aged mice; whereas, acute morphine cardioprotection was lost (Peart and Gross, 2004a). These initial studies also suggested that the robust protection seen with chronic morphine was not blocked by a wide variety of classical preconditioning antagonists which have been shown to block acute morphine protection.

In partial agreement with our results, Kuzume and co-workers (Kuzume et al., 2003) reported that a 24 hour infusion of the δ-opioid receptor agonist, Met^5-enkephalin (ME) produced a significant reduction in infarct size in rabbits which was blocked by sarc and mito K_{ATP} antagonists, HMR1098 and 5-HD, respectively. However, when this same group infused ME for 14 days in mice, no cardioprotection occurred. These divergent results are puzzling but there were many differences between the two studies including mouse background and length of exposure to morphine and ME which might be responsible.

As mentioned above a five day exposure to morphine produces a chronic and robust cardioprotective phenotype. In lieu of these findings, we decided to investigate possible signaling pathways involved in acute (AM) morphine-induced protection versus chronic (CM) morphine-induced protection (Peart and Gross, 2006). The results suggested that CM preconditioning is mediated by a PKC-independent pathway which involved PKA, β_2-

adrenergic receptors and G_s proteins; whereas, AM preconditioning is partially mediated by G_i proteins and PKC.

III. Opioid Effect on Ischemic Arrhythmogenesis

Although the anti- or proarrhythmic effect of opioids has not been clearly elucidated, studies have suggested that the opioidergic system does influence the electrical conductivity of the heart via receptor-dependent and -independent mechanisms (see review by Pugsley (Pugsley, 2002)). Likewise, methadone and levacetylmethadol has been associated with QT interval prolongation and torsade de pointes in animal models of addiction and in opiate users (De Bels et al., 2003; Eikenburg and Stickney, 1979; Kang et al., 2003; Kornick et al., 2003; Krantz et al., 2002; Lipski et al., 1973; Martell et al., 2003). The receptor-independent actions of opioids on arrhythmogenesis are caused by a direct interaction with cardiac ion channels, in particular Na^+ channels and K^+ channels (Boachie-Ansah et al., 1989; Brasch, 1986; Fagbemi et al., 1983; Hung et al., 1998).

There is evidence as well that arrhythmogenesis of the heart is dependent upon opioidergic receptor activity. Our laboratory demonstrated that the δ_1-selective opioid agonist, TAN-67, has a potent antiarrrhythmic and antifibrillatory effect in rat hearts subjected to an ischemia/reperfusion protocol (Fryer et al., 2000). Maslov and investigators (Maslov et al., 1996) demonstrated that rats adapt to stress by synthesizing and/or releasing an increased amount of opioid peptides and showed a decreased severity and incidence of cardiac arrhythmias when subjected to a catecholamine challenge. This effect was abolished by the nonselective opioid antagonist naloxone. Kita and group (Kita et al., 1998) also demonstrated that the mito K_{ATP} blocker 5-HD and opioid antagonist naloxone inhibited ischemic preconditioning-induced suppression of reperfusion arrhythmias in the anesthetized rat heart. Theses results suggest that stimulation of δ-opioid receptors is antiarrhythmic and antifibrillatory. It is not known whether TAN-67's antiarrhythmic effect is receptor-mediated or acts peripherally or centrally.

As far as the κ-receptor is concerned, studies suggest that this opioid receptor may be the subtype that is involved in ischemia-induced arrhythmogenesis (Lee, 1992). It has been demonstrated that stimulation of the κ-opioid receptor by micromolar concentrations of the κ-agonist, U50,448H, abolished β-adrenoceptor-induced arrhythmias and intracellular Ca^{2+} oscillations (Yu et al., 1999b). However, results from the same laboratory (Bian et al., 1998b) have also reported that U50,448H, at lower nanomolar concentrations produced a proarrhythmic effect via a phospholipase C/inositol phosphate-3/Ca^{2+} and pertussis-sensitive pathway. Thus, the effect of activating the κ-opioid receptor may be antiarrhythmic at higher concentrations and proarrhythmic at lower concentrations. These pro- or antiarrhythmogenic dose-dependent effects of κ-opioid receptor stimulation were first observed by Kaschube and Brasch (Kaschube and Brasch, 1991) and Pugsley and co-workers (Pugsley et al., 1992b; Pugsley et al., 1993) such that the arrhythmogenic actions of κ-opioids occur at low doses, due to activation of the opioid receptors and the antiarrhythmic actions occur at higher doses, due to a non-opioid receptor, direct interaction with the cardiac membrane.

IV. Classical Opioid Receptors in Cardiac Function and Heart Failure Progression

A. Opioid Receptors and Modulation of Cardiac Contractility

There is evidence which supports the existence of an endogenous opioid receptor system in the heart that may contribute to functional changes in the normal and diseased myocardium. Activation of δ- and κ-opioid receptors has been shown to directly modulate cardiac contractility (Bian et al., 1998a; Ela et al., 1997; Ela et al., 1993; Laurent et al., 1985; Sheng et al., 1997; Sheng and Wong, 1996; Ventura et al., 1991; Ventura et al., 1992; Wenzlaff et al., 1998). In rat cardiac myocytes, δ-opioid receptor stimulation suppresses intracellular calcium cycling (i.e., alters L-type Ca^{2+} channel current and sarcoplasmic reticulum Ca^{2+} handling), ultimately reducing the contractile state of the heart (Ela et al., 1997; Ventura et al., 1992; Xiao et al., 1993). On the other hand, enkephalins, the endogenous opioid peptides for the δ-opioid receptor, exerted a positive inotropic effect on cultured chick embryo ventricular cells, increasing the contractile state of the cardiomyocytes (Laurent et al., 1985). Also, stimulating the δ_2-opioid receptor enhanced cardiac performance by increasing responsiveness to cytosolic Ca^{2+} despite reduced release of calcium (Fujita et al., 2000). Similarly, dynorphin, an endogenous κ-opioid receptor agonist, caused a transient increase in contractility due to an increase in the free Ca^{2+} available to the heart; however, chronic exposure of dynorphin resulted in a decrease in contractility of the heart because of depletion of intracellular Ca^{2+} pools (Bian et al., 1998a; Sheng et al., 1997; Sheng and Wong, 1996; Ventura et al., 1991; Ventura et al., 1992). In addition, stimulation of δ- and κ-opioid receptors markedly inhibits the β-adrenergic receptor-mediated increase in cardiac contractility which can be reversed by the non-selective opioid antagonist, naloxone (Pepe et al., 2004; Pepe et al., 1997; Xiao et al., 1997; Yu et al., 1998; Yu et al., 1999c). There is evidence that conditions such as chronic hypoxia (Shan et al., 2002) or hypertension (Yu et al., 1999c) modulate the interaction between the opioidergic and β-adrenergic systems, and opioid peptides are produced and co-released with catecholamines from peripheral neuronal terminals in the heart (Wilson et al., 1980). Furthermore, endogenous enkephalins negate sympathetic (β_1-adrenergic) actions on the heart (Pepe et al., 1997; Xiao et al., 1997). Alterations in catecholamine activity and adrenergic signaling are also underlying mechanisms of the pathogenesis of heart failure (Keys and Koch, 2004). Despite a number of interesting observations pertaining to the opioidergic system and cardiac function, the role of opioid receptors in the development and progression of heart failure remains to be determined.

B. Opioidergic System in Heart Failure Development and Progression

Heart failure is a leading cause of death in industrialized nations and can manifest from conditions such as hypertension, coronary heart disease, valve abnormalities and cardiomyopathy (Hwang et al., 2001; Rosamond et al., 2008; Towbin et al., 1999). At its most basic level, heart failure is the inability of the myocardium to pump sufficient blood to meet the metabolic needs of the tissues of the body (Rosamond et al., 2008). In an effort to increase perfusion, the heart hypertrophies or remodels, an early phase of heart failure progression known as the compensatory phase (Erdmann, 2000). These initially beneficial alterations become maladaptive and the heart begins to fail (Erdmann, 2000). Many factors and signaling pathways, including the opioidergic system (Barron, 1999), are known to be altered in heart failure progression with modulation of beta adrenergic receptor function and calcium signaling being the most prominent (Chu et al., 2002; Dorn and Molkentin, 2004; Hajjar et al., 1998; Hwang et al., 2001; Levitsky et al., 1991; Liew and Dzau, 2004; Molkentin and Dorn, 2001; Takano et al., 2003; Thompson et al., 2004; Wilkins and Molkentin, 2004; Zou et al., 2002).

Indications that the opioidergic system could be a factor in cardiac pathology date back as far as 1976 when a patient hospitalized for a heroin overdose subsequently developed acute cardiomyopathy (Paranthaman and Khan, 1976). Since that time, studies have shown that opioid peptide and receptor levels and expression, as well as opioidergic function, are altered in cardiomyopathy and heart failure originating from multiple etiologies, including high blood pressure and genetic mutations. Hypertensive patients were found to have decreased plasma levels of β-endorphin and leu-enkephalin (Kraft et al., 1987; Kraft et al., 1988; Ku, 2006; Zheng et al., 1995), while increased KOR but decreased DOR levels were found in hearts from the spontaneously hypertensive rat (Dumont and Lemaire, 1988; Zimlichman et al., 1996). Studies in our lab showed an attenuated ability of the KOR agonist U50488H to decrease left ventricular systolic pressure in hearts from hypertensive hamsters (Bolte et al., 2009a). Studies in cardiomyopathic hamsters found increased KOR, PDYN, dynorphin B, and PPE (Ouellette and Brakier-Gingras, 1988; Ouellette et al., 1991; Ventura et al., 1998; Ventura and Pintus, 1997). Furthermore, our studies showed enhanced negative inotropic and lusitropic effects of the KOR and DOR in the cardiomyopathic and failing hamster heart (Bolte et al., 2009b), suggesting that the opioid receptor system is altered in the development and progression to heart failure. In addition, patients with heart failure, cardiogenic shock and post-MI were reported to have elevated opioid peptide levels (Fontana et al., 1998a; Fontana et al., 1998b; Oldroyd et al., 1995), but decreased opioids peptide levels were observed in patients with atrial fibrillation or after hemorrhage (Feuerstein and Siren, 1987; Krumins et al., 1985; Lendeckel et al., 2005). However, these studies merely describe alterations in the endogenous opioidergic system but fail to investigate whether these alterations are causative or correlative.

Modulating cardiac growth under physiological or pathological stresses is necessary for proper cardiac function and dysregulation of growth and cardiac remodeling is an underlying hallmark of cardiac dysfunction and heart failure (Erdmann, 2000). The opioid receptor system has been implicated in cardiac growth under stress situations such as development (McLaughlin, 2002; Zagon et al., 1999) or catecholamine activity (Shan et al., 2007; Wang et al., 2004). Administration of the opioidergic antagonist naltrexone (NTX) to pregnant females resulted in increased heart size in the offspring and the EOP met-enkephalin was shown to inhibit cell growth and DNA synthesis across all three germ layers and in the heart, while NTX had the opposite effect, increasing DNA synthesis (McLaughlin, 2002; Zagon et al., 1999). In addition, classical opioidergic signaling, particularly that mediated by the KOR, has been shown to attenuate cardiac hypertrophy following chronic adrenergic stimulation (Shan et al., 2007; Wang et al., 2004). However, whether this is true for other causes of hypertrophy, such as pressure or volume overload, remains to be determined. This is of particular interest because the opioidergic and adrenergic systems belong to the G-protein coupled receptor family with sometimes synergistic or opposing signaling (Holmer and Homcy, 1991; Pepe et al., 2004), and opioidergic signaling has been shown to inhibit adrenergic signaling at concentrations below which they have a physiological effect independently (Pepe et al., 1997; Wong and Shan, 2001; Yu et al., 1998), thus the ability of opioidergic signaling to antagonize adrenergic-induced hypertrophy may be a result of the interaction between the two receptor systems.

Retrospective studies of opiate addicts provide an opportunity to evaluate the influence of chronic opioidergic stimulation in the human; however, these studies have produced conflicting results. One such study showed this population only developed severe coronary artery disease a third as often as the general population (Marmor et al., 2004). Similarly, a longitudinal study in which male heroin addicts were followed for 33 years found that only 11.7% of deaths were due to cardiovascular disease (Hser et al., 2001) compared to one-third of all deaths in the general American population (Rosamond et al., 2008). Accounting for violent deaths and overdoses, cardiovascular disease still amounted to less than 20% of deaths in this population. These studies indicate chronic opiate exposure may protect against cardiovascular disease; however, opiate addicts have also been shown to present with abnormal ECG readings, decreased cardiac output, and pulmonary edema (Ehret et al., 2007; Frishman et al., 2003). Furthermore, another retrospective study found that abusers of heroin or morphine developed acute myocardial congestion and interstitial fibrosis, secondary to increased expression of inflammatory markers in the heart (Dettmeyer et al., 2009), suggesting chronic opiate exposure may be an inducer of the heart failure phenotype. In accordance with these findings, a study using the ADHERE database found that patients with acute decompensated heart failure who received morphine as part of their treatment regimen had greater morbidity and mortality than patients not receiving morphine treatment (Peacock et al., 2008). This population also experienced lengthier hospitalizations, more frequent admission of longer duration to the intensive care unit, increased likelihood of needing ventilation and increased occurrence of dyspnea (Peacock et al., 2008). Thus, the interplay between opioid exposure, whether chronic or acute, and heart failure requires further elucidation and data suggests the opioidergic system may play discordant roles in different etiologies of heart failure. For instance, studies in our own lab have shown that opioid receptor signaling and function is augmented in hypertrophic cardiomyopathy (Bolte et al., 2009b) but blunted in hypertension (Bolte et al., 2009a), two etiologies that can manifest into heart failure.

The above clinical findings and retrospective studies indicate the degree to which the expression of opioidergic peptides and receptors may be altered in cardiovascular disease, but do not address whether these differences result in changes in myocardial opioidergic function. In this regard, several animal studies have suggested therapeutic effects of opioidergic signaling in heart failure development and progression. In a canine right heart failure model, six week administration of the nonselective opioid receptor antagonist, naloxone, improved cardiac contractility, prevented a decline in baroreflex sensitivity, and increased the effects of isoproterenol or forskolin without altering β-adrenergic receptor density, while it had no effect in sham-treated animals (Yatani et al., 1997). In this model, naloxone also increased arterial pressure and peripheral blood flow (Imai et al., 1994). Further evidence indicated DOR-mediated increased opioidergic tone in right heart failure as administration of a selective DOR antagonist produced similar results to naloxone, while more selective MOR antagonists failed to elicit a response (Imai et al., 1994). KOR antagonists, however, were not investigated. Supporting the notion of increased opioidergic tone in heart failure, canines with congestive heart failure had increased plasma BE levels (Liang et al., 1987; Pugsley, 2002), a ligand for both DOR and MOR. Furthermore, treatment with the non-selective opioid receptor antagonist, nalmefene, increased aortic blood pressure, cardiac output, LV contractility, and blood flow to the myocardium, kidney, and skeletal muscle (Liang et al., 1987; Pugsley, 2002). These effects were not seen in sham-treated animals (Liang et al., 1987; Pugsley, 2002). Studies in our own lab showed that hearts from cardiomyopathic hamsters were more sensitive to the negative inotropic and lusitropic effects of the kappa agonist U50,488H and the delta agonist Tan-67, but not the mu agonist, fentanyl, compared to age-matched control hamster hearts (Bolte et al., 2009b), also suggesting a cardiosuppressive tone of opiodergic signaling in heart failure. Furthermore, these effects were receptor-mediated and involved enhanced inhibition of cAMP formation as well as decreased intracellular Ca^{2+} transients vi a G_i protein-mediated mechanism (Bolte et al., 2009b). These opioid effects in failing hearts were blocked by inhibitors specific for the DOR and KOR (Bolte et al., 2009b). Similar to the dog model, patients with acute congestive heart failure had elevated plasma levels of BE, ME, and DYN (Fontana et al., 1993), with the degree of this increase relating to severity of pathology (Pugsley, 2002). Administration of naloxone to patients with severe congestive heart failure increased heart rate and blood pressure (Fontana et al., 1993), suggesting a cardiosuppressive opioidergic tone in these individuals. Right atrial strips from heart failure patients showed decreased negative inotropic effect when incubated with morphine compared to atrial strips from healthy individuals (Llobell and Laorden, 1996). This finding is in contrast to our hamster studies, it is not clear at this time if this represents a species difference or due to tissues taken from different chamber and hemisphere of the heart. Diminished opioidergic function in the diseased myocardium may result from receptor desensitization following a prolonged increase in opioid expression, giving further credence to enhanced opioidergic tone in heart failure. Further studies found similar results using the selective agonists DAMGO, DPDPE, and U50488H; however, the effects were not receptor mediated as they could not be attenuated by antagonist administration (Llobell and Laorden, 1996).

Although the preponderance of evidence suggests increased opioidergic tone is deleterious in the progression to heart failure, data has shown positive cardiac parameters following opioid administration in the failing heart. Acute administration of diamorphine, a morphine derivative with DOR selectivity, prior to a cardiopulmonary exercise test in patients with congestive heart failure increased VO_2 in exercise, tidal volume at rest and in exercise, and aerobic exercise capacity compared to placebo (Williams et al., 2003). Similarly, morphine was found to significantly decrease breathlessness as well as "unload" the heart in patients with NYHA class III or IV heart failure as it causes venodilation, decreases the sensation of dyspnea, reduces sympathetic nervous system activation, and reduces myocardial oxygen demand(Fonarow, 2002; Guntupalli, 1984; Johnson et al., 2002). Neither of these studies used any cardiac parameters as an endpoint.

Results, showing altered opioid expression or that treatment can positively or negatively influence heart failure development and its progression, suggest a relationship between the opioidergic system and this cardiac disease. The described alterations in opioidergic expression and function may be a compensatory mechanism of the disease and observed improvements following opioidergic administration only alleviate symptoms but are not involved in the underlying course of pathology. To date, no studies have been performed involving chronic administration of opioid agonists or antagonists to ascertain whether enhanced opioidergic tone can induce heart failure in population at risk. An initial study by McCloskey and colleagues (McCloskey et al., 2008), using a modified KOR, may gleam information about chronic activation of opioid receptor signaling. These investigators observed that even without OR agonist stimulation that basal constituently active KOR alone induced cardiomyopathy with 90% mortality in transgenic mice by 16 weeks (McCloskey et al., 2008). At the molecular level, these transgenic mice hade decreased sarcoplasmic/endoplasmic reticulum Ca^{2+} ATPase (SERCA) levels, decreased peak cytosolic Ca^{2+} and increased cytosolic Ca^{2+} at diastole (McCloskey et al., 2008). Opioids are commonly abused recreationally and are being prescribed for usage in conditions of chronic pain such as arthritis (Gibofsky and Barkin, 2008; Walker, 2003). If one considers also that opioids are commonly administered for analgesia, including in patients at risk for or suffering from heart failure, determining whether opioids are protective or detrimental to heart failure development and in what etiologies becomes essential.

SUMMARY AND CONCLUSIONS

There is a growing body of evidence to suggest that opioids are able to produce both an acute, delayed and chronic cardioprotective phenotype. Although studies suggest that activation of all three classical opioid receptors (mu, delta, or kappa) may result in cardioprotection depending on the species, age of the animals used and the selectivity of the agonists and antagonists used to tease out the dominant receptor, it appears that the receptor most likely to be the major one responsible for the cardioprotective phenotype is the δ-opioid receptor and met-enkephalin is most likely the major peptide responsible for the cardioprotective phenotype. The recent finding of chronic opioid-induced cardioprotection may provide a new mechanism to extend the window of protection for therapeutic benefit (Gross and Gross, 2007).

Patients suffering from heart failure are commonly prescribed opioids/opiates to alleviate the associated pain, and morphine has been administered to reduce the symptoms of congestion (Timmis et al., 1980). As opioids can act to inhibit normal cardiac function (Barron, 2000; Clo et al., 1985; Ventura et al., 1989; Wenzlaff et al., 1998; Xiao et al., 1993), which is already deficient in patients with heart failure, it may prove counterproductive to administer these agents to individuals who already have compromised cardiac function. Similarly, caution must also be used in prescribing opioid agonists to patients who are devoid of cardiovascular disease risk factors as our laboratory has shown increased sensitivity to a delta agonist, likely mediated by the D_2OR, prior to pathology in hearts at risk for heart failure (Bolte et al., 2009b). If cardiac opioid receptor signaling antagonizes efficient cardiac function and triggers cardiac hypertrophy and failure, then chronic opioid agonist administration could induce a failing phenotype in a previously healthy heart. In addition, little is known as to what effect long-term opioid/opiate usage may have on cardiac function and heart failure progression either in individuals at risk for cardiovascular disease or in those without preexisting risk factors. This represents an important future direction.

There are many questions yet to be answered concerning the effects of opioids in the ischemic/reperfused heart or heart failure. Further research, both basic and clinical, may lead to the development of selective opioid agonists or antagonists which prove useful for cardioprotection against ischemia-reperfusion injury and/or treatment of heart failure.

REFERENCES

Aitchison, K.A., Baxter, G.F., Awan, M.M., Smith, R.M., Yellon, D.M., and Opie, L.H. (2000). Opposing effects on infarction of delta and kappa opioid receptor activation in the isolated rat heart: implications for ischemic preconditioning. *Basic Res. Cardiol. 95*, 1-10; discussion 11.

Barron, B.A. (1999). Opioid peptides and the heart. *Cardiovasc. Res. 43*, 13-16.

Barron, B.A. (2000). Cardiac opioids. *Proc. Soc. Exp. Biol. Med. 224*, 1-7.

Barron, B.A., Oakford, L.X., Gaugl, J.F., and Caffrey, J.L. (1995). Methionine-enkephalin-Arg-Phe immunoreactivity in heart tissue. *Peptides 16*, 1221-1227.

Baxter, G.F., Goma, F.M., and Yellon, D.M. (1995). Involvement of protein kinase C in the delayed cytoprotection following sublethal ischaemia in rabbit myocardium. *Br. J. Pharmacol. 115*, 222-224.

Baxter, G.F., Marber, M.S., Patel, V.C., and Yellon, D.M. (1994). Adenosine receptor involvement in a delayed phase of myocardial protection 24 hours after ischemic preconditioning. *Circulation 90*, 2993-3000.

Bian, J.S., Pei, J.M., Cheung, C.S., Zhang, W.M., and Wong, T.M. (2000). kappa -opioid receptor stimulation induces arrhythmia in the isolated rat heart via the protein kinase C/Na(+)-H(+)exchange pathway. *J. Mol. Cell Cardiol. 32*, 1415-1427.

Bian, J.S., Wang, H.X., Zhang, W.M., and Wong, T.M. (1998a). Effects of kappa-opioid receptor stimulation in the heart and the involvement of protein kinase C. *Br. J. Pharmacol. 124*, 600-606.

Bian, J.S., Zhang, W.M., Xia, Q., and Wong, T.M. (1998b). Phospholipase C inhibitors attenuate arrhythmias induced by kappa-receptor stimulation in the isolated rat heart. *J. Mol. Cell Cardiol. 30*, 2103-2110.

Boachie-Ansah, G., Sitsapesan, R., Kane, K.A., and Parratt, J.R. (1989). The antiarrhythmic and cardiac electrophysiological effects of buprenorphine. *Br. J. Pharmacol. 97*, 801-808.

Bodnar, R.J., and Hadjimarkou, M.M. (2003). Endogenous opiates and behavior: 2002. *Peptides 24*, 1241-1302.

Bolte, C., Newman, G., and Schultz Jel, J. (2009a). Hypertensive state, independent of hypertrophy, exhibits an attenuated decrease in systolic function on cardiac kappa-opioid receptor stimulation. *Am. J. Physiol. Heart Circ. Physiol. 296*, H967-975.

Bolte, C., Newman, G., and Schultz Jel, J. (2009b). Kappa and delta opioid receptor signaling is augmented in the failing heart. *J. Mol. Cell Cardiol. 47*, 493-503.

Brasch, H. (1986). Influence of the optical isomers (+)- and (-)-naloxone on beating frequency, contractile force and action potentials of guinea-pig isolated cardiac preparations. *Br. J. Pharmacol. 88*, 733-740.

Chien, G.L., and Van Winkle, D.M. (1996). Naloxone blockade of myocardial ischemic preconditioning is stereoselective. *J. Mol Cell Cardiol. 28*, 1895-1900.

Chu, G., Haghighi, K., and Kranias, E.G. (2002). From mouse to man: understanding heart failure through genetically altered mouse models. *J. Card. Fail 8*, S432-449.

Clo, C., Muscari, C., Tantini, B., Pignatti, C., Bernardi, P., and Ventura, C. (1985). Reduced mechanical activity of perfused rat heart following morphine or enkephalin peptides administration. *Life Sci. 37*, 1327-1333.

Dawn, B., and Bolli, R. (2002). Role of nitric oxide in myocardial preconditioning. *Ann. N. Y. Acad. Sci. 962*, 18-41.

De Bels, D., Staroukine, M., and Devriendt, J. (2003). Torsades de pointes due to methadone. *Ann. Intern. Med. 139*, E156.

Dettmeyer, R., Friedrich, K., Schmidt, P., and Madea, B. (2009). Heroin-associated myocardial damages--conventional and immunohistochemical investigations. *Forensic. Sci. Int. 187*, 42-46.

Dorn, G.W., 2nd, and Molkentin, J.D. (2004). Manipulating cardiac contractility in heart failure: data from mice and men. *Circulation 109*, 150-158.

Dumont, M., and Lemaire, S. (1988). Increased content of immunoreactive Leu-enkephalin and alteration of delta-opioid receptor in hearts of spontaneously hypertensive rats. *Neurosci. Lett. 94*, 114-118.

Ehret, G.B., Desmeules, J.A., and Broers, B. (2007). Methadone-associated long QT syndrome: improving pharmacotherapy for dependence on illegal opioids and lessons learned for pharmacology. *Expert Opin. Drug Saf. 6*, 289-303.

Eikenburg, D.C., and Stickney, J.L. (1979). Cardiac effects of 1-alpha-acetylmethadol. V. Possible involvement of the autonomic nervous system. *Toxicol. Appl. Pharmacol. 51*, 91-100.

Ela, C., Barg, J., Vogel, Z., Hasin, Y., and Eilam, Y. (1997). Distinct components of morphine effects on cardiac myocytes are mediated by the kappa and delta opioid receptors. *J. Mol. Cell Cardiol. 29*, 711-720.

Ela, C., Hasin, Y., and Eilam, Y. (1993). Opioid effects on contractility, Ca(2+)-transients and intracellular pH in cultured cardiac myocytes. J. *Mol. Cell Cardiol. 25*, 599-613.

Erdmann, E. (2000). The management of heart failure--an overview. *Basic Res. Cardiol. 95 Suppl 1*, I3-7.

Fagbemi, O., Kane, K.A., Lepran, I., Parratt, J.R., and Szekeres, L. (1983). Antiarrhythmic actions of meptazinol, a partial agonist at opiate receptors, in acute myocardial ischaemia. *Br. J. Pharmacol. 78*, 455-460.

Feuerstein, G., and Siren, A.L. (1987). The opioid system in cardiac and vascular regulation of normal and hypertensive states. *Circulation 75*, I125-129.

Fonarow, G.C. (2002). Pharmacologic therapies for acutely decompensated heart failure. *Rev. Cardiovasc. Med. 3 Suppl 4*, S18-27.

Fontana, F., Bernardi, P., Merlo Pich, E., Tartuferi, L., Boschi, S., De Iasio, R., and Spampinato, S. (1998a). Opioid peptides in response to mental stress in asymptomatic dilated cardiomyopathy. *Peptides 19*, 1147-1153.

Fontana, F., Bernardi, P., Pich, E.M., Boschi, S., De Iasio, R., and Spampinato, S. (1998b). Endogenous opioid peptides and mental stress in congestive heart failure patients. *Peptides 19*, 21-26.

Fontana, F., Bernardi, P., Pich, E.M., Capelli, M., Bortoluzzi, L., Spampinato, S., and Canossa, M. (1993). Relationship between plasma atrial natriuretic factor and opioid peptide levels in healthy subjects and in patients with acute congestive heart failure. *Eur. Heart J. 14*, 219-225.

Forster, K., Kuno, A., Solenkova, N., Felix, S.B., and Krieg, T. (2007). The delta-opioid receptor agonist DADLE at reperfusion protects the heart through activation of pro-survival kinases via EGF receptor transactivation. *Am. J. Physiol. Heart Circ. Physiol. 293*, H1604-1608.

Frishman, W.H., Del Vecchio, A., Sanal, S., and Ismail, A. (2003). Cardiovascular manifestations of substance abuse: part 2: alcohol, amphetamines, heroin, cannabis, and caffeine. *Heart Dis. 5*, 253-271.

Fryer, R.M., Hsu, A.K., Eells, J.T., Nagase, H., and Gross, G.J. (1999). Opioid-induced second window of cardioprotection: potential role of mitochondrial KATP channels. *Circ. Res. 84*, 846-851.

Fryer, R.M., Hsu, A.K., Nagase, H., and Gross, G.J. (2000). Opioid-induced cardioprotection against myocardial infarction and arrhythmias: mitochondrial versus sarcolemmal ATP-sensitive potassium channels. *J. Pharmacol. Exp. Ther. 294*, 451-457.

Fujita, S., Smart, S.C., and Stowe, D.F. (2000). Enhanced contractile responsiveness to cytosolic Ca(2+) by delta-2 opioid agonist deltorphin in intact guinea pig hearts. *J. Mol. Cell Cardiol. 32*, 1647-1659.

Gibofsky, A., and Barkin, R.L. (2008). Chronic pain of osteoarthritis: considerations for selecting an extended-release opioid analgesic. *Am. J. Ther. 15*, 241-255.

Gross, E.R., and Gross, G.J. (2007). Ischemic Preconditioning And Myocardial Infarction: An Update and Perspective. D*rug Discov. Today Dis. Mech. 4*, 165-174.

Gross, E.R., Hsu, A.K., and Gross, G.J. (2004). Opioid-induced cardioprotection occurs via glycogen synthase kinase beta inhibition during reperfusion in intact rat hearts. *Circ. Res. 94*, 960-966.

Gross, E.R., Hsu, A.K., and Gross, G.J. (2006). The JAK/STAT pathway is essential for opioid-induced cardioprotection: JAK2 as a mediator of STAT3, Akt, and GSK-3 beta. *Am. J. Physiol. Heart Circ. Physiol. 291*, H827-834.

Gross, E.R., Hsu, A.K., and Gross, G.J. (2007). GSK3beta inhibition and K(ATP) channel opening mediate acute opioid-induced cardioprotection at reperfusion. *Basic Res. Cardiol. 102*, 341-349.

Gross, G.J. (2003). Role of opioids in acute and delayed preconditioning. J. *Mol. Cell Cardiol. 35*, 709-718.

Guntupalli, K.K. (1984). Acute pulmonary edema. *Cardiol. Clin. 2*, 183-200.

Gutstein, H.B., Rubie, E.A., Mansour, A., Akil, H., and Woodgett, J.R. (1997). Opioid effects on mitogen-activated protein kinase signaling cascades. *Anesthesiology 87*, 1118-1126.

Hajjar, R.J., Muller, F.U., Schmitz, W., Schnabel, P., and Bohm, M. (1998). Molecular aspects of adrenergic signal transduction in cardiac failure. *J. Mol. Med. 76*, 747-755.

Holaday, J.W. (1983). Cardiovascular effects of endogenous opiate systems. *Annu. Rev. Pharmacol. Toxicol. 23*, 541-594.

Holmer, S.R., and Homcy, C.J. (1991). G proteins in the heart. A redundant and diverse transmembrane signaling network. *Circulation 84*, 1891-1902.

Howells, R.D., Kilpatrick, D.L., Bailey, L.C., Noe, M., and Udenfriend, S. (1986). Proenkephalin mRNA in rat heart. *Proc. Natl. Acad .Sci. USA 83*, 1960-1963.

Hser, Y.I., Hoffman, V., Grella, C.E., and Anglin, M.D. (2001). A 33-year follow-up of narcotics addicts. *Arch Gen. Psychiatry 58*, 503-508.

Hung, C.F., Tsai, C.H., and Su, M.J. (1998). Opioid receptor independent effects of morphine on membrane currents in single cardiac myocytes. *Br. J. Anaesth. 81*, 925-931.

Hwang, J.J., Dzau, V.J., and Liew, C.C. (2001). Genomics and the pathophysiology of heart failure. *Curr. Cardiol. Rep. 3*, 198-207.

Imagawa, J., Baxter, G.F., and Yellon, D.M. (1997). Genistein, a tyrosine kinase inhibitor, blocks the "second window of protection" 48 h after ischemic preconditioning in the rabbit. *J. Mol. Cell Cardiol. 29*, 1885-1893.

Imai, N., Kashiki, M., Woolf, P.D., and Liang, C.S. (1994). Comparison of cardiovascular effects of mu- and delta-opioid receptor antagonists in dogs with congestive heart failure. *Am. J. Physiol. 267*, H912-917.

Jiang, X., Shi, E., Nakajima, Y., and Sato, S. (2004). Inducible nitric oxide synthase mediates delayed cardioprotection induced by morphine in vivo: evidence from pharmacologic inhibition and gene-knockout mice. *Anesthesiology 101*, 82-88.

Johnson, M.J., McDonagh, T.A., Harkness, A., McKay, S.E., and Dargie, H.J. (2002). Morphine for the relief of breathlessness in patients with chronic heart failure--a pilot study. *Eur. J. Heart Fail 4*, 753-756.

Juhaszova, M., Zorov, D.B., Kim, S.H., Pepe, S., Fu, Q., Fishbein, K.W., Ziman, B.D., Wang, S., Ytrehus, K., Antos, C.L., *et al.* (2004). Glycogen synthase kinase-3beta mediates convergence of protection signaling to inhibit the mitochondrial permeability transition pore. *J. Clin. Invest 113*, 1535-1549.

Kang, J., Chen, X.L., Wang, H., and Rampe, D. (2003). Interactions of the narcotic l-alpha-acetylmethadol with human cardiac K+ channels. *Eur. J. Pharmacol. 458*, 25-29.

Kaschube, M., and Brasch, H. (1991). Negative chronotropic but no antiarrhythmic effect of (+)- and (-)-naloxone in rats and guinea pigs. *Cardiovasc. Res. 25*, 230-234.

Keys, J.R., and Koch, W.J. (2004). The adrenergic pathway and heart failure. *Recent Prog. Horm. Res. 59*, 13-30.

Kin, H., Zatta, A.J., Lofye, M.T., Amerson, B.S., Halkos, M.E., Kerendi, F., Zhao, Z.Q., Guyton, R.A., Headrick, J.P., and Vinten-Johansen, J. (2005). Postconditioning reduces infarct size via adenosine receptor activation by endogenous adenosine. *Cardiovasc. Res. 67*, 124-133.

Kin, H., Zhao, Z.Q., Sun, H.Y., Wang, N.P., Corvera, J.S., Halkos, M.E., Kerendi, F., Guyton, R.A., and Vinten-Johansen, J. (2004). Postconditioning attenuates myocardial ischemia-reperfusion injury by inhibiting events in the early minutes of reperfusion. *Cardiovasc. Res. 62*, 74-85.

Kita, H., Miura, T., Tsuchida, A., Hasegawa, T., and Shimamoto, K. (1998). Suppression of reperfusion arrhythmias by preconditioning is inhibited by an ATP-sensitive potassium channel blocker, 5-hydroxydecanoate, but not by protein kinase C blockers in the rat. *J. Cardiovasc. Pharmacol. 32*, 791-797.

Kodani, E., Xuan, Y.T., Shinmura, K., Takano, H., Tang, X.L., and Bolli, R. (2002). Delta-opioid receptor-induced late preconditioning is mediated by cyclooxygenase-2 in conscious rabbits. *Am. J. Physiol. Heart Circ. Physiol. 283*, H1943-1957.

Kornick, C.A., Kilborn, M.J., Santiago-Palma, J., Schulman, G., Thaler, H.T., Keefe, D.L., Katchman, A.N., Pezzullo, J.C., Ebert, S.N., Woosley, R.L., *et al.* (2003). QTc interval prolongation associated with intravenous methadone. *Pain 105*, 499-506.

Kraft, K., Theobald, R., Kolloch, R., and Stumpe, K.O. (1987). Normalization of blood pressure and plasma concentrations of beta-endorphin and leucine-enkephalin in patients with primary hypertension after treatment with clonidine. *J. Cardiovasc. Pharmacol. 10 Suppl 12*, S147-151.

Kraft, K.S., Behrendt, M., Kolloch, R., and Stumpe, K.O. (1988). Reduced beta-endorphin secretion in young patients with mild essential hypertension at rest and during exercise. *J. Hypertens Suppl. 6*, S381-383.

Krantz, M.J., Lewkowiez, L., Hays, H., Woodroffe, M.A., Robertson, A.D., and Mehler, P.S. (2002). Torsade de pointes associated with very-high-dose methadone. *Ann. Intern. Med. 137*, 501-504.

Krumins, S.A., Faden, A.I., and Feuerstein, G. (1985). Opiate binding in rat hearts: modulation of binding after hemorrhagic shock. *Biochem. Biophys. Res. Commun. 127*, 120-128.

Ku, Y.H. (2006). Role of limbic peptidergic circuits in regulation of arterial pressure, relevant to development of essential hypertension. *Neuropeptides 40*, 299-308.

Kuzume, K., Wolff, R.A., Amakawa, K., and Van Winkle, D.M. (2003). Sustained exogenous administration of Met5-enkephalin protects against infarction in vivo. *Am. J. Physiol. Heart Circ. Physiol. 285*, H2463-2470.

Laurent, S., Marsh, J.D., and Smith, T.W. (1985). Enkephalins have a direct positive inotropic effect on cultured cardiac myocytes. *Proc. Natl. Acad. Sci. USA 82*, 5930-5934.

Law, P.Y., and Loh, H.H. (1999). Regulation of opioid receptor activities. *J. Pharmacol. Exp. Ther. 289*, 607-624.

Lee, A.Y. (1992). Stereospecific antiarrhythmic effects of naloxone against myocardial ischaemia and reperfusion in the dog. *Br. J. Pharmacol. 107*, 1057-1060.

Lee, A.Y., and Wong, T.M. (1986). Naloxone attenuates augmentation of cAMP levels and arrhythmias following myocardial ischaemia and reperfusion in the isolated perfused rat heart. *Clin. Exp. Pharmacol. Physiol. 13*, 707-710.

Lendeckel, U., Muller, C., Rocken, C., Laube, B., Tager, M., Huth, C., Klein, H.U., and Goette, A. (2005). Expression of opioid receptor subtypes and their ligands in fibrillating human atria. P*acing Clin. Electrophysiol. 28 Suppl 1*, S275-279.

Levitsky, D., de la Bastie, D., Schwartz, K., and Lompre, A.M. (1991). Ca(2+)-ATPase and function of sarcoplasmic reticulum during cardiac hypertrophy. *Am. J. Physiol. 261*, 23-26.

Liang, B.T., and Gross, G.J. (1999). Direct preconditioning of cardiac myocytes via opioid receptors and KATP channels. *Circ. Res. 84*, 1396-1400.

Liang, C.S., Imai, N., Stone, C.K., Woolf, P.D., Kawashima, S., and Tuttle, R.R. (1987). The role of endogenous opioids in congestive heart failure: effects of nalmefene on systemic and regional hemodynamics in dogs. *Circulation 75*, 443-451.

Liew, C.C., and Dzau, V.J. (2004). Molecular genetics and genomics of heart failure. *Nat. Rev. Genet. 5*, 811-825.

Lipski, J., Stimmel, B., and Donoso, E. (1973). The effect of heroin and multiple drug abuse on the electrocardiogram. *Am. Heart J. 86*, 663-668.

Liu, X.L., Mok, C.P., Lee, Y.S., and Wong, T.M. (1988). Evaluation of the anti-arrhythmic potency of naltrexone in the isolated ischaemic rat heart. Zhongguo Yao Li Xue Bao *9*, 40-43.

Llobell, F., and Laorden, M.L. (1996). Effects of mu-, delta-and kappa-opioid agonists in the failing human myocardium. *Pharmacol. Rev. Commun. 8*, 299-307.

Marmor, M., Penn, A., Widmer, K., Levin, R.I., and Maslansky, R. (2004). Coronary artery disease and opioid use. *Am. J. Cardiol. 93*, 1295-1297.

Martell, B.A., Arnsten, J.H., Ray, B., and Gourevitch, M.N. (2003). The impact of methadone induction on cardiac conduction in opiate users. *Ann. Intern. Med. 139*, 154-155.

Maslov, L.N., Krylatov, A.V., and Lishmanov Iu, B. (1996). [Participation of endogenous mu- and delta-opiate receptor agonists in mechanisms of the anti-arrhythmia effect of adaptation]. *Biull. Eksp. Biol. Med. 121*, 24-25.

McCloskey, D.T., Turcato, S., Wang, G.Y., Turnbull, L., Zhu, B.Q., Bambino, T., Nguyen, A.P., Lovett, D.H., Nissenson, R.A., Karliner, J.S., *et al.* (2008). Expression of a Gi-coupled receptor in the heart causes impaired Ca2+ handling, myofilament injury, and dilated cardiomyopathy. *Am. J. Physiol. Heart Circ. Physiol. 294*, H205-212.

McLaughlin, P.J. (2002). Exposure to the opioid antagonist naltrexone throughout gestation alters postnatal heart development. *Biol. Neonate 82*, 207-216.

McLaughlin, P.J., and Wu, Y. (1998). Opioid gene expression in the developing and adult rat heart. *Dev. Dyn. 211*, 153-163.

Millington, W.R., Rosenthal, D.W., Unal, C.B., and Nyquist-Battie, C. (1999). Localization of pro-opiomelanocortin mRNA transcripts and peptide immunoreactivity in rat heart. *Cardiovasc. Res. 43*, 107-116.

Molkentin, J.D., and Dorn, I.G., 2nd (2001). Cytoplasmic signaling pathways that regulate cardiac hypertrophy. *Annu. Rev. Physiol. 63*, 391-426.

Oldroyd, K.G., Gray, C.E., Carter, R., Harvey, K., Borland, W., Beastall, G., and Cobbe, S.M. (1995). Activation and inhibition of the endogenous opioid system in human heart failure. *Br. Heart J. 73*, 41-48.

Ouellette, M., and Brakier-Gingras, L. (1988). Increase in the relative abundance of preproenkephalin A messenger RNA in the ventricles of cardiomyopathic hamsters. *Biochem. Biophys. Res. Commun. 155*, 449-454.

Ouellette, M., Dumont, M., Lemaire, S., and Brakier-Gingras, L. (1991). Expression of the preproenkephalin A gene in the heart of cardiomyopathic hamsters. *Biochem. Cell Biol. 69*, 449-453.

Paranthaman, S.K., and Khan, F. (1976). Acute cardiomyopathy with recurrent pulmonary edema and hypotension following heroin overdosage. *Chest 69*, 117-119.

Patel, H.H., Hsu, A.K., and Gross, G.J. (2004). COX-2 and iNOS in opioid-induced delayed cardioprotection in the intact rat. *Life Sci. 75*, 129-140.

Peacock, W.F., Hollander, J.E., Diercks, D.B., Lopatin, M., Fonarow, G., and Emerman, C.L. (2008). Morphine and outcomes in acute decompensated heart failure: an ADHERE analysis. *Emerg. Med. J. 25*, 205-209.

Peart, J.N., Gross, E.R., and Gross, G.J. (2004). Effect of exogenous kappa-opioid receptor activation in rat model of myocardial infarction. *J. Cardiovasc. Pharmacol. 43*, 410-415.

Peart, J.N., Gross, E.R., and Gross, G.J. (2005). Opioid-induced preconditioning: recent advances and future perspectives. *Vascul. Pharmacol. 42*, 211-218.

Peart, J.N., Gross, E.R., Reichelt, M.E., Hsu, A., Headrick, J.P., and Gross, G.J. (2008). Activation of kappa-opioid receptors at reperfusion affords cardioprotection in both rat and mouse hearts. *Basic Res. Cardiol. 103*, 454-463.

Peart, J.N., and Gross, G.J. (2004a). Chronic exposure to morphine produces a marked cardioprotective phenotype in aged mouse hearts. *Exp. Gerontol. 39*, 1021-1026.

Peart, J.N., and Gross, G.J. (2004b). Exogenous activation of delta- and kappa-opioid receptors affords cardioprotection in isolated murine heart. *Basic Res. Cardiol. 99*, 29-37.

Peart, J.N., and Gross, G.J. (2004c). Morphine-tolerant mice exhibit a profound and persistent cardioprotective phenotype. *Circulation 109*, 1219-1222.

Peart, J.N., and Gross, G.J. (2006). Cardioprotective effects of acute and chronic opioid treatment are mediated via different signaling pathways. *Am. J. Physiol. Heart Circ. Physiol. 291*, H1746-1753.

Pepe, S., van den Brink, O.W., Lakatta, E.G., and Xiao, R.P. (2004). Cross-talk of opioid peptide receptor and beta-adrenergic receptor signalling in the heart. *Cardiovasc. Res. 63*, 414-422.

Pepe, S., Xiao, R.P., Hohl, C., Altschuld, R., and Lakatta, E.G. (1997). 'Cross talk' between opioid peptide and adrenergic receptor signaling in isolated rat heart. *Circulation 95*, 2122-2129.

Pugsley, M.K. (2002). The diverse molecular mechanisms responsible for the actions of opioids on the cardiovascular system. *Pharmacol. Ther. 93*, 51-75.

Pugsley, M.K., Penz, W.P., Walker, M.J., and Wong, T.M. (1992a). Antiarrhythmic effects of U-50,488H in rats subject to coronary artery occlusion. *Eur. J. Pharmacol. 212*, 15-19.

Pugsley, M.K., Penz, W.P., Walker, M.J., and Wong, T.M. (1992b). Cardiovascular actions of the kappa-agonist, U-50,488H, in the absence and presence of opioid receptor blockade. *Br. J. Pharmacol. 105*, 521-526.

Pugsley, M.K., Saint, D.A., Penz, M.P., and Walker, M.J. (1993). Electrophysiological and antiarrhythmic actions of the kappa agonist PD 129290, and its R,R (+)-enantiomer, PD 129289. *Br. J. Pharmacol. 110*, 1579-1585.

Pyle, W.G., Lester, J.W., and Hofmann, P.A. (2001). Effects of kappa-opioid receptor activation on myocardium. *Am. J. Physiol. Heart Circ. Physiol. 281*, H669-678.

Pyle, W.G., Smith, T.D., and Hofmann, P.A. (2000). Cardioprotection with kappa-opioid receptor stimulation is associated with a slowing of cross-bridge cycling. *Am. J. Physiol. Heart Circ. Physiol. 279*, H1941-1948.

Rosamond, W., Flegal, K., Furie, K., Go, A., Greenlund, K., Haase, N., Hailpern, S.M., Ho, M., Howard, V., Kissela, B., *et al.* (2008). Heart disease and stroke statistics--2008 update: a report from the American Heart Association Statistics Committee and Stroke Statistics Subcommittee. *Circulation 117*, e25-146.

Saotome, M., Katoh, H., Yaguchi, Y., Tanaka, T., Urushida, T., Satoh, H., and Hayashi, H. (2009). Transient opening of mitochondrial permeability transition pore by reactive oxygen species protects myocardium from ischemia-reperfusion injury. *Am. J. Physiol. Heart Circ. Physiol. 296*, H1125-1132.

Schultz, J.e.-J., Hsu, A.K., Nagase, H., and Gross, G.J. (1998). TAN-67, a delta 1-opioid receptor agonist, reduces infarct size via activation of Gi/o proteins and KATP channels. *Am. J. Physiol. 274*, H909-914.

Schultz, J.E., and Gross, G.J. (2001). Opioids and cardioprotection. *Pharmacol. Ther. 89*, 123-137.

Schultz, J.E., Hsu, A.K., and Gross, G.J. (1996). Morphine mimics the cardioprotective effect of ischemic preconditioning via a glibenclamide-sensitive mechanism in the rat heart. *Circ. Res. 78*, 1100-1104.

Schultz, J.E., Rose, E., Yao, Z., and Gross, G.J. (1995). Evidence for involvement of opioid receptors in ischemic preconditioning in rat hearts. *Am. J. Physiol. 268*, H2157-2161.

Schultz, J.J., Hsu, A.K., and Gross, G.J. (1997). Ischemic preconditioning and morphine-induced cardioprotection involve the delta (delta)-opioid receptor in the intact rat heart. *J. Mol. Cell Cardiol. 29*, 2187-2195.

Shan, D., Wang, H., Su, Y., Jing, Y., and Wong, T.M. (2007). kappa-opioid receptor stimulation inhibits cardiac hypertrophy induced by beta1-adrenoceptor stimulation in the rat. *Eur. J. Pharmacol. 555*, 100-105.

Shan, J., Yu, X.C., Fung, M.L., and Wong, T.M. (2002). Attenuated "cross talk" between kappa-opioid receptors and beta-adrenoceptors in the heart of chronically hypoxic rats. *Pflugers Arch. 444*, 126-132.

Sheng, J.Z., Dong, H., and Wong, T.M. (1994). Effects of U-50 488H, a kappa-agonist, on action potentials of isolated ventricular papillary muscle of guinea pigs. Zhongguo Yao Li Xue Bao *15*, 36-38.

Sheng, J.Z., Wong, N.S., Wang, H.X., and Wong, T.M. (1997). Pertussis toxin, but not tyrosine kinase inhibitors, abolishes effects of U-50488H on [Ca2+]i in myocytes. *Am. J. Physiol. 272*, C560-564.

Sheng, J.Z., and Wong, T.M. (1996). Chronic U50,488H abolishes inositol 1,4,5-trisphosphate and intracellular Ca2+ elevations evoked by kappa-opioid receptor in rat myocytes. *Eur. J. Pharmacol. 307*, 323-329.

Springhorn, J.P., and Claycomb, W.C. (1989). Preproenkephalin mRNA expression in developing rat heart and in cultured ventricular cardiac muscle cells. *Biochem. J. 258*, 73-78.

Springhorn, J.P., and Claycomb, W.C. (1992). Translation of heart preproenkephalin mRNA and secretion of enkephalin peptides from cultured cardiac myocytes. *Am. J. Physiol. 263*, H1560-1566.

Standifer, K.M., and Pasternak, G.W. (1997). G proteins and opioid receptor-mediated signalling. *Cell Signal 9*, 237-248.

Takano, H., Hasegawa, H., Nagai, T., and Komuro, I. (2003). Implication of cardiac remodeling in heart failure: mechanisms and therapeutic strategies. *Intern. Med. 42*, 465-469.

Thompson, R.B., Rungwerth, K., and Koch, W.J. (2004). Gene therapy for heart failure. *Ann. Med. 36 Suppl 1*, 106-115.

Timmis, A.D., Rothman, M.T., Henderson, M.A., Geal, P.W., and Chamberlain, D.A. (1980). Haemodynamic effects of intravenous morphine in patients with acute myocardial infarction complicated by severe left ventricular failure. *Br. Med. J. 280*, 980-982.

Towbin, J.A., Bowles, K.R., and Bowles, N.E. (1999). Etiologies of cardiomyopathy and heart failure. *Nat. Med. 5*, 266-267.

Ventura, C., Bastagli, L., Bernardi, P., Caldarera, C.M., and Guarnieri, C. (1989). Opioid receptors in rat cardiac sarcolemma: effect of phenylephrine and isoproterenol. *Biochim. Biophys. Acta 987*, 69-74.

Ventura, C., Capogrossi, M.C., Spurgeon, H.A., and Lakatta, E.G. (1991). Kappa-opioid peptide receptor stimulation increases cytosolic pH and myofilament responsiveness to Ca2+ in cardiac myocytes. *Am. J. Physiol. 261*, H1671-1674.

Ventura, C., Maioli, M., Pintus, G., Posadino, A.M., and Tadolini, B. (1998). Nuclear opioid receptors activate opioid peptide gene transcription in isolated myocardial nuclei. *J. Biol. Chem. 273*, 13383-13386.

Ventura, C., and Pintus, G. (1997). Opioid peptide gene expression in the primary hereditary cardiomyopathy of the Syrian hamster. III. Autocrine stimulation of prodynorphin gene expression by dynorphin B. *J. Biol. Chem. 272*, 6699-6705.

Ventura, C., Pintus, G., Vaona, I., Bennardini, F., Pinna, G., and Tadolini, B. (1995). Phorbol ester regulation of opioid peptide gene expression in myocardial cells. Role of nuclear protein kinase. *J. Biol. Chem. 270*, 30115-30120.

Ventura, C., Spurgeon, H., Lakatta, E.G., Guarnieri, C., and Capogrossi, M.C. (1992). Kappa and delta opioid receptor stimulation affects cardiac myocyte function and Ca2+ release from an intracellular pool in myocytes and neurons. *Circ. Res. 70*, 66-81.

Walker, J.S. (2003). Anti-inflammatory effects of opioids. *Adv. Exp. Med. Biol. 521*, 148-160.

Wang, G., Wang, H., Yang, Y., and Wong, T.M. (2004). Kappa-opioid receptor stimulation inhibits growth of neonatal rat ventricular myocytes. *Eur. J. Pharmacol. 498*, 53-58.

Wang, G.Y., Wu, S., Pei, J.M., Yu, X.C., and Wong, T.M. (2001). Kappa- but not delta-opioid receptors mediate effects of ischemic preconditioning on both infarct and arrhythmia in rats. *Am. J. Physiol. Heart Circ. Physiol. 280*, H384-391.

Weil, J., Eschenhagen, T., Fleige, G., Mittmann, C., Orthey, E., and Scholz, H. (1998). Localization of preproenkephalin mRNA in rat heart: selective gene expression in left ventricular myocardium. *Am. J. Physiol. 275*, H378-384.

Weil, J., Zolk, O., Griepentrog, J., Wenzel, U., Zimmermann, W.H., and Eschenhagen, T. (2006). Alterations of the preproenkephalin system in cardiac hypertrophy and its role in atrioventricular conduction. *Cardiovasc. Res. 69*, 412-422.

Wenzlaff, H., Stein, B., and Teschemacher, H. (1998). Diminution of contractile response by kappa-opioid receptor agonists in isolated rat ventricular cardiomyocytes is mediated via a pertussis toxin-sensitive G protein. *Naunyn Schmiedebergs Arch. Pharmaco*l. *358*, 360-366.

Wilkins, B.J., and Molkentin, J.D. (2004). Calcium-calcineurin signaling in the regulation of cardiac hypertrophy. *Biochem. Biophys. Res. Commun. 322*, 1178-1191.

Williams, S.G., Wright, D.J., Marshall, P., Reese, A., Tzeng, B.H., Coats, A.J., and Tan, L.B. (2003). Safety and potential benefits of low dose diamorphine during exercise in patients with chronic heart failure. *Heart 89*, 1085-1086.

Wilson, S.P., Klein, R.L., Chang, K.J., Gasparis, M.S., Viveros, O.H., and Yang, W.H. (1980). Are opioid peptides co-transmitters in noradrenergic vesicles of sympathetic nerves? *Nature 288*, 707-709.

Wong, T.M., and Huang, X.D. (1988). [Opioid peptides, opioid receptors and ischemia-reperfusion induced cardiac arrhythmias]. Sheng Li Ke Xue Jin Zhan *19*, 357-359.

Wong, T.M., and Shan, J. (2001). Modulation of sympathetic actions on the heart by opioid receptor stimulation. *J. Biomed. Sci. 8*, 299-306.

Wu, S., Li, H.Y., and Wong, T.M. (1999). Cardioprotection of preconditioning by metabolic inhibition in the rat ventricular myocyte. Involvement of kappa-opioid receptor. *Circ. Res. 84*, 1388-1395.

Xiao, R.P., Pepe, S., Spurgeon, H.A., Capogrossi, M.C., and Lakatta, E.G. (1997). Opioid peptide receptor stimulation reverses beta-adrenergic effects in rat heart cells. *Am. J. Physiol. 272*, H797-805.

Xiao, R.P., Spurgeon, H.A., Capogrossi, M.C., and Lakatta, E.G. (1993). Stimulation of opioid receptors on cardiac ventricular myocytes reduces L type Ca2+ channel current. *J. Mol. Cell Cardiol. 25*, 661-666.

Yatani, A., Imai, N., Himura, Y., Suematsu, M., and Liang, C.S. (1997). Chronic opiate-receptor inhibition in experimental congestive heart failure in dogs. *Am. J. Physiol. 272*, H478-484.

Younes, A., Pepe, S., Barron, B.A., Spurgeon, H.A., Lakatta, E.G., and Caffrey, J.L. (2000). Cardiac synthesis, processing, and coronary release of enkephalin-related peptides. *Am. J. Physiol. Heart Circ. Physiol. 279*, H1989-1998.

Yu, X., Zhang, W., Bian, J., and Wong, T.M. (1999a). Pro- and anti-arrhythmic effects of a kappa opioid receptor agonist: a model for the biphasic action of a local hormone in the heart. *Clin. Exp. Pharmacol. Physiol. 26*, 842-844.

Yu, X.C., Li, H.Y., Wang, H.X., and Wong, T.M. (1998). U50,488H inhibits effects of norepinephrine in rat cardiomyocytes-cross-talk between kappa-opioid and beta-adrenergic receptors. *J. Mol. Cell Cardiol. 30*, 405-413.

Yu, X.C., Wang, H.X., Pei, J.M., and Wong, T.M. (1999b). Anti-arrhythmic effect of kappa-opioid receptor stimulation in the perfused rat heart: involvement of a cAMP-dependent pathway. *J. Mol Cell Cardiol. 31*, 1809-1819.

Yu, X.C., Wang, H.X., Zhang, W.M., and Wong, T.M. (1999c). Cross-talk between cardiac kappa-opioid and beta-adrenergic receptors in developing hypertensive rats. *J. Mol. Cell Cardiol. 31*, 597-605.

Zagon, I.S., Wu, Y., and McLaughlin, P.J. (1999). Opioid growth factor and organ development in rat and human embryos. *Brain Res. 839*, 313-322.

Zatta, A.J., Kin, H., Yoshishige, D., Jiang, R., Wang, N., Reeves, J.G., Mykytenko, J., Guyton, R.A., Zhao, Z.Q., Caffrey, J.L., *et al.* (2008). Evidence that cardioprotection by postconditioning involves preservation of myocardial opioid content and selective opioid receptor activation. *Am. J. Physiol. Heart Circ. Physiol. 294*, H1444-1451.

Zhang, W.M., Jin, W.Q., and Wong, T.M. (1996). Multiplicity of kappa opioid receptor binding in the rat cardiac sarcolemma. *J. Mol. Cell Cardiol. 28*, 1547-1554.

Zhang, W.M., and Wong, T.M. (1998). Suppression of cAMP by phosphoinositol/Ca2+ pathway in the cardiac kappa-opioid receptor. *Am. J. Physiol. 274*, C82-87.

Zheng, X., Zhang, T., Ding, H., and Wang, C. (1995). Plasma levels of beta-endorphin, leucine enkephalin and arginine vasopressin in patients with essential hypertension and the effects of clonidine. *Int. J. Cardiol. 51*, 233-244.

Zimlichman, R., Gefel, D., Eliahou, H., Matas, Z., Rosen, B., Gass, S., Ela, C., Eilam, Y., Vogel, Z., and Barg, J. (1996). Expression of opioid receptors during heart ontogeny in normotensive and hypertensive rats. *Circulation 93*, 1020-1025.

Zou, Y., Takano, H., Akazawa, H., Nagai, T., Mizukami, M., and Komuro, I. (2002). Molecular and cellular mechanisms of mechanical stress-induced cardiac hypertrophy. *Endocr. J. 49*, 1-13.

In: Biochemistry Research Updates
Editor: Simon J. Baginski
ISBN 978-1-61209-700-8

Chapter 5

ANGIOTENSIN CONVERTING ENZYME INHIBITORS: A CLASS OF POTENT ANTIHYPERTENSIVE AGENTS

Sharad Kumar Panday**, Jagdish Prasad and Manohar Bhushan Pathak

Department of Applied Chemistry,
Faculty of Engineering and Technology,
M. J. P. Rohilkhand University, Bareilly, U. P., India

1. ABSTRACT

Hypertension, a major cause of heart-diseases is caused by malfunction of one or more of the complex set of mechanisms controlling blood pressure. In last four decades specific agents interacting with these regulatory mechanisms have become available. Introduction in early eighties of Captopril, a rationally designed potent inhibitor of Angiotensin converting enzymes (ACE), heralded a new era in antihypertensive therapy. Since then rapid strides have been made in understanding of the active site and the spatial requirements for ACE inhibitors. This chapter has been aimed to report the existing knowledge and to enable researchers all over the world to exploit further with an objective to develop better therapeutic agents.

Bio receptors being chiral, interactions with organic molecules are strongly influenced by the chirality of the substrate. The two enantiomers of a drug are some times known to elicit differential and in some cases even opposite responses, amply heightened by work on ACE inhibitors. The present trend, in drug research, is to design and synthesize molecules with designated chirality. ACE inhibitors represent a class of rationally designed drugs based on the comprehensive knowledge of the enzyme and its binding sites and had led to great revolution in antihypertensive therapy. ACE is a Zn^{++} containing metalloenzyme similar to carboxypeptidases, whose binding sites have been studied properly and a hypothetical model incorporating all the binding sites have been developed. Based on the hypothetical model for ACE binding sites and the role which it plays in living organism, attempts have been made to develop potent inhibitors of ACE with an objective to control blood pressure. Early researches on ACE inhibitors started with snake venom peptides, showing antihypertensive activity. Though these could not be

* E-mail: drsharadpandey@yahoo.com.

used as drugs due to high toxicity but certainly opened an avenue to look for development of antihypertensive agents on these lines. Further researches coupled with structure activity relationship (SAR) studies led to development of various ACE inhibitors and ultimately in early eighty's first rationally designed drug "Captopril" was introduced in market. Since then various other drugs acting through same mechanism and controlling blood pressure effectively have come out.

The chapter entitled "Angiotensin converting enzymes inhibitors: A class of potent antihypertensive agents" is based on literature reports and intended to summarized the available knowledge and to give an update picture of major advances towards the design and synthesis of angiotensin converting enzymes inhibitors and their role in controlling blood pressure and shall undoubtedly be valuable to readers working in the area. This chapter gives information about the designing and developing of ACE inhibitors reported till date with future prospectives to achieve further progress as for as designing and development of even better ACE inhibitors is concerned.

2. INTRODUCTION

Hypertension is a major health risk factor and its management by drugs has reduced the incidence of stroke, Heart failure and renal damage. Though described by a single term, hypertension is a consequence of various disorders and increase in peripheral vascular resistance is the hallmark of the all types of hypertension.

Some of the factors responsible for producing hypertension are as follows:

1. Mechanical factors: Constriction or obstruction of blood flow.
2. Neural mechanism:
   i. adrenergic-increased CVS out put
   ii. para sympathatic- reduced CVS output
   iii. catecholamines:
      a increased circulatory levels/production
      b altered metabolism
      c altered release
      d altered central CVS input
3. Renopressor mechanism:
   i. increased renin production/ release
   ii. increased production of renin substrate
   iii. altered conversion of Angiotensin I to Angiotensin II
4. Renal parenchymal function: Altered Na+/K+ balance, water balance, ionic balance
5. Hormonal factors: Adrenal cortical hormone (cortisone/aldosterone) vasopressin, growth hormone, seratonin.
6. Reduced participation of depressor factors: Histidine, kallikrein kinin system, prostaglandlins

Though there has been great advancement in understanding and regulating neural mechanisms, renopressor mechanisms have also been understood in quite detail. Of all these better understood conversion of Angiotensin I to Angiotensin II catalyzed by angiotensin converting enzyme (ACE) had been the target of rational drug design. The Angiotensin converting enzyme inhibitors are presently being regarded as drugs of first choice as monotherapy or combination therapy. ACE inhibitors possess several advantages over conventional antihypertensive therapy including the lack of interaction with the central and autonomic nervous systems, thus drowsiness and impotence are not the indicated side effects nor is the postural hypertension. Investigations also pointed out the possible role of ACE inhibitors [1a] in reversing blood vessel damage.

3. Angiotension Converting Enzyme (ACE) Inhibitors and Renin–Angiotensin System

The beginning of 1980s may well be remembered as a turning point in the treatment of hypertension, one of the most common of man's afflications [1b]. In 1980 and 1981, the first antihypertensive agent designed to interact specifically with one of the physiological system involved in the blood pressure control, became available through out the world. This drug Captopril was the first orally active inhibitor of ACE. Following this intensive researches in many laboratories have led to newer agents, of which many are undergoing clinical trials, and the widening use of these drugs, would surely lead to a better understanding of therapy of hypertension.

The development of orally active ACE inhibitors is an outstanding achievement in the elucidation of Renin-Angiotensin system [3]. Studies on molecular details of this mechanism for the blood pressure control began in 1940s and 1950s through the efforts of research groups in North and South America and Europe[4]. Renin one of the key components of the system, is a highly specific protease synthesized in the juxtaglomerular apparatus of kidney from which it is released to blood stream. Renin is also found in the brain vascular tissue, amniotic fluid and the submaxillary glands. The action of circulating renin on a globular plasma protein known as Angiotensinogen, cleaves a single peptide bond to release a N-terminal decapeptide (Figure 1). This decapeptide, called Angiotensin-I, is devoid of any biological activity but the action of a second membrane bound endopeptidase, angiotensin converting enzyme[5] (ACE) cleaves a dipeptide unit from the C-terminal of Angiotensin-I (A-I) to yield an octapeptide called Angiotensin-II (A-II) which is to date the most potent endogenous vasoconstructive substance.

Angiotensin- II alters the blood pressure via at least two mechanisms.

1. By direct vasoconstructive action on blood vessels.
2. By stimulating adrenal gland to release aldosterone which helps in the retention of sodium and water.

A-II is linked to renin release in a negative feed back loop, but its invivo half life is very short due to hydrolysis by several inactivating enzymes collectively called angiotensinases. ACE is also known as kininase II[6], because it removes a C-terminal dipeptide from bradykinin, a potent vasodepressor nonapeptide. Thus ACE playing a vital role in two ways via generation of a vasopressor agent (A-II) and inactivation of a vasodepressor agent bradykinin, elevates the blood pressure.

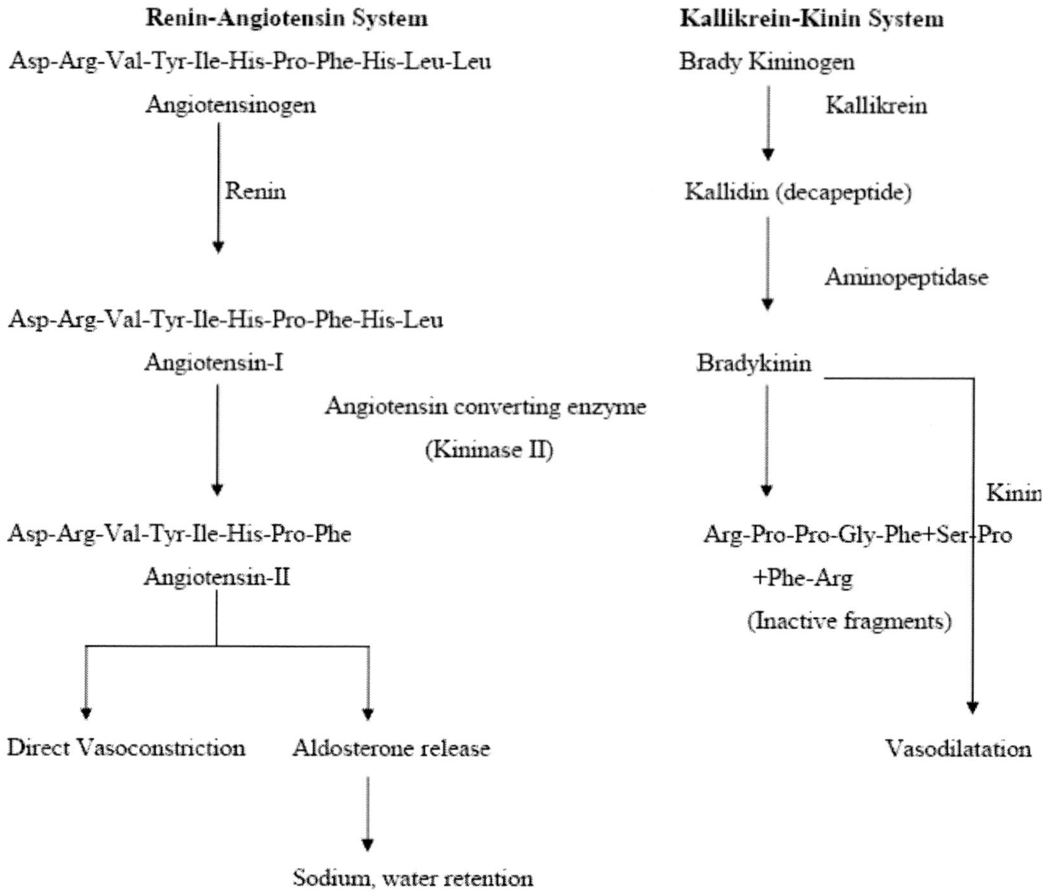

Figure 1. The role of Angiotensin Converting Enzyme in the Renin-Angiotensin and Kallikrein-Kinin System.

4. ANGIOTENSION CONVERTING ENZYME (ACE)

Angiotensin converting enzyme (EC 3.4.15.1) is a dipeptidyl carboxypeptidase. Though ACE has yet not been crystallized, its properties and specificity has already been studied in quite detail in recent years[7][8]. ACE a glycoprotein seems to be similar in rats, rabbits, dogs, pigs and humans, and is present in large number of tissues including lung, kidney, plasma, brain and in the endothelial linings of vasculature.

The most abundant form of ACE has molecular weights in the range of 130000-140000 KDa of which about 25% is carbohydrate by weight. There is one atom of Zn^{++} per molecule of enzyme and its removal by EDTA, inactivates the enzyme. Heavy metal salts also inhibit the enzyme[7][8]. Taken together these data strongly suggested that Angiotensin converting enzyme is a Zn^{++} containing metalloenzyme and shares similarities with other well studied metalloproteases such as carboxypeptidase A and thermolysin.

5. An Overview of Early Work on ACE Inhibitors

Researches on Ranin-Angiotensin system have been carried out in various laboratories all over the world in the last few decades, and lot of work has been reported. Studies on ACE inhibitors and their structure activity relationship (SAR) studies have also been reviewed[9-15].

Earlier leads in this area emerged when scientists, due to the hypotensive effect of snake poisons, digested it to smaller fragments and some of these fragments obtained fom the digest of Bathrops jaraca[16-20] and Agkistrodon halys blmhoffi[21][22], specially $BPP_{5\alpha}$ **1**[23] showed remarkable inhibition of ACE (Table 1).

Table 1. Analogs of BPP $_{5\alpha}$

No.	Structure	IC_{50} (µg/ml)
1.	P-Glu-Lys-Trp-Ala-Pro BPP $_{5\alpha}$	0.05
2.	P-Glu-Lys-Phe-Ala-Pro	0.05
3.	Phe-Ala-Pro	1.4
4.	Ala-Pro	50

Table 2. Dipeptide inhibitors of ACE

No.	Structure	IC_{50} (µg)
5.	Gly-Tro	30
6.	Gly-Pro	450
7.	Gly-Phe	630
8.	Gly-Lys	5400
9.	Gly-Glu	5400
10.	Val-Gly	1100
11.	Ile-Gly	1200
12.	Arg-Gly	1200
13.	Try-Gly	2000
14.	Ala-Gly	2500
15.	Lys-Gly	3200
16.	Pro-Gly	17000
17.	Val-Trp	1.6
4.	Ala-Pro	230*

However, none of these fragments were active by oral route. These findings provided considerable encouragement to researchers involved in the designing of small orally active ACE inhibitors, and led to synthesis and evaluation of ACE inhibitory potential of a large number of peptides. The most active of these peptides, Val-Trp, 17 (Table2) was a competitive inhibitor[24] of ACE (K_i=$3x10^{-7}$ M). However this compound failed to inhibit the contractile response of A-I on guinea pig ileum presumably because of difficulties in transport or metabolism. These early studies on peptides as ACE inhibitors, through not very rewarding in the development of an orally active inhibitor, yet these were of immense help in understanding the active site of the enzyme and as a result of these early studies a hypothetical model for active site encompassing later development also, could be proposed.

6. Hypothetical Model for ACE Binding Site

ACE a Zn^{++} metalloenzyme is similar to carboxypeptidase A[25], carboxypeptidase B[26] and thermolysin[27], whose active site structures were determined by x-ray crystallography[28-30]. Although ACE could not be available for X-ray analysis yet, the structural and mechanistic similarities between ACE and these better studied carboxypeptidases, was the key factor in the development of a hypothetical active site model for ACE which has been further modified based on extensive structure activity relationship studies on various classes of inhibitors. Structure activity relationship studies reported for the newer inhibitors greatly facilitated the interpretation of the hypothetical model, when role of each putative binding site was determined[10-15].

Figure 2 shows the active site model model with a generalized peptide inhibitor bound to it. Sites marked S_2, S_1, S_1^1, S_2^1 are hydrophobic pockets in the hypothetical model for ACE binding site. Studies on larger peptide inhibitors had shown that mainly three subsites designated S_1, S_1^1, S_2^1 are responsible for the binding of peptide substrates and inhibitors to ACE

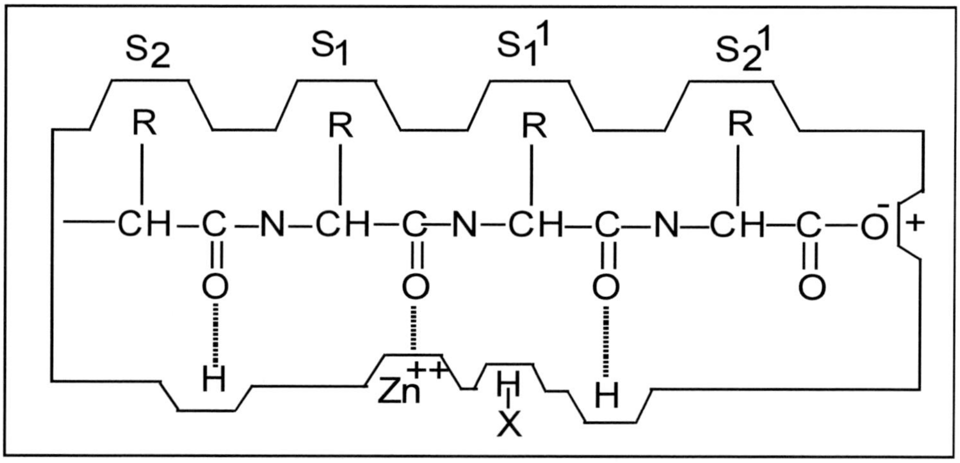

Figure 2. Hypothetical model for ACE binding site showing only those functional groups implicated so far in binding of inhibitors.

Subsites S_2^1

Subsite S_2^1 strongly affects the binding of inhibitors, but has little effect on the enzymes hydrolytic specificity[31]. ACE may hydrolyze substrates bearing virtually any L-Amino acid except Glu or Asp in the S_2^1 position[5]. Apparently the requirements for binding to S_2^1 depends on the nature of other interactions between the inhibitors and the enzyme.

Subsites S_1^1

The specificity on subsites S_1^1 has a moderate effect on the binding of substrates and inhibitors[31]. Peptides containing a proline residue or a D-amino acid at S_1^1 are not hydrolyzed by ACE. In the dipeptide series Gly, Pro, bind most poorly to S_1^1 but in the hippuryl dipeptide series Pro binds best. Again the binding requirements for S_1^1 must be different depending on the size of inhibitors, Val, Ala, Arg, Lys all bind well to S_1^1.

Subsites S_1

The ease with which ACE hydrolyzes peptide bond is greatly affected by the specificity of this subsite[31]. Studies with carboxyalkydipeptides show that phenylalkyl and aminoalkyl substituent positioned for hydrophobic interaction with S_1 enhance the activity of these compounds [32]. Figure 2 also indicates the possible presence at the ACE active site, of functional groups which may be involved in the binding of substrate and inhibitors, and which probably participate in the hydrolytic mechanism of the enzyme [33].

Figure 2 shows a cationic site represented by a (+) symbol. The Zn++ ion is shown binding to the carbonyl of scissile amide bind. The proton donor which protonates the nitrogen of the departing peptide fragment is represented by X-H. This group is probably a tyrosine-OH as in carboxypeptidase[33]. In addition to these, this model also incorporates symbols for proton doner site (H) which could bind to amide carbonyls and acceptor sites (∩), which could bind to N-H groups of the substrate peptide backbone.

7. The Design of Captopril

Captopril the first rationally designed drug introduced in the market in 1980, was developed by Squibb group [12][13][34][35]. Its development was based on the presumption that ACE is related structurally and mechanistically to carboxy peptidaseA. Reports[36][37] by Byers and Wolfenden that D-2-benzyl succinic acid was a potent inhibitor of caroxypeptidase A (ki =4.5x10-7 M), led to application of the same analogy for the synthesis of ACE inhibitors by Sqibb group. .A proline terminus and methyl substitutions were used like Ala- Pro grouping, whose contribution to high potency was evidenced during their snake venom inhibitor studies. Though the most potent of the carboxyalkanoyl inhibitors 20 (Table 3) had K_i of 8x10-7M, these unlike to snake venom fragments, were orally active. Substitution of a weak Zn^{++} chelator, COOH, by a strong thiol chelator led to thousand fold increase in the inhibitory potency of the resultant compounds 21 and 22 (Table 3).

The better of the two 22 was designated Captopril (SQ, 14, 225), ki $=1.7\times10^{-9}$M. Captopril was more potent than teprotide and was reported to be a quite specific inhibitor even in respect to other Zn^{++} metallopeptidases. Its interaction with the active site of ACE was proposed as shown in figure 3.

Table 3. Carboxyalkyl and mercaptoalkanoyl derivatives of proline

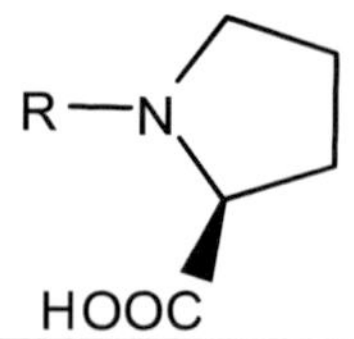

No.	R	IC_{50} (μM)	K_i(μM)
18.	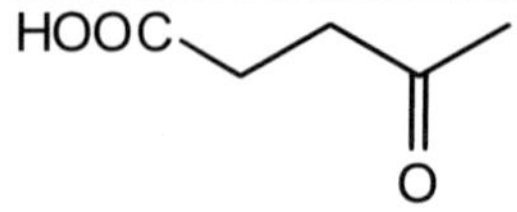	330	-
19.	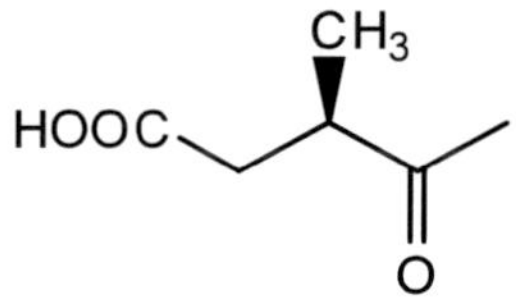	22	2.5
20.	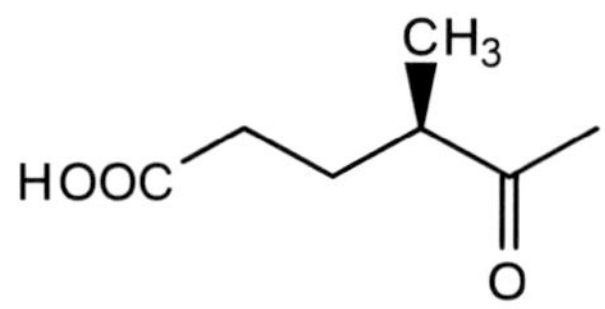	4.9	0.8
21.	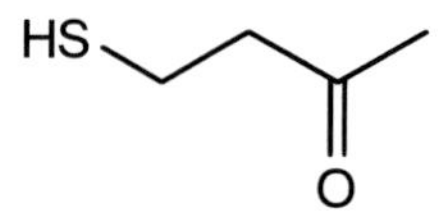	0.20	0.012
22.	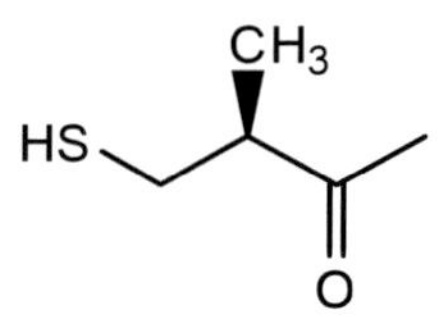	0.023	0.0017
23.	Teprotide (SQ,20,881)	0.55	0.10

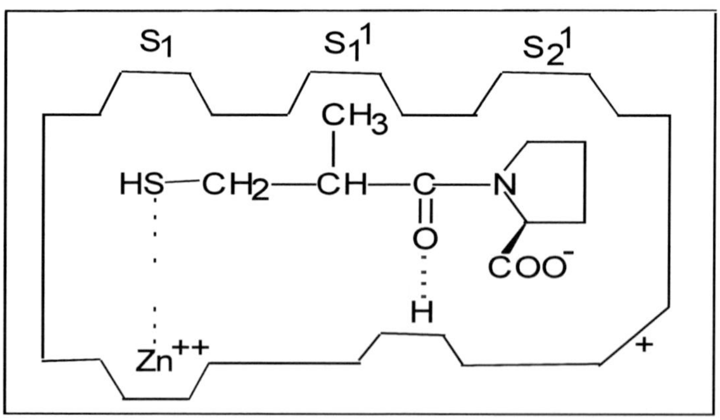

Figure 3. The binding interactions of Captopril with the active site of ACE as hypothesized by Petrillo and Ondetti.[10].

8. THE N-CARBOXYALKYL DIPEPTIDE DEVELOPMENT

Patchett and associates tried to design a non-thiol ACE inhibitor, with a view that this compound could possibly be more stable than Captopril and side effects of Captopril like lack of interaction with the central and automatic nervous systems, drowsiness and rashes which were related to thiol may also be avoided[32]. Like Captopril, the N-Carboxyalkyl dipeptide inhibitors originated from the same Bayers and Wolfenden byproduct design. Based on this rational, numerous Carboxyalkyl dipeptides have been synthesized and tested (Table 4)[15]. Of these, the phenylethyl analog 26 showed excellent activity[32] indicating that this group interacts with S_1 subsite of ACE (figure 4).

The phenylethyl compound was selected for detailed pharmacological studies based on its good intravenous inhibition of Angiotensin-I. Diastereomers were separated and the more potent isomer 27 was confirmed by X-ray crystallography[38] and therefore an independent synthesis for the compound having S,S,S- configuration was achieved. Unfortunately this compound 27 (MK-422, Enalaprilat) did not have satisfactory oral activity in rats, dogs and men[39]. However its ethyl ester 28 (Enalapril, MK-421) showed excellent oral activity. The dominant effect of esterification could possibly be to lower the basicity of the NH group (Enalapril, Pka=5.5, EnalaprilatPka =7.6) and also to remove the charges of carboxylate. Thus Enalapril is a prodrug requiring the action of esterases in the body to liberate the highly potent ACE inhibitor Enalaprilat.

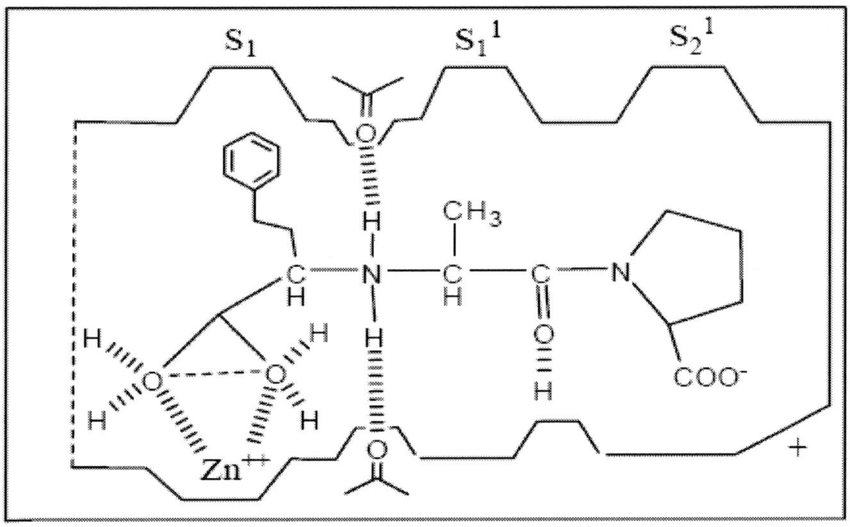

Figure 4. Proposed binding interactions between Enalaprilat (MK-422) and angiotensin converting Enzyme [32].

To overcome the oral activity deficiency of Enalaprilat, lysyl proline derivatives 29 (Lisinopril, MK 5210 was synthesized. This compound was more slowly and less completely absorbed than Enalapril. The in vitro inhibitory activities of a number of close analogs of Enalapril and Lisinopril have been published by Merck group[40-42]. Apart from high in vitro activity Lisinopril also exhibited excellent oral activity in animal models and eventually humans without a requirement of prodrug design[32][42-44].

Table 4. ACE inhibition by carboxyalkanoyl prolines and N-carboxyalkyl dipeptides

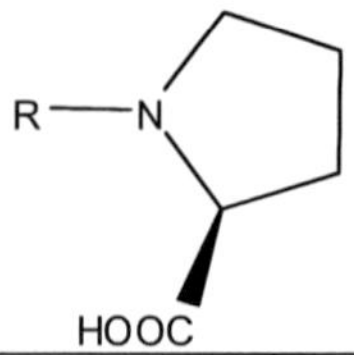

No.	R	IC_{50}(nM)
20.	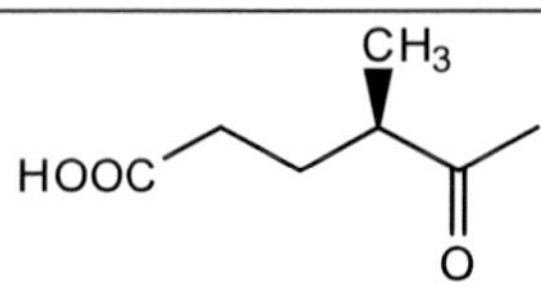	4900
24.	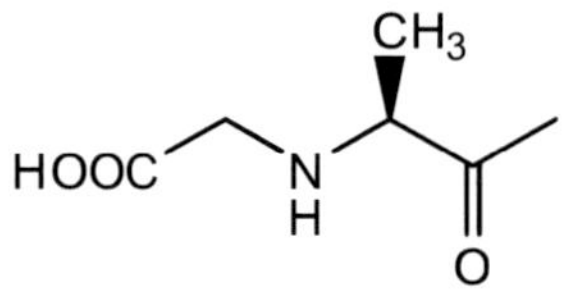	2400
25.	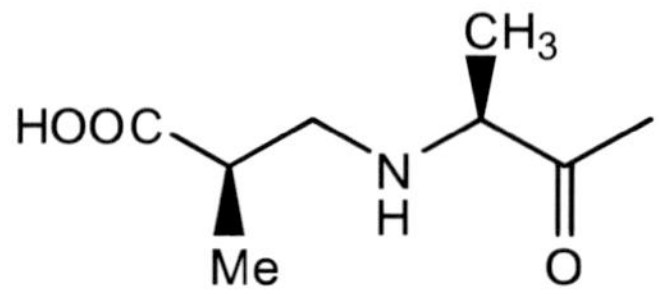	92
26.	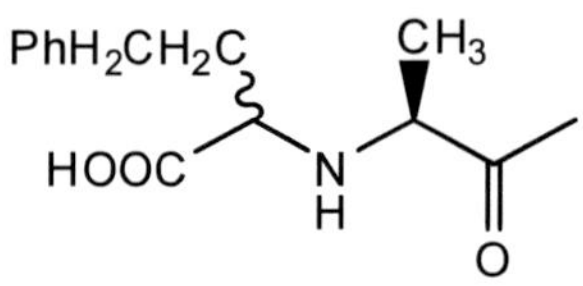	3.8
27.	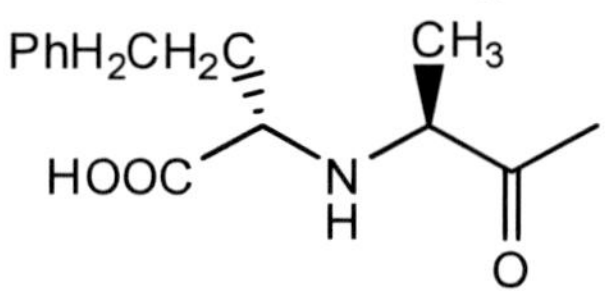	1.2
28.	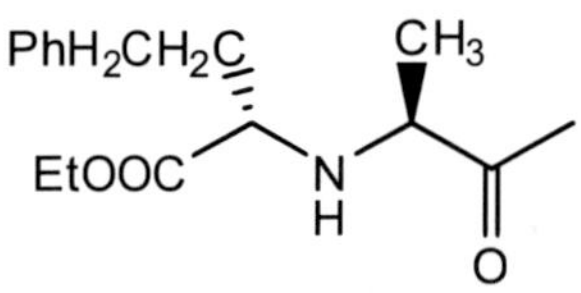	1200
29.	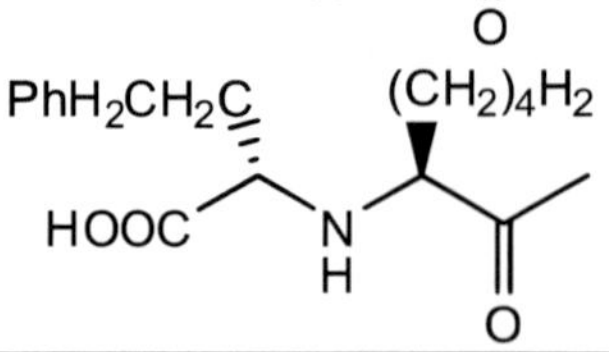	1.2

9. Structure Activity Relationship Studies on Various Classes of Molecules by Variations at the Different Site of ACE

On the basis of structure activity relationship studies on various classes of molecules it is possible to asses the role that each putative binding site in the ACE active site model, might play in binding to inhibitors as well as substrates.

Table 5. ACE inhibition by variations at the cationic site

No.	R	IC_{50}(μM)
22.	Captopril	0.023
30.	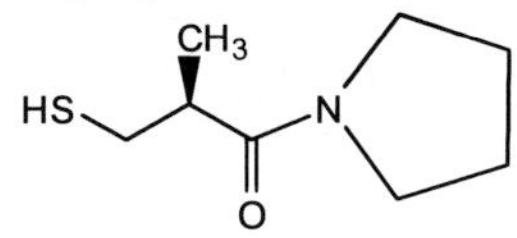	240
31.	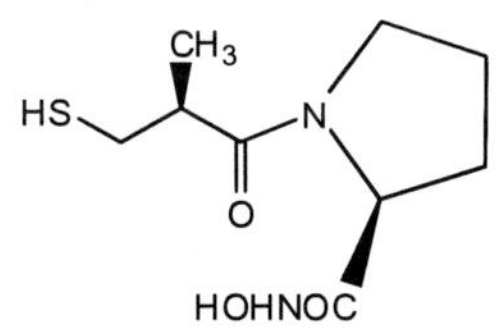	0.022
32.	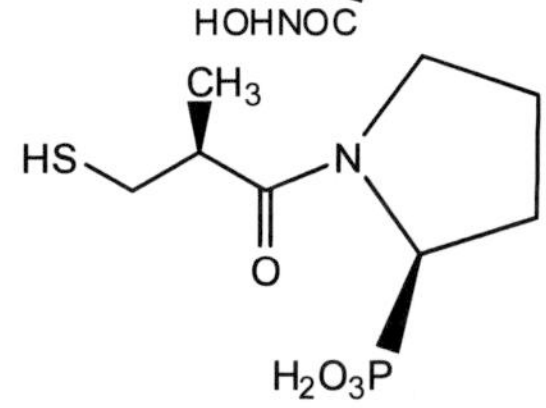	1.7
33.	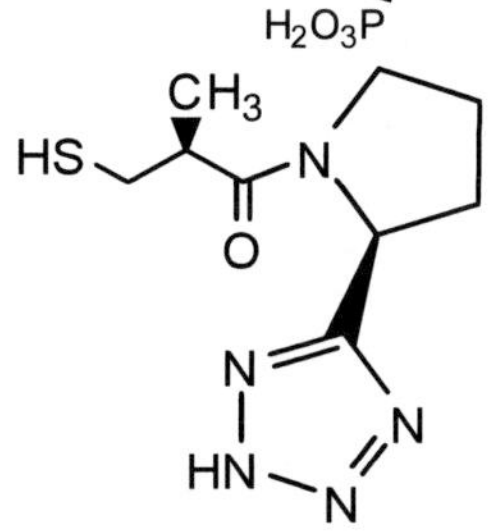	0.26
34.	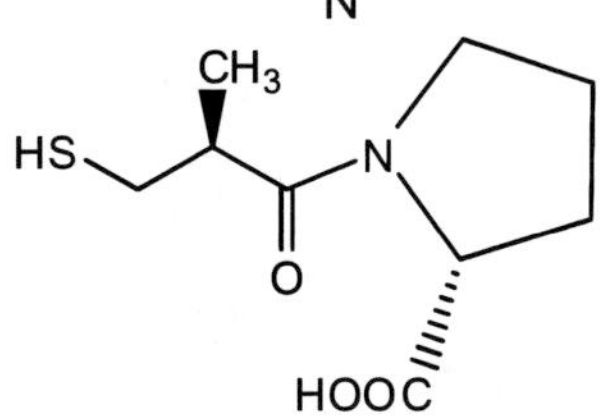	1800
35.	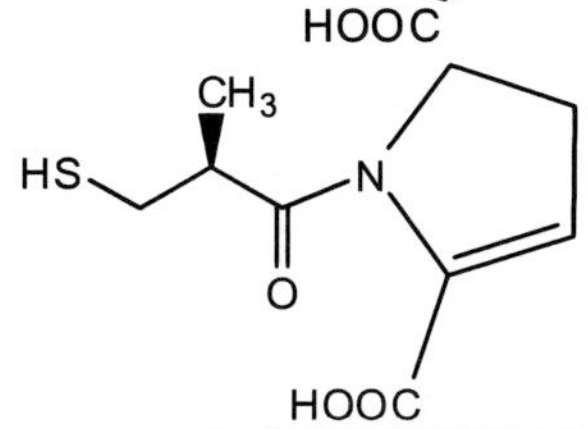	0.65

9.1. Variations at the Cationic Site

Comparison of Captopril 22 with its des-carboxy analog 30 (Table 5) shows the critical nature of this interaction. A similar comparison of $BPP_{5\alpha}$ vs its des-carboxy analog[12] reveals even greater requirement for this interaction. Thus, the presence of a strong anionic group which could bind to the cationic site seems to be a necessity for any effective ACE inhibitor.

Attempts in various laboratories have been made to modify the cationic site interactions by replacing the carboxylic group with various substituents. Various compounds with different anionic residues have been synthesized and their SAR studies have been reported. (Table-5)

The synthesis and biological activity of carboxylic acid replacement analogues of Captopril was investigated[45]. Replacement of carboxylic group with –CONHOH resulted in compound 31 which was equipotent to Captopril invitro. Replacement of –COOH with PO_3H_2 as in compound 32 resulted in somewhat reduced activity. Compound 33 showed activity greater than 32 but less as compared to 31.

Alacipril, a carboxy derived derivative of captopril was synthesized as a prodrug which showed activity equal to Captopril but with a longer duration of action[46-49].

As already stated, only L(S)-aminoacids are tolerated at the cationic site. This is exemplified in compound 34, a D-proline analog of Captopril which is about 75 times less active than Captopril[50]. However this critical interaction can apparently be satisfied by planar geometry of the dehydro derivative[50] 35.

9.2. Variations at the Site S_2^1

The relatively flexible binding requirements of S_2^1 subsite of ACE have permitted the preparation of many novel analogs of Captopril and Enalapril in which the proline residue has been replaced by various groups such as heterocyclic acids, substituted glycines, pyroglutamates and 4-substituted proline serrogates, resulting in the development of various new classes of ACE inhibitors.

(a) Replacement of Proline Part with Heterocyclic Acids

Heterocyclic acid derived ACE inhibitors in particular have received much attention. Numerous heterocyclic compounds have been synthesized and tested for their ACE inhibitory activity (Table 6).

1,3-Thiazolidine-4-carboxylic acid derivative 36 (K_i =6x10^{-9}M) exhibited invitro and invivo inhibitory activities of an order similar to that of Captopril[51-53]. A C-2 derivative of 36 having a 2-hydroxyphenyl substituent 37, designated as SA-446 and DE-040 (Fentiapril) was about 3-5 times as potent as Captopril invitro[54-56]. The isomeric C-2 derivative is at least 1000 fold less active[55]. Disubstitution at this position drastically reduced the activity. Contrary to Captopril's structure activity profile, an added methyl substituent in the S_1^1 position of 37 slightly decreased the activity. Isomeric 3-hydroxy and 4-hydroxyphenyl derivatives were less active than 37[56]. The invivo properties of Fentiapril varried considerably depending upon the animal model[54, 57-60]. In stroke prone spontaneously hypertensive rats, compound 37 was equally effective as an antihypertensive as Captopril, when given orally despite a much weaker inhibition of plasma ACE[58].

Replacement of proline part of Captopril with 1,3-oxazolidine-4-carboxylic acid[61] resulted in compound 38, which was five times less active than Captopril. Compound 39, a product of benzofusion on C-4,5 of proline residue of Captopril was found to be 3-7 times more potent invitro than Captopril and was more effective in reducing blood pressure in spontaneous hypertensive rats[62-64]. Both tetrahydroquinoline 40 and tetrahydroisoquiniline 41 derivatives were comparable to Captopril as ACE inhibitors[63-65]. However the isomeric tetrahydroisoquinoline derivative 42 was much less active[63]. As with 37, inhibitors 40 and 41 did not require S_1^1 methyl substituent for high in vitro activity. In invivo studies, 40 and 41 were less effective than indoline derivative 39[63].

Synthesis and ACE inhibitory activity of 3-acyl-1-arylenemethyl-2-oxo imidazolidine-4-carboxylic acid derivatives was also reported[66]. Resultant compound 43 was found to be the much potent inhibitor invitro studies comparable to Enalaprilat.

Benzothiazepine derivatives were also studied extensively as ACE inhibitors[67-70]. R-3-[(3)-1-carboxy-1,5-benzothiazepine-5-acetic acid (CV-5975) 44 was reported to be a potent inhibitor of ACE. Various saturated bicyclic amino acids[71] have been synthesized and evaluated for ACE inhibitory activity. The best known compound appeared to be 45 whose invitro activity was comparable to Enalaprilat.

Nonsulphydril compound quinapril (CI 906,22) 46 was synthesized[72] and found to be more active invivo although it was less active as compared to Enalaprilat in invitro studies.

Some bicyclic compounds[73] have been designed and tested for ACE inhibitory activity. Of these the Cilazapril 47 was found to be the much potent inhibitor. It was designed as prodrug like Enalapril which on hydrolysis gave the active inhibitor 48. 10-Aryl sulphonamide substituted N-acyl-(1-carboxy-3-phenylpropyl)-l-alanyl-L-proline 49 was also found to be potent ACE inhibitor[74].

(b) Glycine Analogs

The value of α-amino acids in the S_2^1 binding site of ACE is well documented[14]. Several years back novel type of Captopril analogs were described in which the amino acid used in place of proline was N-substituted glycine[75]. One of these compounds 50 (RHC-3454) was reported to be equipotent to Captopril invitro. Other alicyclic and aromatic substitutions in place of cyclopentyl were also explored[75, 76]. Out of these glycine analogs the pivaloyl thioester of 50 (RHC-3659, pivalopril) was selected for clinical trials[77- 79] though it was less potent than Captopril.

Various modified Enalapril analogs having glycine as amino acid have been developed[80]. One of these N-[N(S)-1-ethoxycarbonyl-3-phenylpropyl-L-alanyl]-(N-indan-2-yl) glycine 51 (CV-3317) was found to be more potent than Captopril in invitro.

(c) Pyroglutamic Acid and 5-Thioxoproline Derived ACE Inhibitors

As described earlier site S_2^1 of ACE is capable of accepting a wide variety of amino acids, which provided researchers a great deal of flexibility in modifying proline residue by replacement with pyroglutamic acid and 4-substituted prolines.

With an aim to increase the hydrophobicity of the proline molecule one of the research group made an attempt to replace the proline part of Captopril with pyroglutamic acid 52 (Table 7) resulting in the ACE inhibitory activity enhanced to the greater extent[81]. Similar results were also obtained when proline part was substituted with 5- thioxo prolines 53[82]. N-(1(S)-(ethoxycarbonyl)-3-phenylpropyl-(S)-alanyl)-S-pyroglutamic acid[83] 54 was synthesized as modified analog which showed activity about half of Enalapril. Pyroglutamic acid analogs having hetero atom at position-4 were also synthesized and tested for their ACE inhibitory activity[81]. Of these compounds 55 with a oxygen atom at position-4 showed pronounced activity. (Table 6). N-(3'mercaptoalkanoyl)-4-α(S)-(phenyl methyl) pyroglutamic acid 56 showed 60% inhibition of angiotensin converting enzyme at 30Mg/μg in invivo studies[84].

(d) 4-Substituted Proline Analogs

Various analogs of captopril with replacement of the proline part of the Captopril or Enalapril with 4-substituted prolines (Table 6) were designed with an assumption that C-4 substitution at proline with hydrophobic groups will go into the S_2^1 subsite of ACE and in turn this tighter binding with S_2^1 will result in better ACE inhibitory activity . With this basic logic Zofenopril 57 (SQ, 26, 991) having a thiophenyl group at position-4 was synthesized[85- 88] by the Squibb group. As expected it was the much potent inhibitor of ACE both in invitro as well as in vivo studies.

With the encouraging results obtained from the discovery of Zofenopril, various 4-substituted proline analogs have been synthesized in Captopril as well as Enalapril based inhibitor series. Both 4-α and 4-β-substituted proline derived ACE inhibitors (58- 65)(Table-6) have been reported[89-91] along with their IC_{50} values. All these compounds showed promising results providing considerable amount of evidence for C-4substitution of proline to interact with the S_2^1 subsite of the enzyme.

(e) Aromatic compounds

Compound 66 where proline part of captopril was replaced by p-hydroxy phenylmethyl group with simultaneous modification of the mercapto chain, showed ACE inhibitory activity much less as compared to captopril[92].

(f) Pyrrolidine Analogs

Replacing proline part with 4-sunstituted pyrrolidine[93] 67 or 4, 5 substituted pyrrolidine[94] 68 resulted in hypotensive activity higher as compared to Captopril when matched with virtual screening.

(g) Glycosides

Few glycosides have also been found to have potent ACE inhibitory activity. Compound 69 where o-glycosidic linkage was established with substituted aromatic group[95] showed activity much higher than that of captopril. Similarly compound **70** where glycosidic linkage was established with pyrrolidine showed similar results[95].

Table 6. Replacement of proline part of Captopril with heterocyclic acids, glycine analogs

No.	R	IC_{50} (nM)
22.	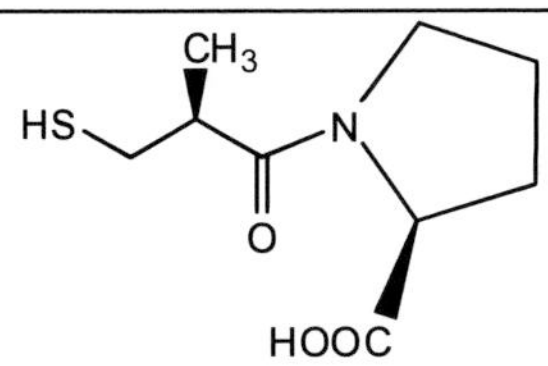	23
36.	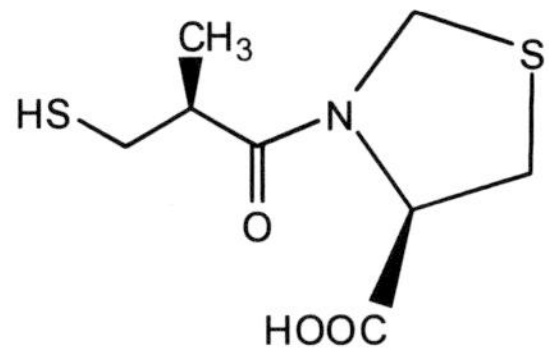	9.5-260
37.	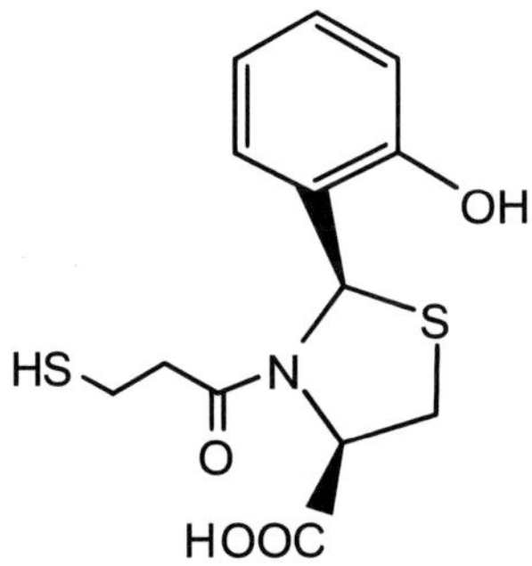	6
38.	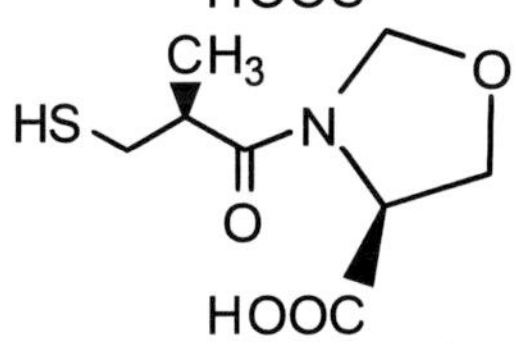	1/5 OF CAPTOPR IL
39.	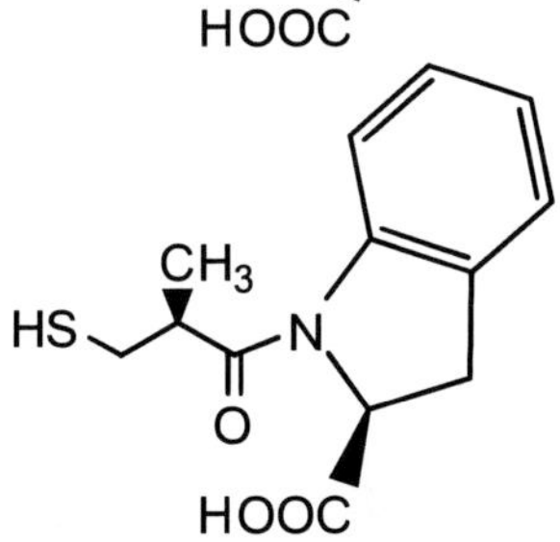	3.7
40.	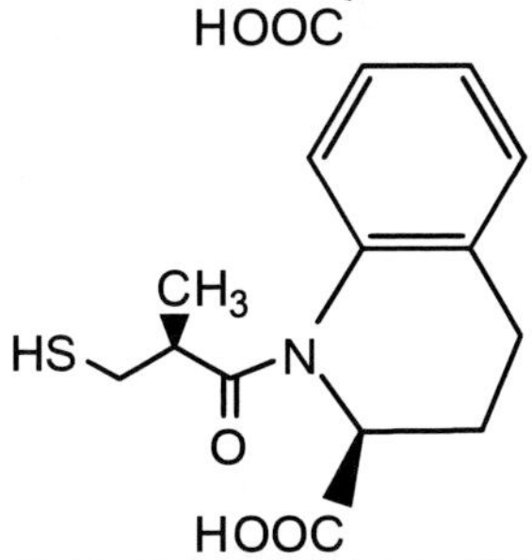	25[a]

Table 6. Replacement of proline part of Captopril with heterocyclic acids, glycine analogs (Continued)

No.	R	IC_{50} (nM)
41.	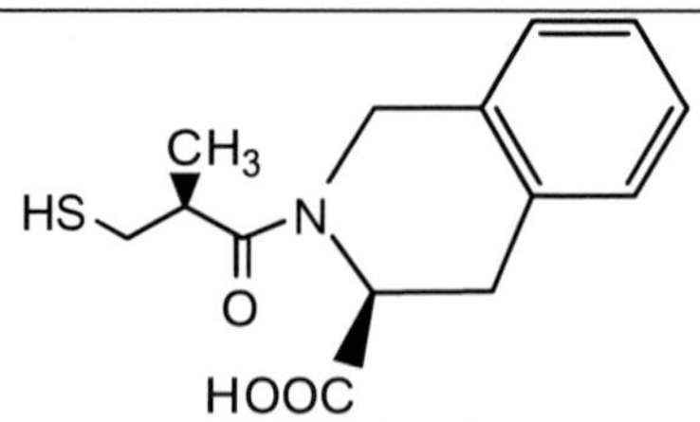	20[a]
42.	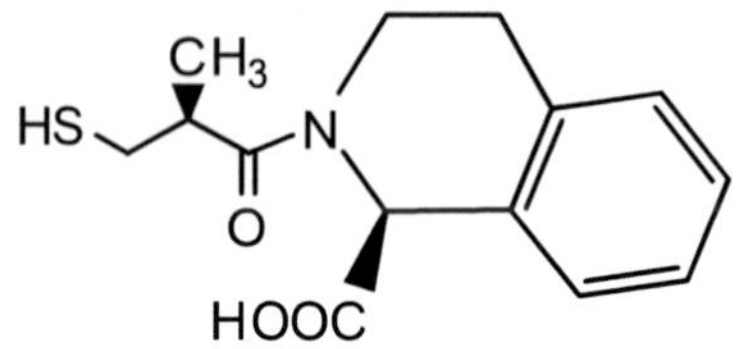	1800[a]
43.	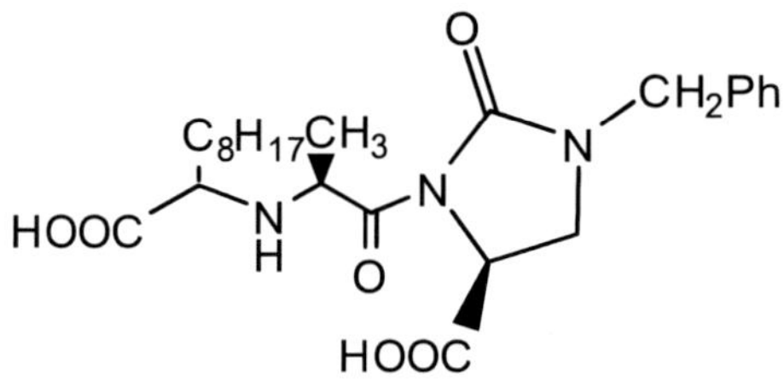	15
44.	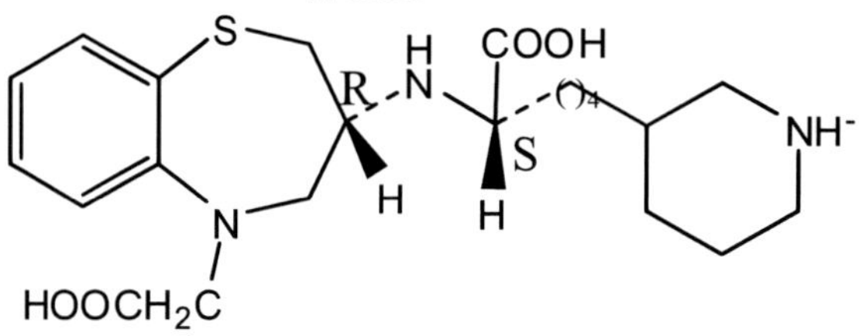	-
45.	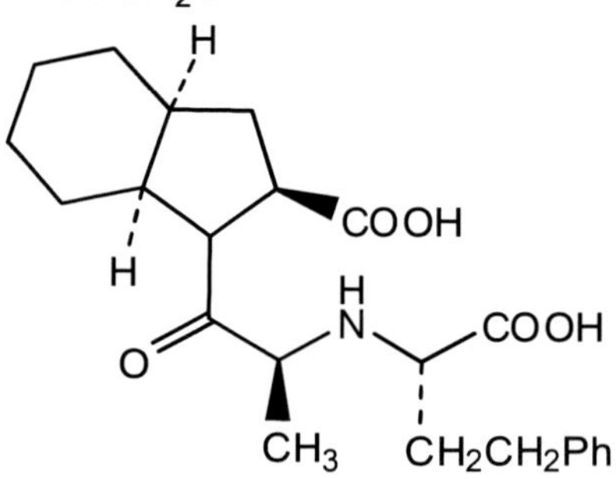	24
46.	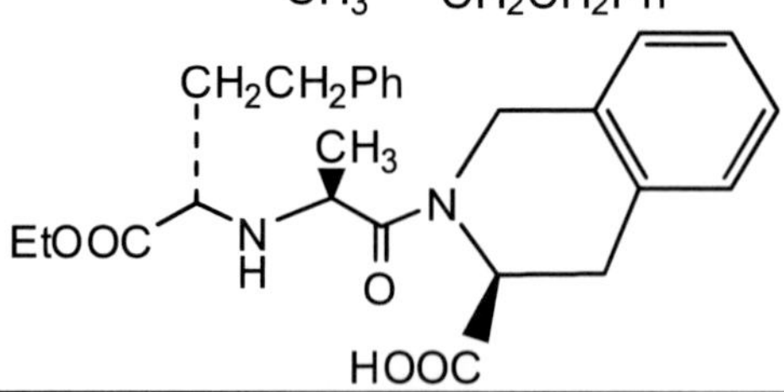	83

No.	R	IC$_{50}$ (nM)
47.	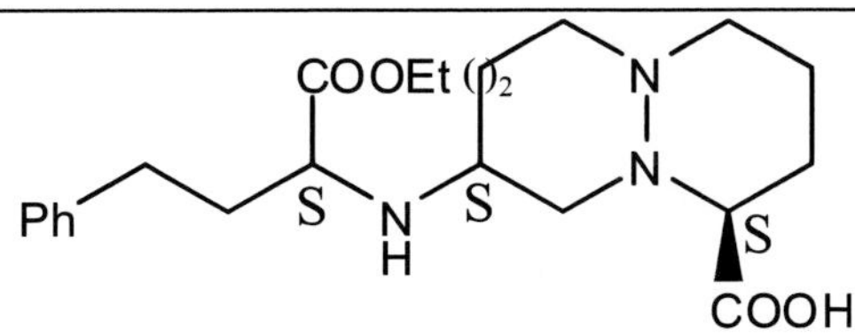	15
48.	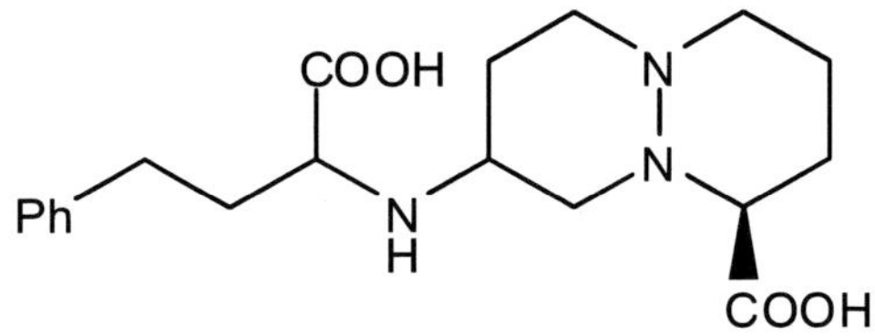	1.6
49.	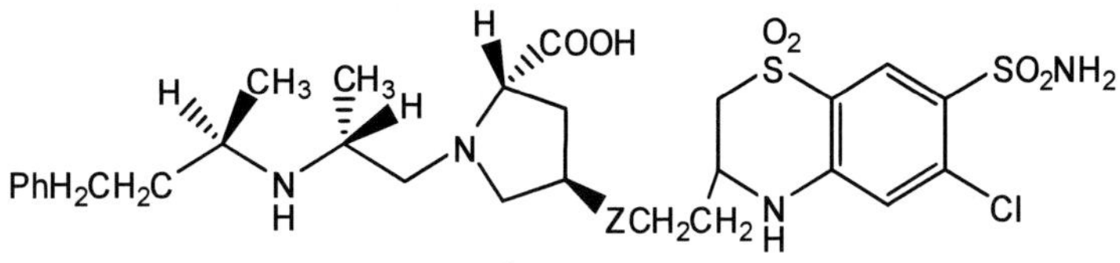	-
50.	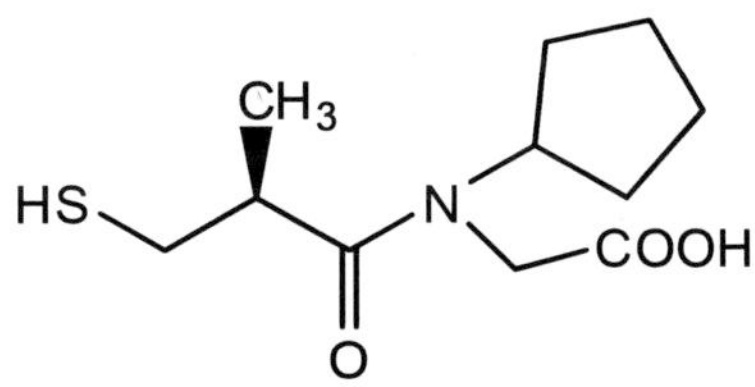	16
51.	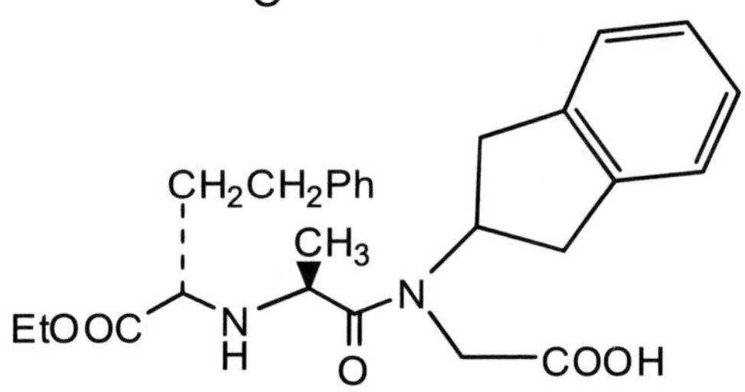 (CV=3317)	-
52.	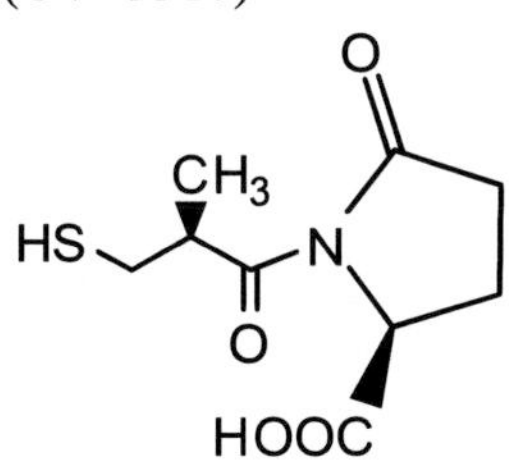	3.6
53.	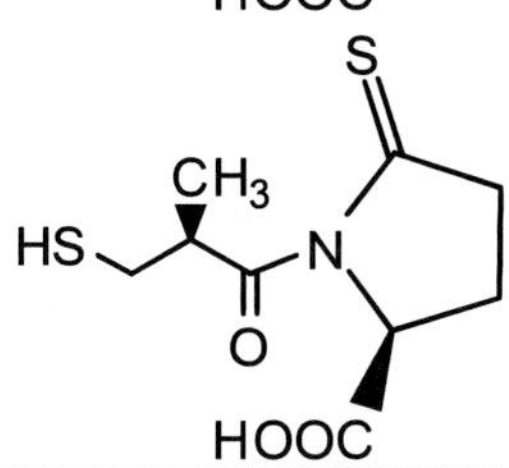	-

Table 6. Replacement of proline part of Captopril with heterocyclic acids, glycine analogs (Continued)

No.	R	IC_{50} (nM)
54.	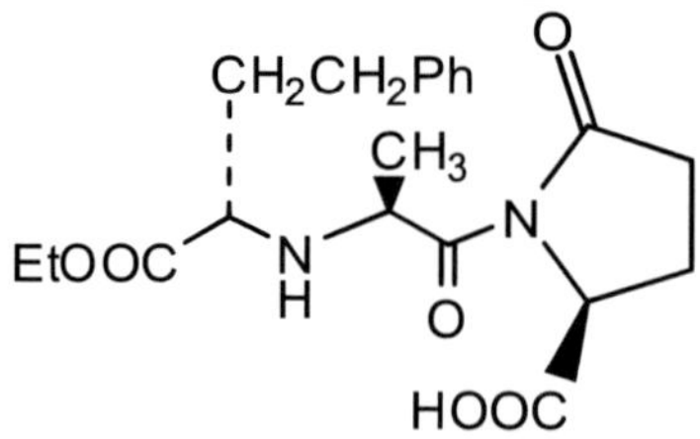	1/2 of Enalapril
55.	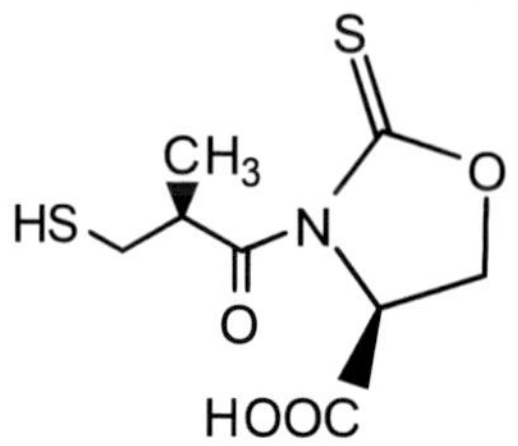	6.4
56.	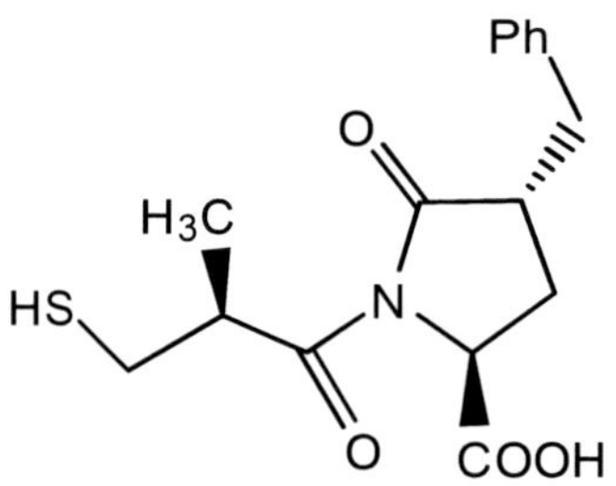	--
57.	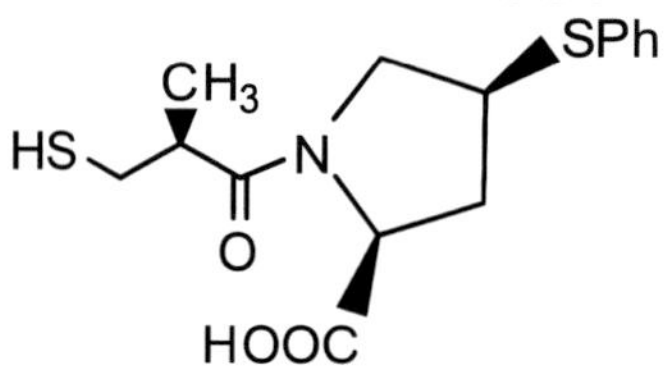	8
58.	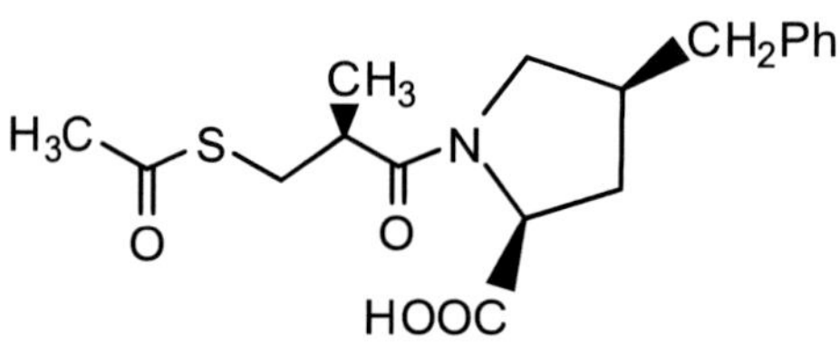	2.0
59.	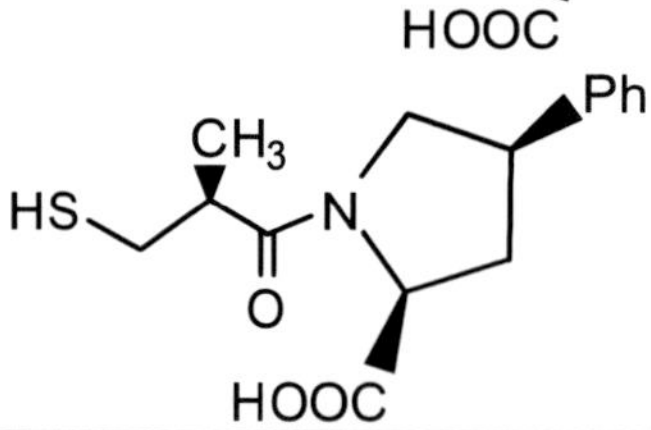	30

No.	R	IC_{50} (nM)
60.	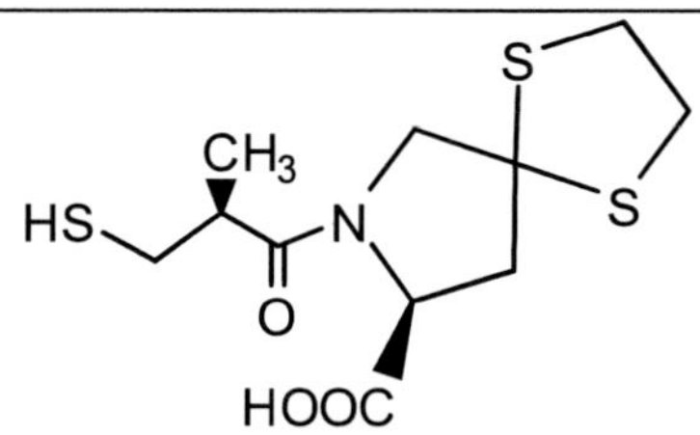	6.5
61.	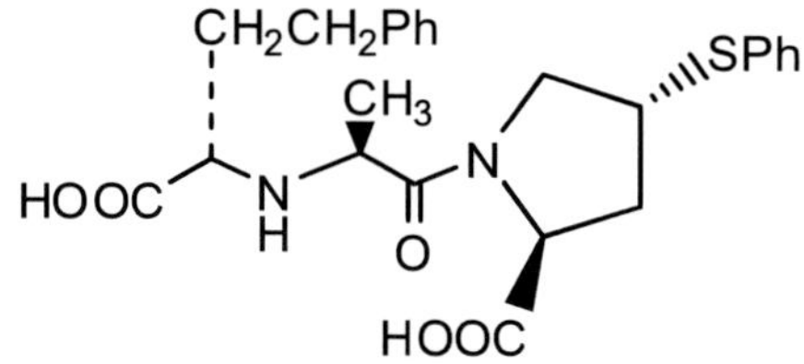	2.0
62.	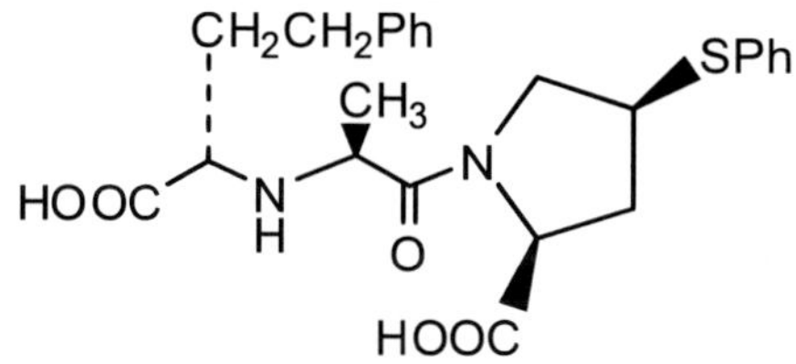	2.0
63.	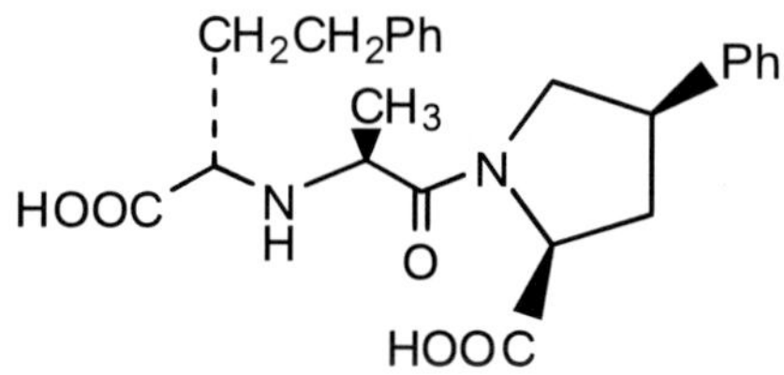	5.0
64.	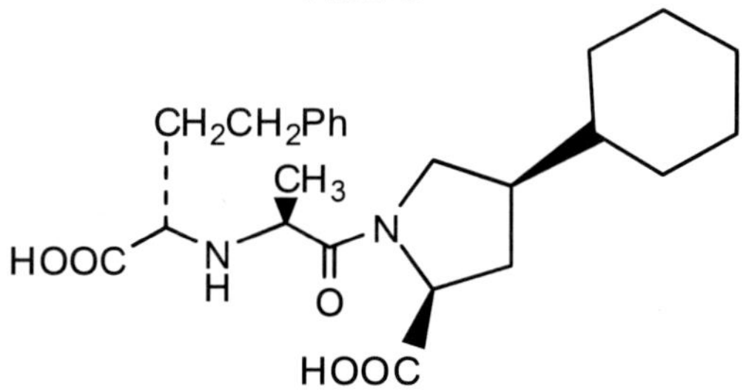	5.0
65.	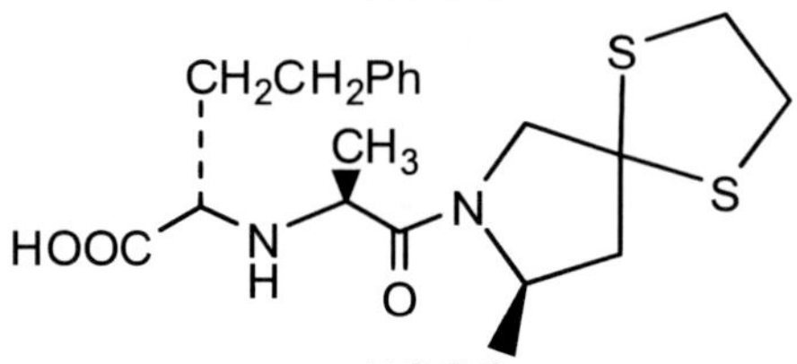	2.0
66.	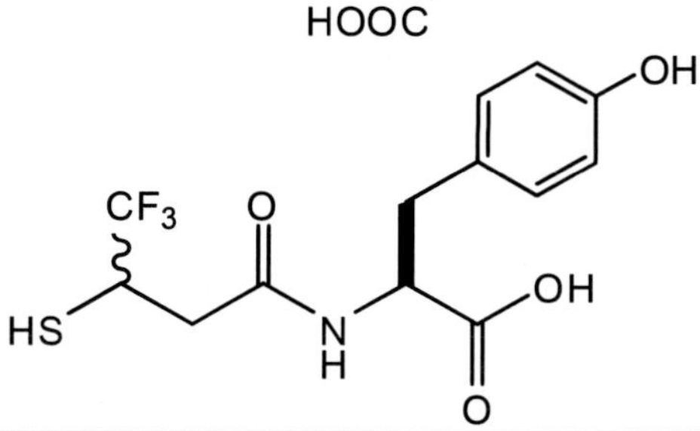	160

Table 6. Replacement of proline part of Captopril with heterocyclic acids, glycine analogs (Continued)

No.	R	IC_{50} (nM)
67	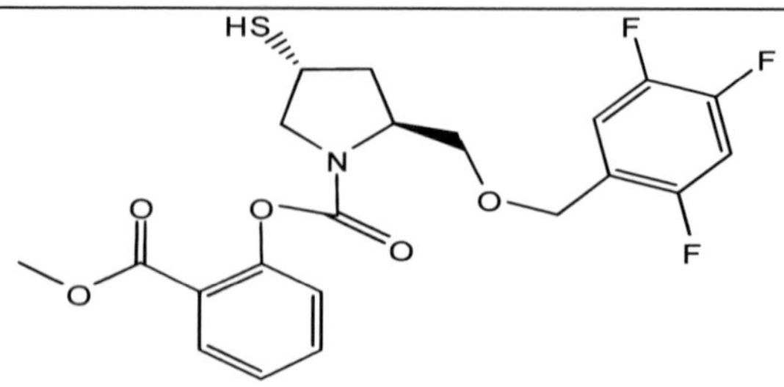	higher than Captopril
68	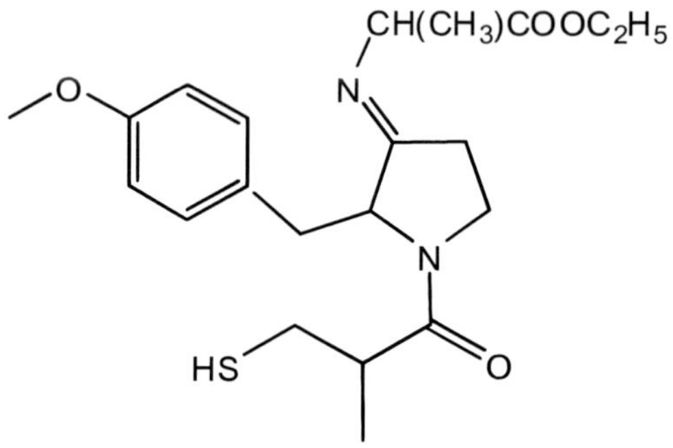	higher than Captopril
69	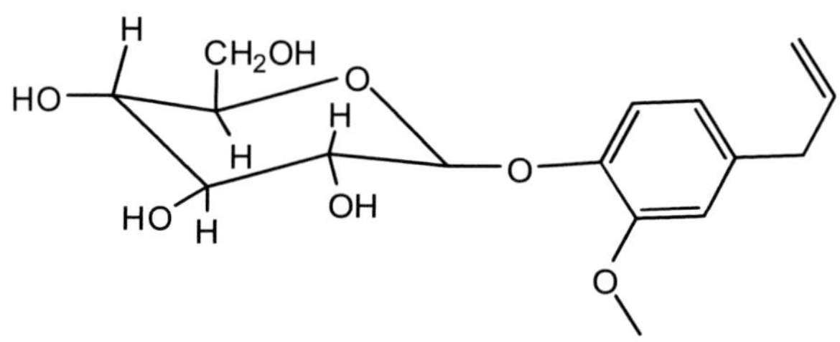	0.5
70	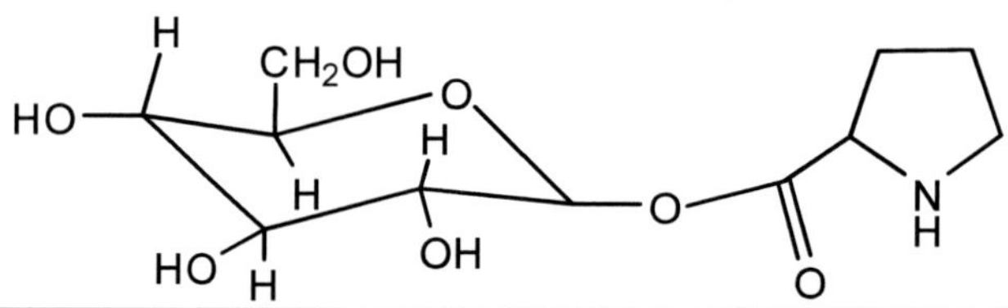	1.7

9.3. Variations at the Site S_1^1 of ACE

(a) Variation of the Methyl Group at S_1^1

The S_1^1 subsite of ACE seems to be different from S_2^1. For example proline is poorly tolerated in this position and hydrophobic groups larger than $-CH_3$ do not significantly enhance the activity of inhibitors. The later point is illustrated in the Table 7, where compounds 71, 72, 74-76 showed this effect[32, 15]. The substituent in this position must be of the L-configuration (72 vs. 73). The activity enhancing properties may quite likely have a conformational origin. Interest in the S_1^1 residue was further hightened when Lisinopril 29 (Table-4) as discussed earlier was found to be much more effective as compared to Enalapril in invivo studies.

Table 7. Replacement of methyl group of N-acyl side chain with other groups

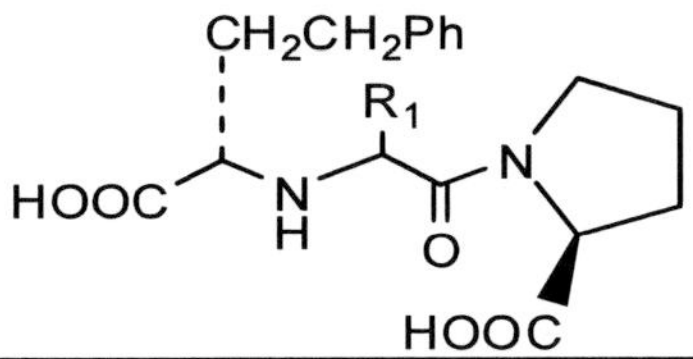

No.	R_1	IC_{50} (nM)
71.	-H	230
72.	$-CH_3(S)$	3.8
73.	$-CH_3(R)$	2500
74.	$-CH_2F$	3.6
75.	$-CH_2CH_3$	5.0
76.	$-CH_2CH_2CH_3$	8.0

Table 8. Role of amide carbonyl

No.	R_1	IC_{50} (nM)
77.	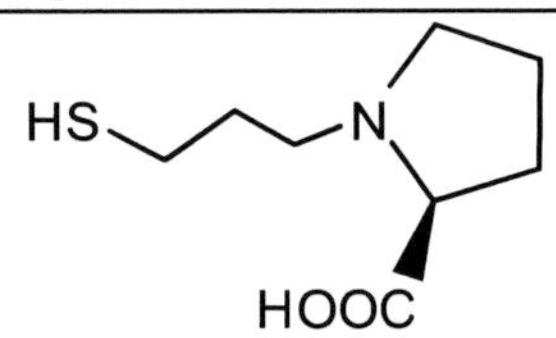	240 22
21.	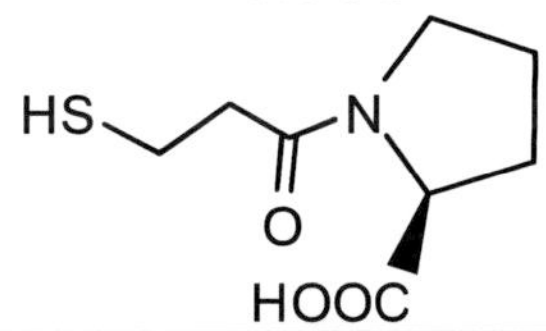	0.20

(b) The Role of Amide Carbonyl

The role of S_1^1 as a hydrogen bond donor site was established for mercaptoacyl amino acids[96] and it was found that inhibitor's amide carbonyl is important for activity as shown by the example given in Table 8. Compound 77 which does not contain an amide carbonyl is about 1000 times less active than 21 which contains an amide carbonyl. Thus from this example it may be concluded that a carbonyl group capable of accepting a hydrogen bond from this site on enzyme must be present.

(c) Modifications in the Thiolchain of Captopril

Extensive work has been carried out in various laboratories to carry out changes at site S_1^1 by modifying the methylmercapto chain of Captopril as shown in table 9. Various compounds have been developed and tested for their ACE inhibitory activity. A series of aroyl analogs have been reported[97] compound 78 (CL, 242,817) is one of the most potent ACE inhibitor in this series. The more active diastereomer has the R-configuration at the S-acetyl group. Compound 78 was found to be equipotent orally with Captopril in inhibiting the A-I pressor responses in normotensive rats and dogs[98].

Table 9. Modifications in the N-acyl side chain of captopril

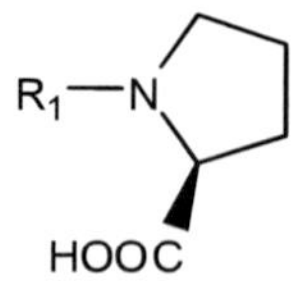

No.	R_1	IC_{50} (nM)
78.	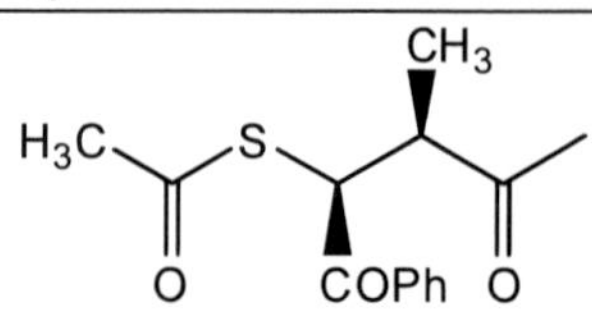	55
79.	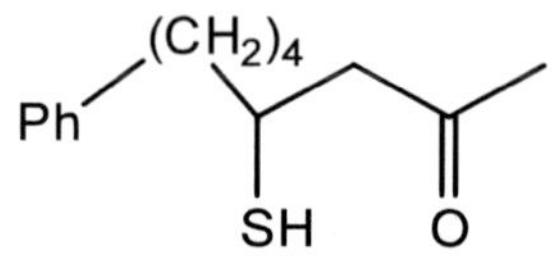	620
80.	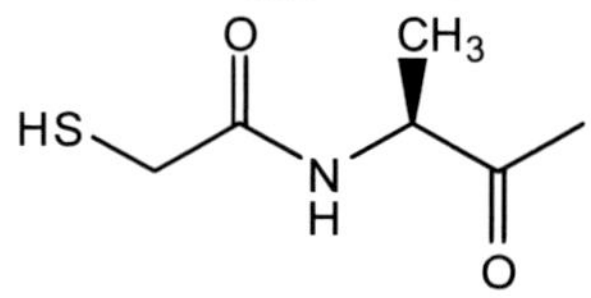	60
81.	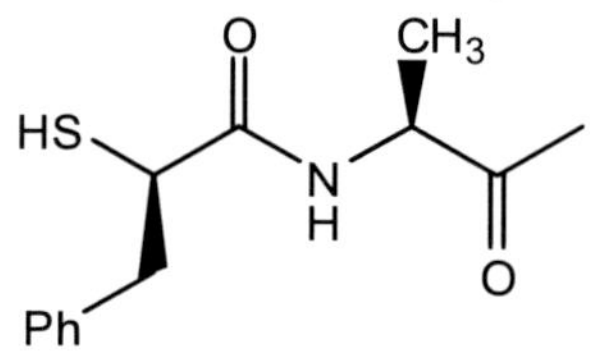	50
82.	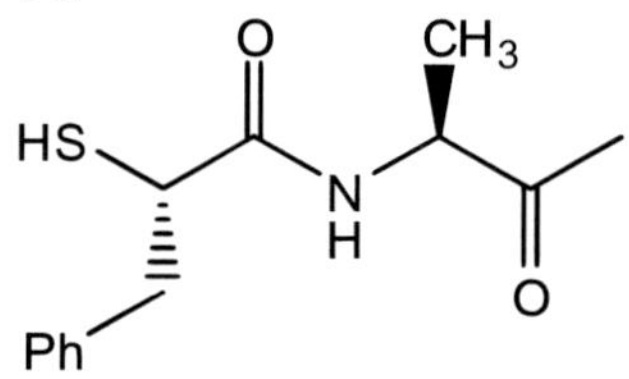	60
83.	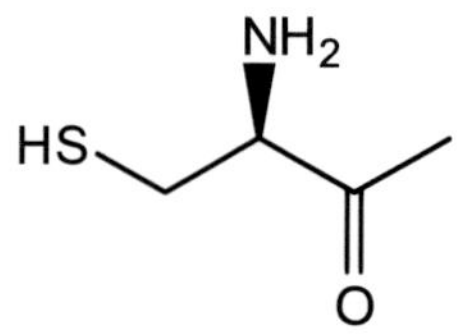	5.7
84.	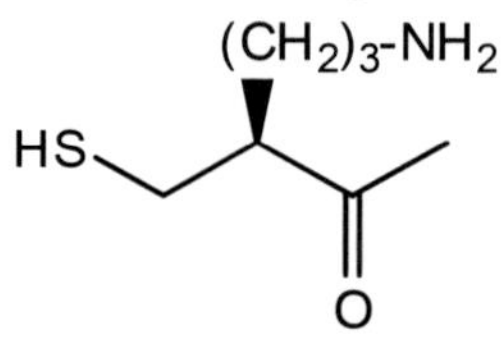	69
85.	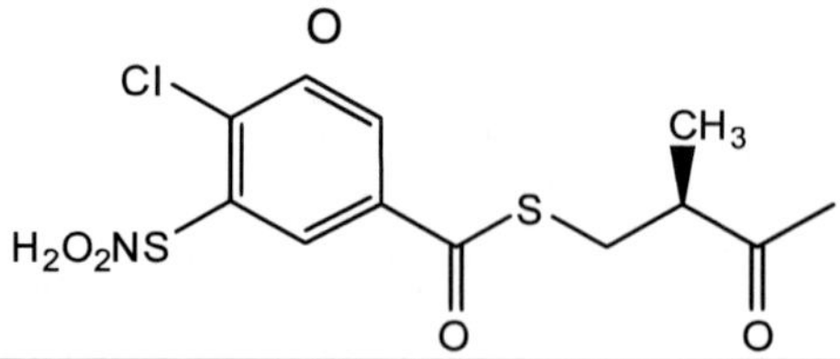	-

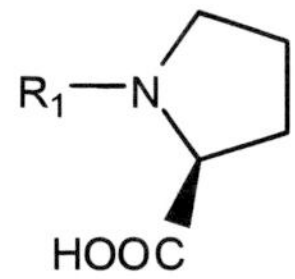

No.	R_1	IC_{50} (nM)
86.	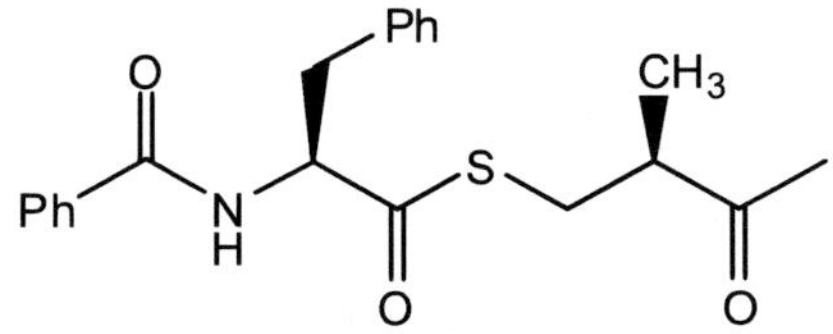	30
87	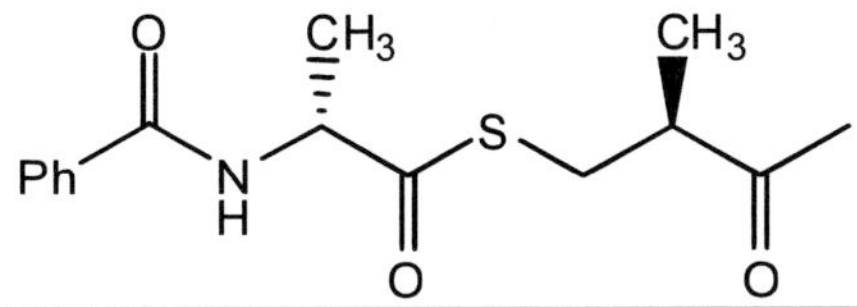	260

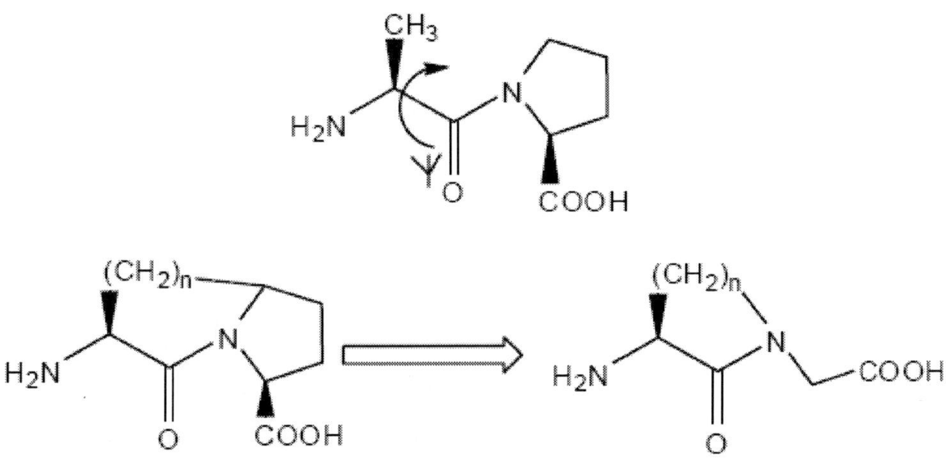

Figure 5.

Compound 79 was reported to have slightly decreased binding to the enzyme[99]. Likewise the introduction of a lipophilic substituent (81 and 82) into 80 did not influence the activity.

D-cysteinyl-L-proline derivative 83 was found to be a good inhibitor of ACE with a Ki of $5.7x10^{-7}$M (ph 8.3)[100]. Compound 84 with an amino propyl group also found to be effective in replacing α-methyl substituent in the $S_1{}^1$ position. These data, however do support the conclusion that $-CH_3$ effect of group in Captopril is largely to augment potency by conformational effect[35].

Thioester, 85 (Y-108) was reported to have different hypotensive mechanism than Captopril[101]. Similarly the N-benzoyl-L-phenylalanine thioester derivative 86 showed invitro activity comparable to that of Captopril[102]. Additional analogs of 86 containing D-alanine residue, 87 (MC-838) was found to be the most active analog in the series invivo and was having a longer duration of action than Captopril[103].It was more effective as compared to Enalapril as well in invivo studies.

Table 10. Conformationally restricted analogs of Captopril

No.	R_1	IC_{50} (μM)
88.	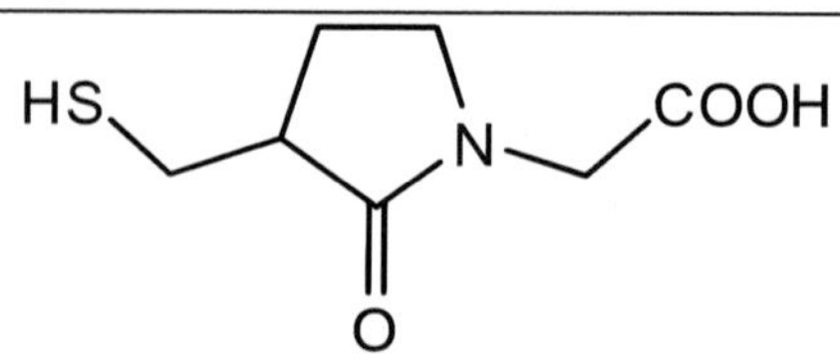	3.7A[a]
89.	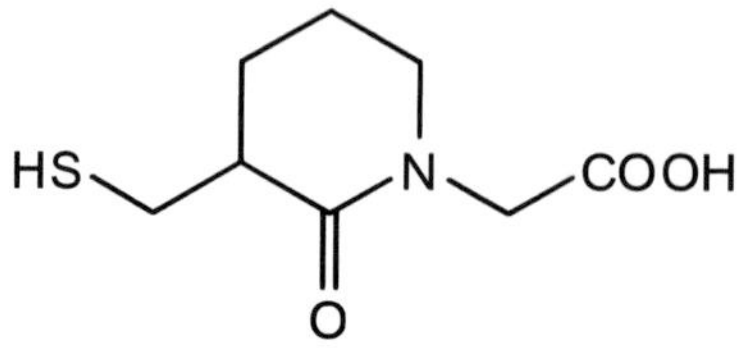	3.4[a]
90.	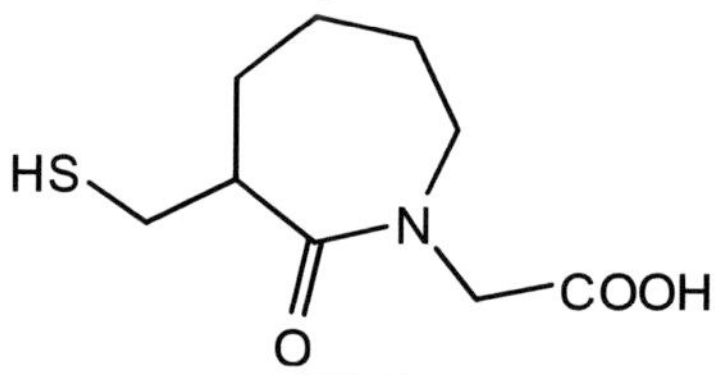	0.17[a]
91.	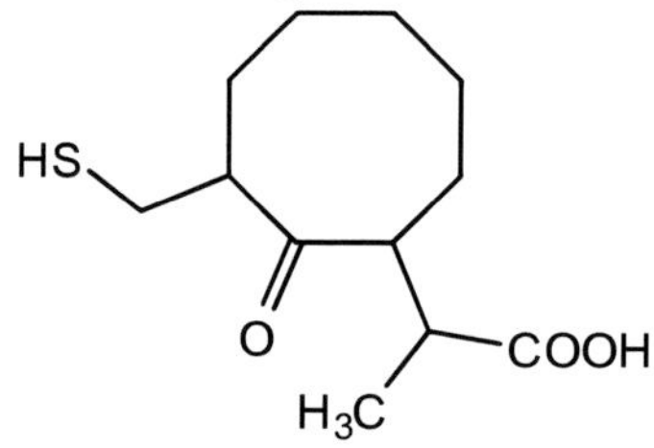	0.053[a]
92.	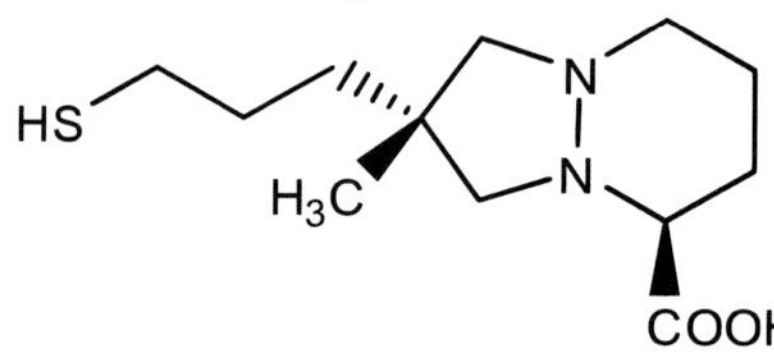	0.10[a]
93.	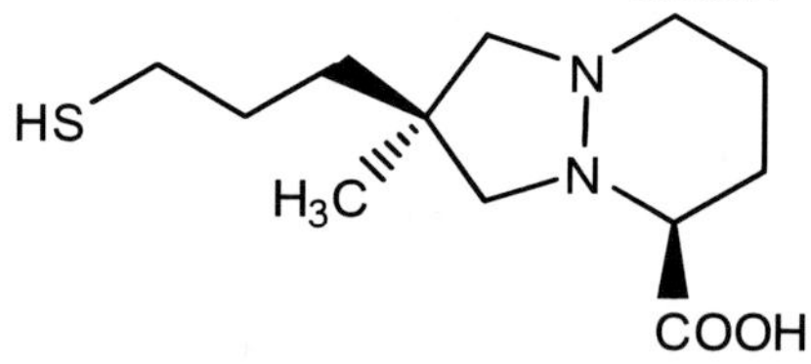	43[a]
94.	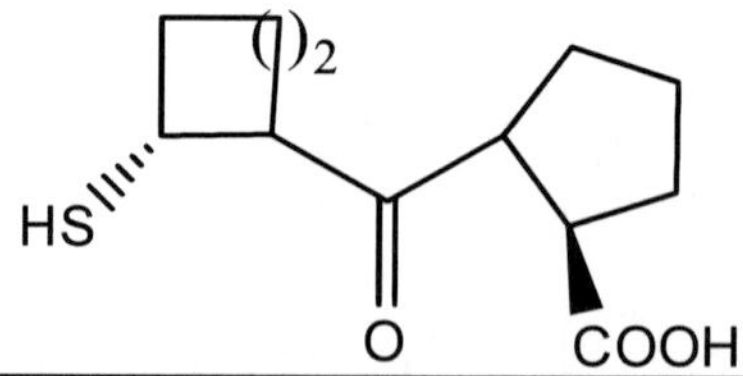	0.58

(d) Conformationally Restricted Inhibitors of ACE

Thorsett and coworkers have examined in some detail the conformational aspects of the sites $S_1{}^1$ and $S_2{}^1$ in the Captopril and Enalapril designs[104- 106]. Spectroscopic studies on L-alanyl-L-proline indicated that trans rotamer with respect to the amide bond was slightly preferred over cis[107, 108]. However spectroscopic studies on Enalaprilat showed that trans rotamer was heavily preferred (4-7/1) in the pH range 3.5-11[38]. Since Enalaprilat was a potent inhibitors of ACE, it was hypothesized that it might not undergo major conformational changes upon binding to the enzyme and thus binds as its trans rotamer, if so the conformation of L-analyl-L-proline part of Enalaprilat could solely be determined by ψ as is illustrated in the Figure 5.

Conformationally restricted analogs of L-analyl-L-proline were then created by bridging a carbon chain between the alanylmethyl group and the C-5 position of the proline ring, for simplicity C-3 and C-4 were initially vomited as in fig 5. The angle ψ is a function of lactam size. Thorsett predicted that 5- and 6-membered lactams would not be good mimics of L-alanyl-L-proline in various inhibitors designs. However they predicted that medium sized lactams (7, 8 and 9) would make good approximations of low energy conformation of Ala-Pro.

(i) Conformationally Restricted Analogs of Captopril

As shown in table 10, 5- and 6-membered lactams 88 and 89 respectively were significantly less effective as compared to Captopril[105, 109, 110]. However, as predicted 7- and 8-membered lactams 90 and 91 were much more effective[105]. In fact compound 91 which had a part of proline ring restored as a methyl substituent, was equipotent with Captopril. The initial study verified the assumption that trans rotamer is present in the bioactive conformation of Captopril. Independently later on other lactams eg. 92, 93 and 94 were also developed as conformationally restricted analogs of Captopril[63, 111, 112].

(ii) Conformationally Restricted Analogs of Enalaprilat

Numerous conformationally restricted analogs of Enalaprilat have been synthesized and reviewed[15]. As expected, medium sized ring inhibitors 97, 98 and 99 displayed excellent inhibition ACE activity as shown in table 11.

Table 11. Conformationally restricted analogs of enalaprilat

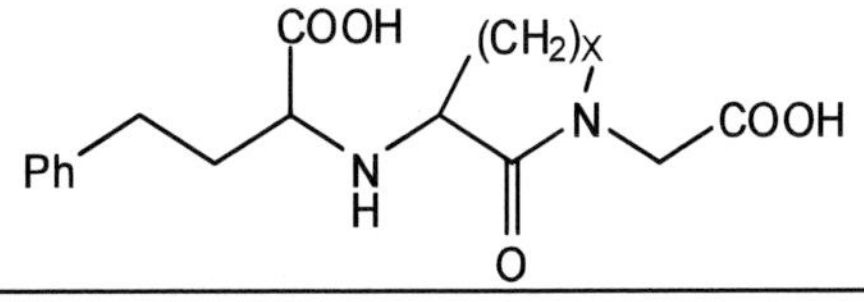

No	Ring size (X)	IC_{50} (nM)[a]
95.	5	5300
96.	6	430
97.	7	19
98.	8	4.8(1.9)[b]
99.	9	8.1

[a] tested as recemic mixture.

[b] tested as the resolved diastereomer.

Table 12. Phosphorous- based ACE inhibitors

No.	Ring size	IC50 (nM)
100.	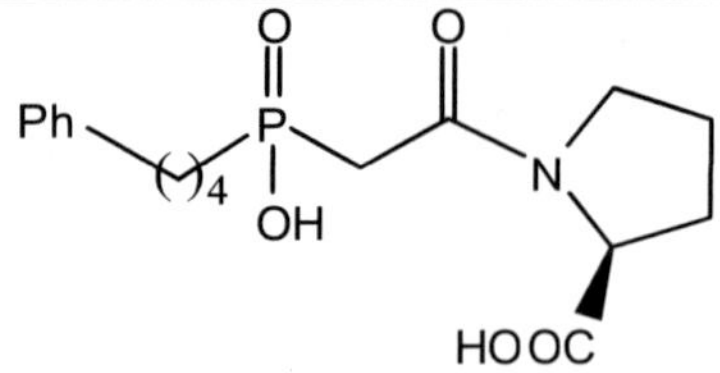	18
101.	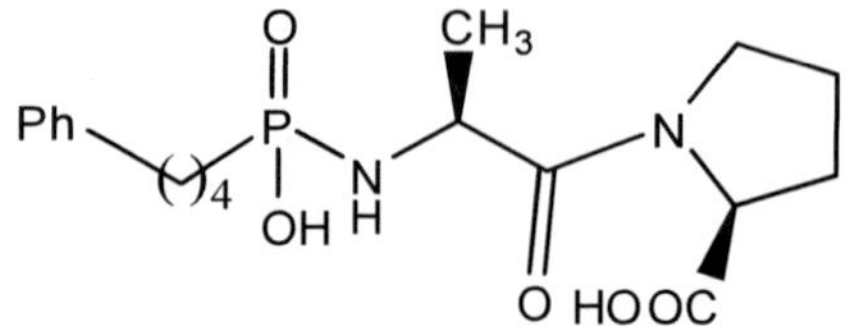	12
102.	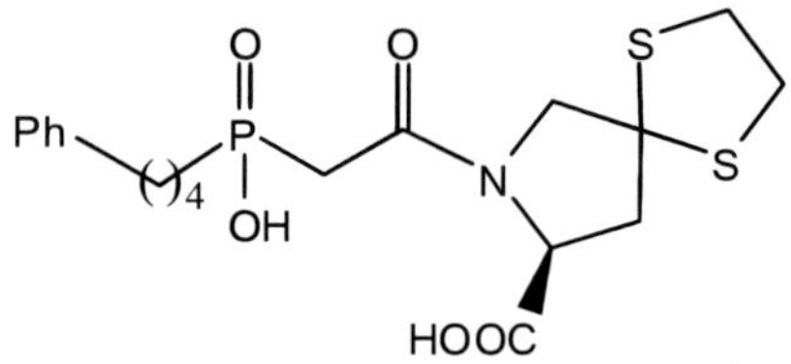	4
103.	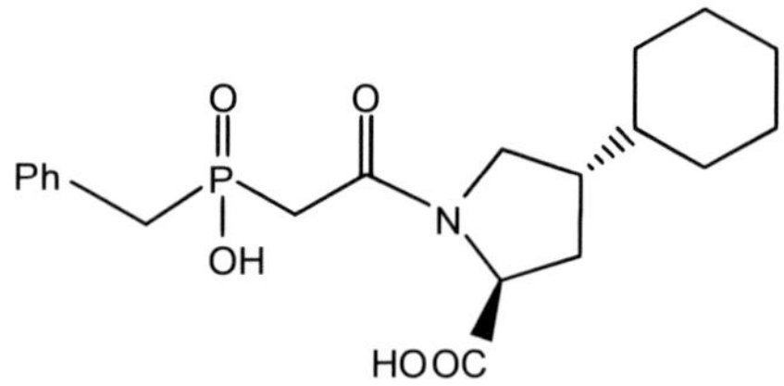	12
104.	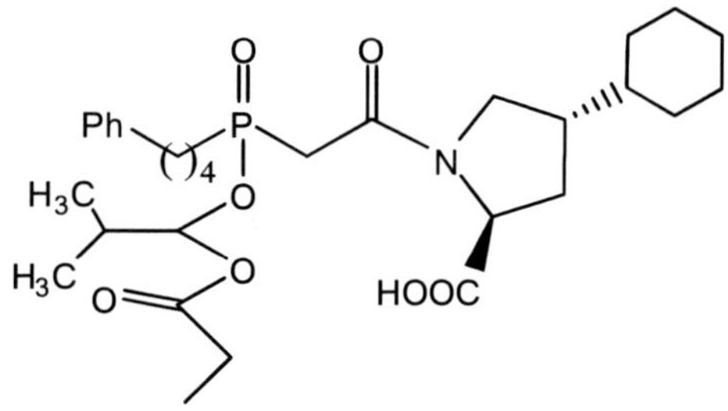	-
105	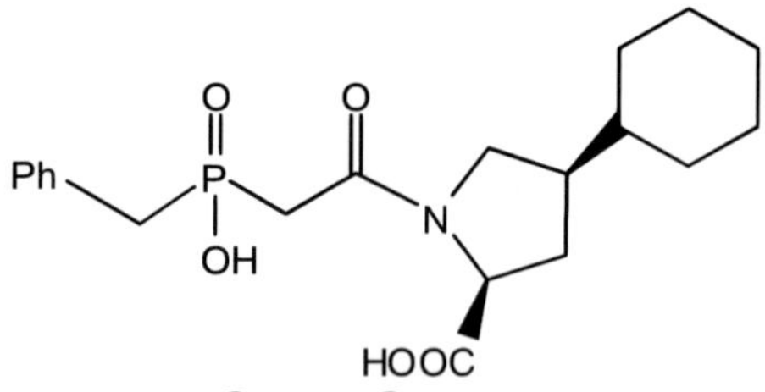	7.0
106.	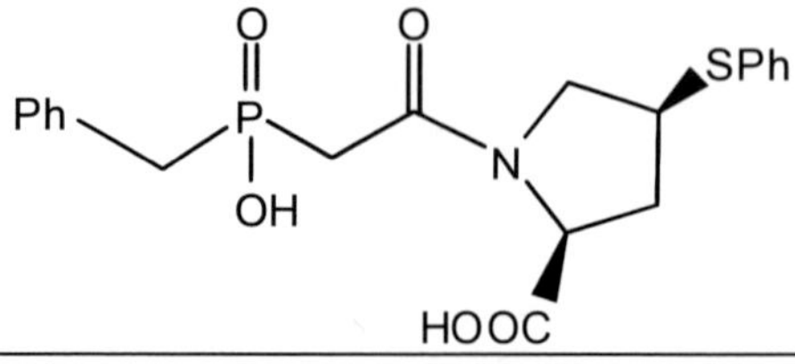	20

No.	Ring size	IC50 (nM)
107.	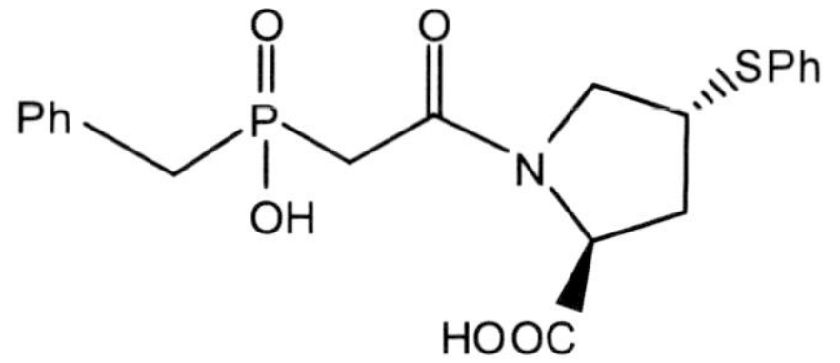	17
108.	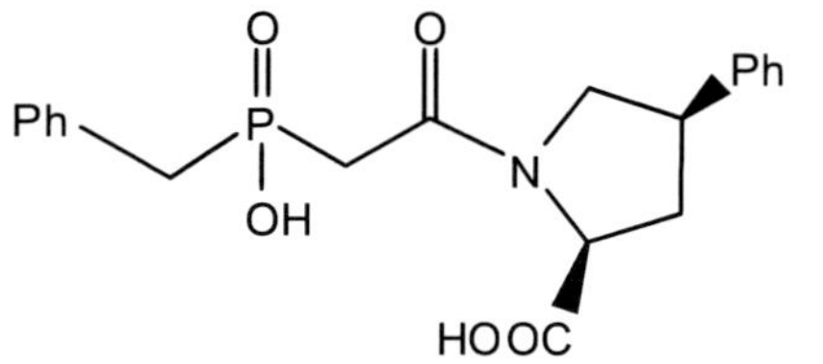	7.0
109.	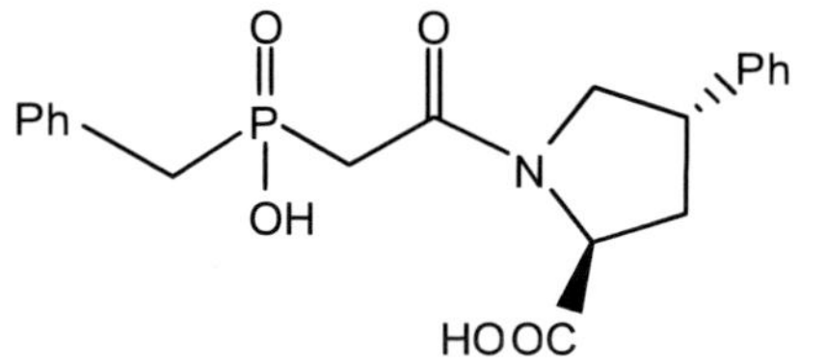	120
110.	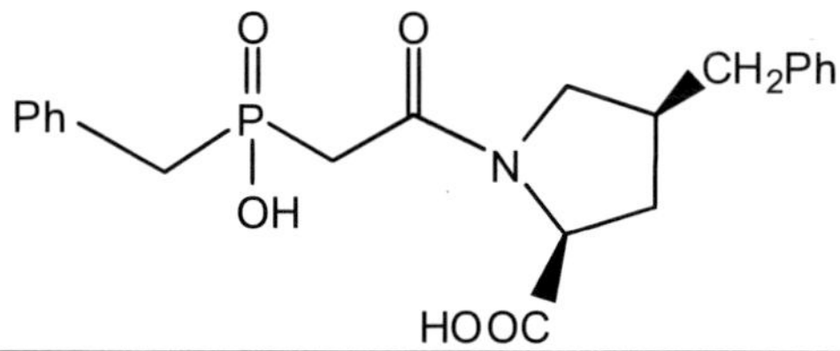	10

(e) Phosphorous- Based Inhibition Designs

The isolation and structure determination of phosphoramidon[N-(α-Lrhamnopyra-nosyloxydroxyphosphosphinyl)-L-leucyl-L-tryptophan] was announced in 1972 and its potent inhibition of Zn^{++} metalloenzyme thermolycin was reported[113, 114] in 1973. Thus based on this model phosphorous based inhibitors were also designed.

Various phosphorous based inhibitors have been synthesized in last few decades and have been summarized[15]in table 12

Compound 100 with phosphinyloxy acyl group was found to be the potent inhibitor of ACE[115] (Table 12). Phophinyloxy acylamino acids have also been reported as inhibitors of ACE[116]. Out of all these, compound 101 was the most potent inhibitor. It was the novel orally active compound, which showed activity greater than Captopril with a longer duration of action. Later on various N-phophinyloxyacyl-4-substituted prolines, e.g., 102 and 103 with modifications at both S_2^1 and S_1^1 subsites were synthesized and reported to have enhanced potency as compared to simple prolines, but these suffered due to the major drawback of poor oral activity. The best compound appeared to be Fosfopril **104** (SQ, 28,555)[117-119]. Fosfopril was taken for clinical trials and results indicated that it was the much potent inhibitors of ACE in humans with a duration of action greater than Captopril[120]. Various other N-phophinyloxyacyl 4-α and 4-β-substituted proline derived ACE inhibitors **105-110**, have been synthesized and reported as potent ACE inhibitors[89-91].(Table 12)

Table 13. Analogs with elongated N-acyl side chain of Enalaprilat

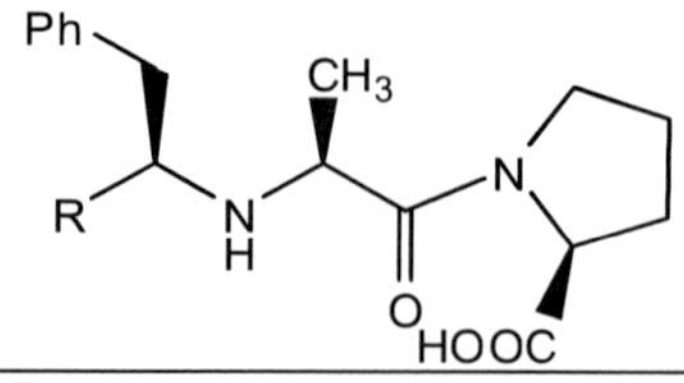

No.	R-	IC_{50} (nM)
111.	-H	330
112.	$-NH_2$	1.4
113.	-NHCOPh	2.7
114.	p-Glu-Lys-NH-	0.05

9.4. Variations at the Site S_1 And S_2 of Angiotensin Converting Enzyme

Development of carboxyalkyl dipeptides (table 4) revealed that phenylethyl group of Enalapril, due to its substitution pattern and stereochemistry might interact with the S_1 site of ACE as shown in fig 4, resulting in the increased binding with the enzyme and thereby increased ACE inhibitory activity. Based on this background associated with the development of Enalapril, the analogs of the N-carboxyalkyl dipeptides which could interact with site S_1 and additional sites S_2 or more as well, had been reported[16, 121]. Compound 114 shown in table 13 is many fold more potent than 111. The p-Flu-Lys-fragment in 114 interacts with the S_2 or S_3 binding sites of ACE leading to enhancement in binding over the tripeptide.

Some other new classes of modified peptides which could possibly interact with S_1 as well as S_2 binding sites of ACE have also been reported as shown in table 14. One such modified analog 5(S)-Benzamido-4-oxo-6-phenylhexanoyl-L-proline 115 was developed as potent ACE inhibitor[122]. It was active than Captopril in invitro studies. Similarly a tetrapeptide derivative 116 showed 50% inhibition of ACE at 37μM[123]. Synthesis and ACE inhibitory activities of N-carboxymethyl dipeptides with a ω-(4-piperidyl)alkyl group were reported[124, 125]. The best known compound among these was found to be alanyl-proline derivative 117, with S,S,S-cofiguaration. Further investigations led to the discovery of peptides[126, 127] as ACE inhibitors having various organic groups. The best compound was the pyrimidine derivative 118, though it was less active than Captopril.

Although extensive work has been performed in various laboratories to develop compounds with increased bindings with S_1 and S_2 sites of enzyme, however looking at overall SAR on various classes of molecules, it could easily be concluded that site S_2 or below that e.g. S_3 are less important and binding with these has not much effect on the overall ACE inhibitory activity

Table 14. Peptides as ACE inhibitors

No.	Ring size	IC_{50}
115.	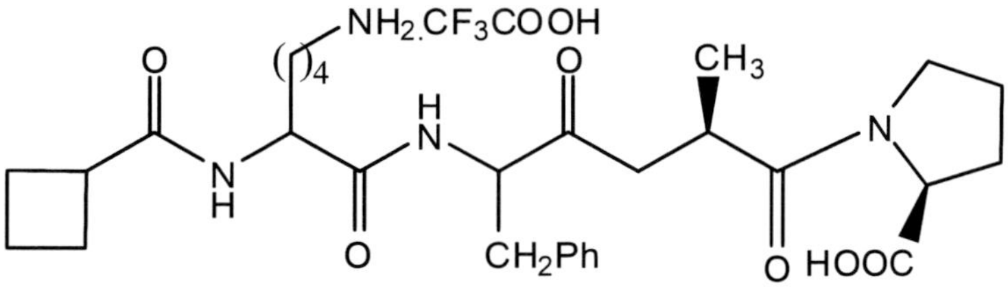	3.0nM
116.	H-Lys-Thr-Ala-Pro-OH	37μM
117.	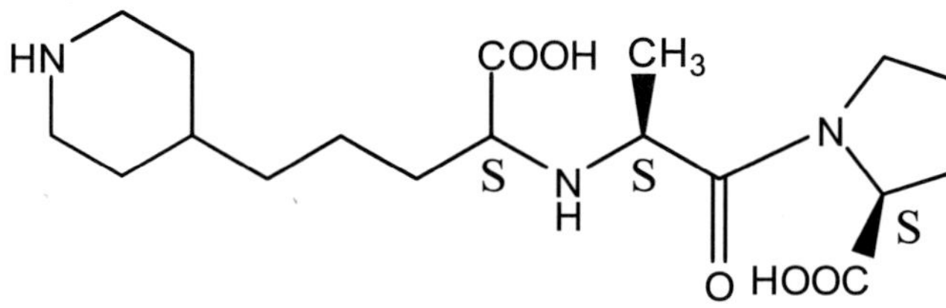	0.5nM
118.	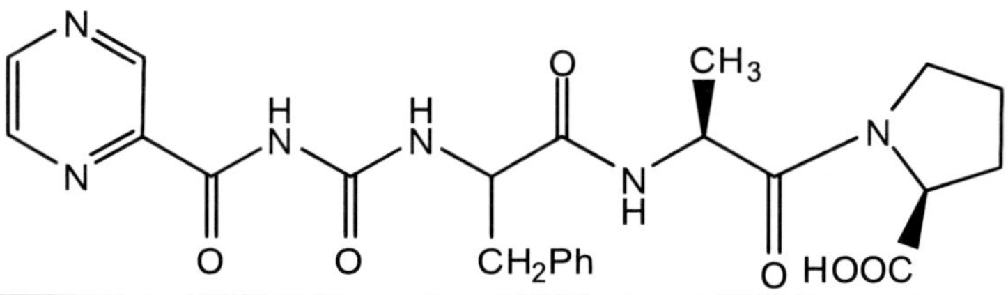	0.85 μM

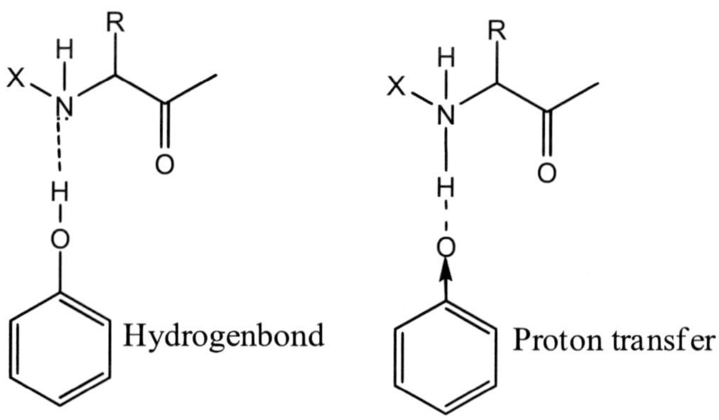

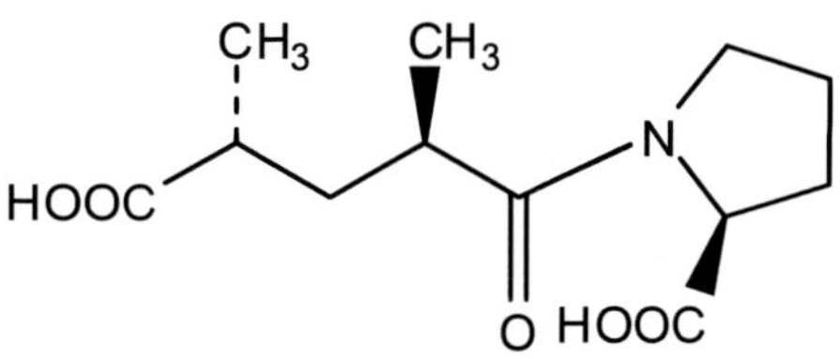

IC_{50}= 4.8 M
119

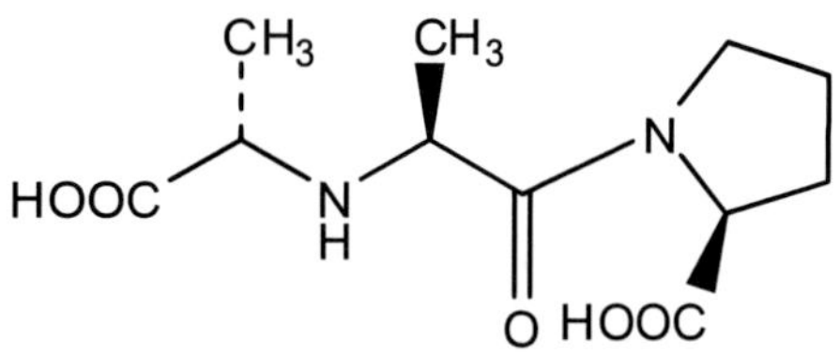

IC_{50}= 0.09 M
120

Figure 6. Role of proton donor site.

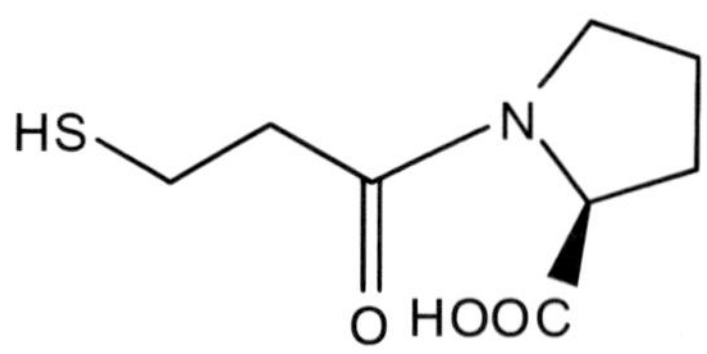

IC_{50}= 0.20 μM

21

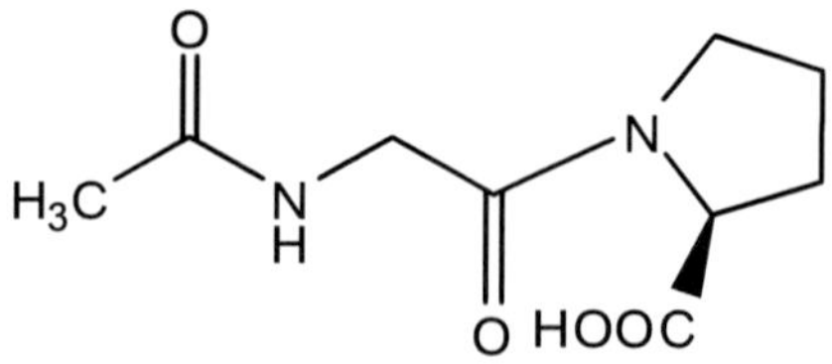

140,000 times less active

121

Figure 7. The role of Zn^{++} ion.

9.5. Proton Donor Site

This site in the hypothetical model for ACE also plays a vital role in the inhibition of ACE inhibitory activity. The proton donor site can be implicated in the three types of inhibitions carboxyalkyl dipeptides, simple dipeptides and phosphoryl dipeptides. The first two types possess fully basic amino group and the nitrogen in the third type is also somewhat basic. The lone electron pair on these functional groups could accept a hydrogen bond (proton) from the enzyme proton donor site as in figure 6.

9.6. The Critical Nature of Zn^{++} Metal Ion

The critical nature of Zn^{++} has also been well determined. The importance of Zn^{++} is known from the fact that when Mercaptan group is replaced by other groups as shown in fig 7, the activity is effectively reduced. Compound 21 containing mercaptan group 140,000 times more active time 121 which has only carbonyl group and has a much weaker binding with Zn^{++}.

Alternatively a proton could be transferred completely from the enzyme to the inhibitor nitrogen resulting in an ionic bond. Such an interaction may or not be involved in the binding of peptide inhibitors or substrate since compounds do not possess the required basic nitrogen atom.

Conclusion Based on Structure Activity Relationship Studies

Based on the SAR studies for various classes of molecules and with a complete study available for each and every binding site of ACE it may concluded that an inhibitor which interacts with the maximum number of sites of ACE in a positive way would be a better inhibitor. As it is generally the case that an inhibitor which occupies and interacts with maximum possible binding sites of enzyme in an effective manner would be a much potent inhibitor. Based on hypothetical model for ACE binding sites and structure activity Relationship (SAR) studies available for various classes of molecules, it could be ascertained that essential requirements in an inhibitor enlisted below must be incorporated while designing an ACE inhibitor with a view to develop a better therapeutic agent.

1. It requires a -COOH group in proline natural state (L-configuration) to interact with the Arg-residue of the enzyme.
2. Hydrophobic pockets to provide stability to enzyme- inhibitor complex by binding to site S_2^1 and S_1^1.
3. An electron rich chelating group as –SH in captopril which could form chelate with Zn^{++} ion of ACE for better binding and in turn potent ACE inhibitory activity.
4. A methyl or equivalent group in the N-acyl side chain to interact with the site S_1^1.
5. An amide carbonyl group in the N-acyl side chain which forms hydrogen bond with the proton donor site.
6. A hydrophobic pocket in the proline residue at C-4 or C-5 which could bind with the site S_2^1 of the enzyme and in turn leads to enhancement of ACE inhibitory potential.
7. The sites S_1, S_2 and so on are of lesser significance as it is evidenced from the fact that larger peptides designed to interact with the site S_1, S_2 or more sites have not shown any remarkable enhancement of ACE inhibitory activity.

ACKNOWLEDGMENT

The authors are thankful to University Grants Commission, New Delhi, for financial assistance in form of a major research project and to Central Drug Research Institute, Lucknow for extending Library facilities to us.

REFERENCES

[1] a. Saxena, P. R., Manintveld, A. J. Trends in pharmacological Sciences, 12, 239, (1991)
b. Laragh, J. H., in Captopril and Hypertension, Case, D.B., Ed. Plenum, New York, 1980 p. 173

[2] Heel, R. C., Brogden, R. N., Speight, T. M. and Avery, G. S., *Drugs*, 20, 409, (1980).

[3] Peach, M. J., *Physiol. Rev.*, 57, 313, (1977).

[4] Fasciolo, J. C., in Hypertension, Genest, J., Koiw, E. and Kuchel, O., Eds., McGraw-Hill, New YorK, 1977, p. 134.

[5] Soffer, R. L., *Ann. Rev. in Biochem.*, 45, 73, (1976).

[6] Yang, H. Y. T., Erdos, E. G. and Levin, Y., *Biochem. Biophys. Acta.*, 214, 374, (1970).

[7] Ondetti, M. A. and Cushman, D. W., *Ann. Rev. in Biochem.*, 51, 283, (1982).

[8] Soffer, R. L., in Biochemical regulation of blood pressure, Soffer, R. L., Ed. Wiley, New YorK., 1981, p. 123.

[9] Wiecek, A. and Grzeszczak, W., *Poly. Arch. Med. Wewn.*, 76 (5-6/11-12) 291-7, (1986), (Pol.)

[10] Petrillo, E. W. Jr. and Ondetti, M. A., *Med. Res. Rev.*, 2, 1, (1982).

[11] Ksander, G. M., Abstracts. 19th national Med. Chem. Symposium, Tucson, Az, June 17-21, (1984), p. 20.

[12] Cushman, D. W. and Ondetti, M. A.; in progress in Med. Chem., Ellis, G. P. and West, G. B., Eds., Elsevier/ North Holland, Amsterdam, 17, 41, (1979).
[13] Ondetti, M. A. and Cushman, D. W., in Biochemical regulation of blood pressure Soffer, R. L., Ed. Wiley, New YorK. (1981), p. 165.
[14] Ondetti, M. A. and Cushman, D. W., *CRC Crit Rev. Biochem*, 16, 381, (1984).
[15] Wyvratt, M. A. and Patchett, A. A., *Med. Res. Rev.*, 5, 483, (1985).
[16] Ferreira, S. H., Bartelt, D. C. and Greene, L. J. *Biochemistry* 9, 2583, (1970).
[17] Greene, L. J., Stewart, J. M. and Ferreira, S. H., *Pharmacol. Res. Commun.* 1, 159, (1969).
[18] Stewart, J. M., Ferreira, S. H. and Greene, L. J., *Biochem. Pharmacol.* 20, 1557, (1971).
[19] Ondetti, M. A., Williams, N. J., Sabo, E.F., Pluscec, J., Weaver, E. R. and Kocy, O., *Biochemistry*, 10, 4033, (1971).
[20] Ondetti, M. A., Williams, N. J., Sabo, E. F., Pluscec, J., Weaver, E. R. and Kocy, O., in progress in peptide research, Proceeding 2nd American peptide symposium, Lande, S., Ed. Gordon and Breach, New York, (1972), p. 249.
[21] Kato, H. and Suzuki, T., *Experientia*, 25, 694, (1969).
[22] Kato, H. and Suzuki, T., *Biochemistry*, 10, 972, (1971)
[23] Cushman, D. W., Pluscec, J., Williams, N. J., Weaver, E.R., Sabo, E. F., Kocy, O., Cheung, H. S. and Ondetti, M. A., *Experientia,* 29, 1032, (1973).
[24] Cheung, H. S., Wang, F. L., Ondetti, M. A., Sabo, E. F and Cushman, D. W., *J. Biol. Chem.*, 255, 401, (1980).
[25] Hartsuck, J. A. and Lipscomb, W. N., in the Enzymes, Boyer, P. D., Ed. Academic, New York, 3, 1, (1971).
[26] Schmid, M. E. and Herriot, J. R., *J. Mol. Biol.*, 103, 175, (1976).
[27] Matthews, B. W., Weaver, L. H. and Kestov, W. R., *J. Biol. Chem.*, 249, 8030, (1974)
[28] Lipscomb, W. N., *Tetrahedron*, 30, 1725, (1974).
[29] Kester, W. R. and Matthews, B. W., *Biochemistry*, 16, 2506, (1977)
[30] Weaver, L. H ., Kester, W. R. and Matthews, B. W., *J. Mol. Biol.*, 114, 119, (1977).
[31] Rohrbach, M. S., Williams, E. B., and Rolstad, R. A. *J. Biol. Biol.*, 256, 225, (1981).
[32] Patchett, A. A., Harris, E. W., Tristram, E. W., Wyvratt, M. J., Wu, M. T., Taub, D., Peterson, E. R., Ikeler, T. J., Broeke, J. L., Payne, L. G., Ondeyka, D. L., Thorsett, E. D., Greenlee, W. J., Lohr, N. S., Hoffsommer, R. D., Joshua, H., Ruyle, W. V., Rothrock, J. W., Aster, S. D. Maycock, A. L., Robinson, F. M., Hirschmann, R., Sweet, C. S., Ulm, E. M., Groos, D. M., Vassil, T. C. and Stone, C.A., *Nature,* 288, 280, (1980)
[33] Bunning, P., Holmquist, B. C., and Riordan, J. F., *Biochem. Biophys. Res. Comm.*, 83, 1442, (1978).
[34] Ondetti, M. A., Rubin,B. and Cushman, D. W. *Science*, 196, 441, (1977).
[35] Cushman, D. W., Cheung, H. S., Sabo, E. F., and Ondetti, M. A., *Biochemistry*, 16, 5484, (1977)
[36] Byers, L. D., and Wolfenden, R. *J. Biochem.,* 247, 606, (1972).
[37] Byers, L. D. and Wolfenden, R., *Biochemistry*, 12, 2070, (1973)
[38] Wyvratt, M. J Tristram, E. W Ikeler, T. J Lohr, N. S Joshua, H., Springer, J. P.,Arison, B. H. and Patchett A. A., *J. Org. Chem.*, 49, 2816, (1984).

[39] Biollaz, J., Burnier, M., Turini, G. A., Brunner, D. B., Porchet, M., Gomez, H. J., Jones, K. H., Ferber, F., Abrams, W. B., Gavras, H. and Brunner, H. R., *Clin. Pharmacol., Ther.*, 29, 665, (1981).

[40] Patchett A. A, in Hypertension and the Angiotensin System, Therapeutic approaches, Raven New York, (1984), p.155

[41] Patchett A. A., *Br. J. Clin. Pharmacology*, 18, (suppl.2), 201S, (1984)

[42] Sweet, C. S., Patchett A. A Ulm, E. H., Gross, D. M., in proceedings of 2nd Sci/RSC, Med. Chem. Symposium, Cambridge, sept.12-14, (1983), Emmett. J. C., Ed. Burligton House, London. p.36.

[43] Rotmensch, H. N., Vlasses, P. H., Swanson, B. N., Irvin, J. D., Harris K. E., Merill, D. G. and Ferguson, R. K., *Am. J. Cardio.*, 53, 116, (1984).

[44] Rotmensch, H. N., Vincent, M., Vlasses, P. H., Swanson, B. N., Irvin, J. D., Hichens, M., Harris K. E. and Ferguson, R. K., *Fed. Proc.*, 43, 1333, (1984)

[45] Almquist, R. G., Chao, W. R. and White, C. J., J. Med. Chem. 28 (8), 1067, (1985)

[46] Sawayama, T., Kinugasa, H., Nishimura, H., Takeyama, K., and Hosoki, K., *U. S. Patent*, 4, 248, 883, (1981)

[47] Hosoki, K Takeyama, K., Okamoto, H., Minato, H. and Kadokawa, T., *Jpn. J. Pharmacol.,* 31, (suppl.) abs. 0-107, (1981)

[48] Kadokawa, T., Hosoki, K and Takeyama, K., (9th International Congress of Pharmacology, London, July, (1984), abst., 1899P and 1900P

[49] Kono, T., *Jpn. Arch. Intern. Med.*, 30, 415, (1983)

[50] Natarajan, S., Condon, M. E., Cohen, M. S., Reid, J., Cushman, D. W., Rubin, B and Ondetti, M. A., in proceeding of 6th American peptide Symposium, Gross. E., Meienhofer, J., Ed. North-Holland, Amsterdam, (1979), p.463

[51] Funae, Y., Komori, T., Sasaki, D. and Yamamoto, K., *Biochem. Pharmacol.*, 29, 1543, (1980).

[52] Mita, I., Iwao, J., Oya, M., Chiba, T. and Iso, T., *Chem. Pharm. Bull.*, 26, 1333, (1978)

[53] Unger, T., Rockhold, R. W., Schaz, K., Vescei, P., Bonner,G., Rascher, W. and Ganten, D., *Clin. Sci.,* 57, 1575, (1979)

[54] (a). Iso, T., Yamauchi, H., Suda, H., Nakata, K., Nishimura, K. and Iwao, J., *Jpn. J. Pharmacol.,* 31, 875, (1981)
(b). Yamauchi, H., Nishimura, K., Nakata, K., Suda, H., Iso, T., Shimizu, M. and Hiramatsu, Y., *Arzneimforsch.*, 37-1 (2), 157, (1987)

[55] Oya, M., Kato, E., Iwao., J., and Yasuoka, N., *Chem. Pharm. Bull.*, 30, 484, (1982)

[56] Oya, M., Baba, T., Kato, E., Kawashima, Y. and Watanabe, T., *Chem. Pharm. Bull.*, 30, 440, (1982)

[57] Iso, T., Yamauchi, H., Iwatani, T., Kito, H. and Horiuchi, M., *Clin. Exp. Hy.*, 4, 321, (1982)

[58] Unger, T., Yukimura, T., Marin-Grez, M., Lang, R. E., Rascher, W. and Ganten, D., *Eur. J. Pharmacol.*, 78, 411, (1982).

[59] Takata, Y., Di-Nicolantonio, R., Mendelsohn, F.A.O., Hutchinson, J.S. and Doyle, A. E., *Clin. Exp. Pharmacol. Physicol.*, 10, 131, (1983).

[60] Abe, Y., Okahara, T., Miura, K., Yukimura, T., Takada, T., Iwatani, T., Iso, T. and Yamamoto, K., *Naunyn Schmie Debergs Arch Pharmacol.*, 325, 356, (1984).

[61] Vasella, A., Voeffray, R., Plass, J. and Huguenin, R., *Helvetica Chemica Acta*, 66, 1241, (1983).

[62] Kim, D. H., Guinosso, C. J., Buzby, G. C., Herbst, D. R., McCaully, R. J., Wicks, T. C. and Wendt, R. L., *J. Med. Chem.*, 26, 394, (1983).

[63] Stanton, J. L., Gruenfeld, N., Babiarz, J.E., Ackerman, M.H., Friedman, R. C., Yuan, A.M. and Macchia, W., *J. Med. Chem.*, 26, 1267, (1983).

[64] Dinish, J. L., Mitz, N. K., Lappe, R. W., Kim, D.H. and Wendt, R. L., *Pharmacologist,* 25, 102, (1983).

[65] Hayashi, K., Ozaki, Y., Nunami, K., Uchida, T., Kato, J., Kinashi, K. and Yoneda, N., *Chem. Pharm. Bull.* 31, 570, (1983).

[66] Hayashi, K., Nunami, K., Kato, J., Yoneda, N., Kubo, M., Ochiai, T. and Ishida, R., *J. Med. Chem.*, 32, 289, (1989).

[67] Kori, M., Itoh, K. and Sugihara, K., *Chem. Pharm. Bull.* 35, 2319, (1987).

[68] Itoh, K. Kori, M., Inada, Y., Nishikawa, K., Kawamatsu, Y. and Sugihara, K., *Chem. Pharm. Bull.* 34, (3), 1128, (1986).

[69] Itoh, K. Kori, M., Inada, Y., Nishikawa, K., Kawamatsu, Y. and Sugihara, K., ibid, 34, (5), 2078, (1986).

[70] Itoh, K., Kori, M., Inada, Y., Nishikawa, K., Kawamatsu, Y. and Sugihara, K., ibid, 34 (9), 3747, (1986).

[71] Blankley, C. J., Kaltenbronn, J. S., Dejohn, D. E., Werner, A, Benett, L. R., Babowski, G., Krolls, U., Johnson, D. R., Pearlman, W. M., Hoefle, M.L., Essenburg, A. D., Cohen, D. M. and Kaplan, H. R., *J. Med. Chem.*, 30 (6), 992, (1987).

[72] Klutchko, S., Blankley, C. J., Fleming, R. W., Hinkley, J. M., Werner, A.E., Nordin, I., Holmes, A., Hoefle, M. L., Cohen, D. M., Essenberg, A. D. and Kaplan, H. R., *J. Med. Chem.*, 29 (10), 1953, (1986).

[73] Attwood, M. R., Hassall, C.H., Krohn,A., Lawton, G. and Redshaw,S., *J. Chem. Soc.*, P.1,6, 1011, (1986).

[74] Mencel, J. J., Regan, J. R., Barton, J., Menard, P. R., Bruno, J. G., Calvo, R. R., Kornberg, B. E., Schwab, A., Neiss, E. S. and Suh, J. T., *J. Med. Chem.*, 33, 1606, (1990).

[75] a. Schwab, A., Weinryb, I., Macerata, R., Rogers, W., Suh, J. and Khandwala, A., *Biochem. Pharmacol.*, 32, 1957, (1983).
b. Menard, P. R., Suh, J. T., Jones, H., Loev, B., Neiss, E. S., Wilde, J., Schwab, A. and Mann, W. S., *J. Med. Chem.*, 28, 328, (1985).

[76] Suh, J. T., Skiles, J. W., Williams, B. E., Youssefyeh, R. D.,Jones, H., Loev, B., Neiss, E.S., Schwab, A., Mann, W. S., Khandwala, A., Wolf, P. S. and Weinryb, I., *J. Med. Chem.*, 28, 57, (1985).

[77] Burnier, M., Turini, G. A., Brunner, H. R., Porchet, M., Kruithof, D., Vukovich, R. A. and Gavras, H., *Br. J. Clin. Pharmacol.*, 12, 893, (1981).

[78] Burnier, M., Bioallaz, J., Brunner, H. R., Turini, G. A., and Gavras, H., *Am. J. Cardiol.*, 49, 1550, (1982).

[79] Wolf, P. S., Mann, W. S., Suh, J. T, Loev, B. and Smith, R.D., *Fed. Proc.*, 43, 1332, 1984.

[80] Miyake, A., Itoh, K. and Oka, Y., *Chem. Pharm. Bull.*, 34(7), 2852, (1986).

[81] Imaki, K., Sakuyama, S., Okada, T., Toda, M., Hayashi, M., Miyamoto, T., Kawasaki, A. and Okegawa, T., *Chem. Pharm. Bull.*, 29, 2210, (1981).

[82] Ondetti, M. A., Gordon, E.M. and Ryono, D.E., U. S. Patent, 4, 234, 489; *Chem. Abst.*, 94, P 103837, (1981).

[83] Johnson, A. L., Price, W. A., Wong, P.C., Vavala, R. F. and Stump, J. M., *J. Med. Chem.*, 28, 1596, (1985).

[84] Panday, S. K., Dikshit, M. and Dikshit, D. K. Med. Chem. Res. DOI 10.1007/s00044-008-9150-z, published on line: 07 Jan., (2009), in press.

[85] Swiss, G. F., Smith, E. M., Neustadt, B. R., Gold, E. H., Chiu, P. J. S., Brown, A. D. and Sommer, J., Abstracts 184th National meeting American Chemical Society, Kanas City, MO September, (1982), MEDI. 77.

[86] Powell, J. R., Rubin, B., Cushman, D. W., Krapcho, J. and Antonaccio, M.J., International society of Hypertension, Ist European Meeting on Hypertension, Milan, Italy, May (1983), Abstract 355.

[87] Dean, A. V., Kriplani, K. J. and Migdalof, B. H., *Fed. Proc.*, 43, 349, (1984).

[88] Deforrest, J. M., Waldron, T. L., Krapcho, J., Turk, C., Rubin, B., Powell, J.R., Caushman, D. W. and Petrillo, E. W., *J. Cardiovas. Pharmacol.*, 13(6), 887, (1989).

[89] a. Krapcho, J., Turk, C., Cushman, D. W., Powell, J. R., Deforrest, J. M., Spitzmiller, E.R., Karanewsky, D. S., Duggan, M., Rovnyak, G., Schwartz, J., Natrajan, S., Godfrey, J. D., Ryono, D. E., Neubeck, R., Atwal, K. S. and Petrillo, E. W. Jr., *J. Med. Chem.*, 31(6), 1148, (1988).
b. Smith, E. M., Swiss, G. F., Neustadt, B. R., McNamara, P., Gold, E. H., Sybertz, E. J. and Baum, T., *J. Med. Chem.*, 32, 1600, (1989) and references cited therein.

[90] Smith, E. M., Swiss, G. F., Neustadt, B. R., Gold, E. H., Sommer, J. A., Brown, A. D., Chiu, P. J. S., Moran, R., Sybertz, E. J. and Baum, T. J. *Med. Chem.*, 31, 875, (1988).

[91] Alexandrou, E. N. and Liakopoulou-Kyriakides, M., Biochem. Int., 21, 271, (1990).

[92] Olimpieri, F., Tambaro, S., Fustero, S., Lazzari, P., Sanchez-Roselló, M., Pani,, L., Volonterio, A., Zanda, M. *Bio. and Med. Chem. Letters*, DOI:10.1016/j.bmcl.2009.06.064, (2009),

[93] Wei, W.; ShengRong, S.; FengQin, F.; GuoQing, H.; ZhanLi, W. *Sci. China, Ser. B Chem.* 51, 786, (2008)

[94] Ismail, M. A. H., Aboul-Enein, M. N. A., Abouzid, K. A M., Ella, D. A. A. E, Ismail, N. S.M., *Bio. and Med. Chem.*, 17, 3739, (2009).

[95] Lohith, K., Vijayakumar, G. R., Somashekar, B. R., Sivakumar, R., Divakar, S. *Eur. J. Med. Chem.*, 41, 1059, (2006).

[96] Condon, M.E., Petrillo, E.W., Ryono, D. E., Reid, J. A., Neubeck, R., Puar, M., Heikas, J. E., Sabo, E.F, Losee, K. A., Cushman, D. W. and Ondetti, M. A., *J. Med. Chem.* 25, 250, (1982).

[97] McEvoy, F. J., Lai, F. M. and Albright, J. D., *J. Med. Chem.*, 26, 381, (1983).

[98] Lai, F. M., Cervoni, P., Tanikella, T., Shepherd, C., Quirk, G., Herzlinger, H. and Stubbs, C. S., *Druv Dev. Res.*, 3, 261, (1983).

[99] Walker, E. R. H., *Brit. U. K. Patent Appl.* GB. 2, 092, 574, (1982).

[100] Hariss, R. B., Ohlsson, J.T. and Wilson, I. B., *Arch. Biochem. Biophys.*, 206, 105, 1981.

[101] Prous, J. R., *Ed. Ann. Drug Data Rep.*, 4, 220, (1982)

[102] Ryan, J. W. and Chung, A., *Adv. Exp. Med. Biol.* 156(B), 1133, (1983).

[103] Kuromaru, S., Tanaka, S., Fujimura, H., Matsunaga, K., Hinohara, Y., Kawamura, N., Nabata, H. and Aono, R., 104th Annual Meeting, Pharmaceutical Society of Japan, Sendai, Japan, March, (1984), Abstr, 29 A, 9-5.

[104] Thorsett, E. D., Hariss, E. E., Aster, S., Peterson, E. R., Patchett, A. A., Ulm, E. H. and Vassil, T.C., Abstracts, 184th National Meeting Americal Chemical Society, Kanas City; MO Sep (1982), MEDI. 78.

[105] Thorsett, E. D., Hariss, E. E., Aster, S., Peterson, E. R., Taub, D., Patchett, A. A., Ulm, E. H. and Vassil, T.C., *Biochem. Biophys. Res. Commun.*, 111, 166, (1983).

[106] Thorsett, E. D., Hariss, E. E., Aster, S. D, Peterson, E. R., Tristram, E. W., Suyder, J. P., Springer, J. P. and Patchett, A. A., in peptides, Structure and function proceeding 8th American peptide symposium, Hruby, V. J. and Rich, D. H., Eds. Pierce Chemical Co. Rocford, IL, (1983), p.555.

[107] Grathwohl, C. and Wuthrich, K. *Biopolymers*, 15, 2043, (1976).

[108] Blomberg, F., Prneterjans, H., Lintner, K., Toma, F. and Fermandjcan, S., *Org. Magn. Reson.*, 11, 598, (1978).

[109] Klutchko, S., Hoefle, M. L., Smith, R. D., Essenberg, A. D., Parker, R. B., Nemeth, V. L., Ryan, M. J., Dugan, D. H. and Kaplan, H. R., *J. Med. Chem.*, 24, 104, (1981).

[110] Condon, M. E., Petrillo, E.W., Ryano, D. E., Reid, J. A., Neubeck, R., Puar, M., Heikes, J. E., Sabo, E. F., Losee, K. A., Cushman, D. W. and Ondetti. M. A. *J. Med. Chem.,* 25, 250, (1982).

[111] Wetthey, J. W. H., Gavin, T. and Desai, M., *J. Med. Chem.,* 27, 816, (1984).

[112] a. Hassall, C. H., Krohn, A., Moody, C. J. and Thomas, W. A., *FEBS Lett.*, 147, 175, (1982).

b. Ciabatti, R., Padova, G., Bellasio, E., Tarzia, G., Depoli, A., Battaglia, F., Cellentani, M., Barone, D. and Baldoli, E., *J. Med. Chem.*, 29, 411, (1986).

[113] Umezawa, S., Tatsuta, K., Izawa, O. and Tsuchiya, T. *Tetrahedron Lett.*, 13, 97, (1972).

[114] Suda, H., Aoyagi, T., Takeuchi, T. and Umezawa, H., *J. Antibiot.*, 26, 621, (1973).

[115] Petrillo, E. W., Cushman, D, W., Duggan, M. E., Heikes, J. E., Karanewsky, D. S., Ondetti, M. A., O'Reilly, B., Rovnyak, G. C., Schwartz, J., Spitzmiller, E. R. and Wang, N. in peptides, structure and function, Proceeding 8th American Peptide Symposium, Hruby, V. J. and Rich, D. H., Eds. Pierce Chemical Co., Rockford, IL, (1983), p. 541.

[116] a. Karanewsky, D. S., Badia, M. C., Cuchman, D. W., Deforrest, J. M., Dejeneka, T., Loots, M. J., Perri, M. G., Petrillo, E. W. Jr. and Powell, J. R. *J. Med, Chem.*, 31, 204, (1988).

b. Karanewsky, D. S., Badia, M. S., Cuchman, D. W., Deforrest, J. M., Dejeneka, T., Lee, V. G., Loots, M. J. and Petrillo, E. W., *J. Med. Chem.*, 33, 1459, (1990).

[117] Petrillo, E. W., 18th Middle Atlantic Regional Meeting, American Chemical society, New york, N. J., May (1984), Abs., 41.

[118] Powell, J. R., Deforrest, J. M., Cushman, D. W., Rubin, B. and Petrillo, E. W., *Fed. Proc.*, 43, 733, (1984).

[119] Powell, J. R., Deforrest, J. M., Cushman, D. W., Rubin, B. and Petrillo, E. W., 9th International Congress of Pharmacology, London, July (1984), abs. 1257.

[120] Duchin, B. I., Lambe, R., Nugent, H. and Darragh, A., 10th Scientific Meetinf International Society of Hypertension, Interlaken, Switzerland, June (1984) Abs. 312.

[121] Greenlee, W. J., Allibone, P. L., Parlow, D. S., Petchett, A. A., Ulm, E. H. and Vassil, T. C., abstracts, 187th Natinal meeting, *American Chemical Society*, St. Louis, MO April (1984), MEDI 17.

[122] Almquist, R. G., Jennings-White, C., Chao, W. R., Steeger, T., Wheeler, K., Rogers, J. and Mitoma, C. J. *Med. Chem.*, 28, 1062, (1985).

[123] Komura, M., Nio, N. and Ariyoshi, Y., Jpn Kokai. Tokkyo Koho, Jp. 01, 226, 898,[89,226,898] 112, P 119954x, (1990).

[124] Waga, T., Matsui, S., Saito, S., Watanabe, M., Kajiwara, y., Shirota, M., Iijima, M. and Kitabatake, K., Arzneim-Forsch, 40, 407, (1990).

[125] Satio, S., Matsui, S., Watanabe, M., Waga, T., Kajiwara, Y., Shirota, M., Iijima, M. and Kitabatake, K., Arzneim-forsch, 40, 257, (1990).

[126] Kayahara, H., Kawabata, H., Murokata, Y., Imai, S., Kurosawa, S., Tomida, I. and Tadasa, K., *Agri. Biol. Chem.*, 52, 289, (1988).

[127] Kayahara, H., Ohashi, A., Tadasa, K. and Kato, S., *Agri. Biol. Chem.*, 54, 1325, (1990).

In: Biochemistry Research Updates
Editor: Simon J. Baginski

ISBN 978-1-61209-700-8

Chapter 6

PROTEIN SUPERFAMILIES BASED PHYLOGENOMIC ANALYSIS OF ARCHAEAL DOMAIN

P. Chellapandi,[1*] and S. Sivaramakrishnan[2]

[1]Department of Bioinformatics, School of Life Sciences, Bharathidasan University, Tiruchirappalli, Tamil Nadu, India

[2]Department of Biotechnology, School of Life Sciences, Bharathidasan University, Tiruchirappalli, Tamil Nadu, India

ABSTRACT

Since a metabolic pathway with similar function has similar evolutionary relationship, the operation of metabolic pathways is essential to the survival of organisms, and metabolism related genes usually constitute only 10–20% of the total number of genes, metabolic pathway-based analysis can minimize the effect of the gene number, and thus allow more precise investigation of evolutionary relationships. Herein, we have shown that protein superfamilies of archaea were diverged and metabolically converged together within the genomes of this domain as a consequence of long evolutionary and diverged metabolic events. However, a close phylogenetic proximity between archaea and bacteria was observed for sharing metabolic functions by protein family divergence. A close phylogenetic boundary between methanogens and methylotrophs was revealed the reverse methanogenesis hypothesis. Energetic metabolism of methanogens are shared their ancestral behaviors with primitive proteobacteria. A mosaic nature of some functional domains of protein superfamilies was observed between mesophilic methanogens and thermophilic archaea/halophilic archaea, which may be resulted due to selection pressures. Overall results revealed that the phylogenetic analysis of protein superfamilies obtained from archaeal domain provided a standpoint for the evolutionary history of some key metabolic modules among methanogens. Therefore, molecular evolutionary hypothesis from this finding would provide a new horizon for growth of research in advancement of metabolomics and metabolic engineering in methanogens.

[*] Tel: +91-431-2407071 Fax: +91-431-2407045 Email: pchellapandi@gmail.com.

Keywords: Archaea, Protein superfamily, Phylogeny, Metabolic modules, Methanogens, Molecular evolution.

Abbreviations

BA: Betaine aldehyde
CO:Carbon monoxide
CoP: Cobalt-precorrin-3C
DB: Dethiobiotin
DHL: Dihydrolipoamide
G-6-P: Glucosamine-6-phosphate
HN: Hydroxyneurosporene
CoHD: CoB-CoM heterodisulfide
AMO: Ammonia monooxygenase
MH: N_5, N_{10}-methylenetetrahydromethanopterin
PU: Pseudouridylate
RUBISCO: Ribulose-5-phosphate carboxylase

Introduction

The extremophilic nature of many archaea has stimulated intense efforts to understand the physiological adaptations for living in extreme environments and to probe the potential biotechnological applications of their stable cellular components. Specific archaeal metabolites have been purified and characterized, and some of them have potential industrial uses (Schiraldi et al., 2002; Alqueres et al., 2007). Thus, sequencing archaeal genomes has led to insights into the mosaic nature of relationships between prokaryotic proteins, particularly among metabolic, transport, and DNA processing proteins (Deppenmeier et al., 2002; Hong et al., 2004a; Kalyuzhnaya et al., 2005). The different metabolic aspects of archaeal domain have been inferred by phylogenetic analyses based on structural and sequence similarities. These insights in turn have led to a deeper molecular understanding of differences and similarities between organisms in the major archaeal phenotypes (Lake et al., 1985; Cambillau and Claverie, 2000; Hong et al., 2004b; Clementino et al., 2007).

About 85% extremophiles are included in archaea in the prokaryotic world (Ciaramella et al., 1995). Among species of archaea, there are a variety of metabolic regimes differing greatly from the better-known metabolic pathways of bacteria and eukaryotes (Brown and Doolittle, 1997; Apic et al., 2008; Ebenhoh et al., 2005). A key to modeling and understanding the evolutionary process is the identification and characterization of the constraints that the evolution perceives as protein diverges (Thorne et al., 1996; Naylor and Brown, 1997; Goldman et al., 1996; 1998). The limitations of widely used models of sequence evolution often prevent more refined and informative questions from being addressed. Protein sequences are more conserved than nucleotide sequences, which in turn to rather than protein structure. A protein superfamily is a protein class composed of one or more protein families that has low sequence identity exhibiting 5-10% matching in their sequences. The structural and functional aspects of these proteins that can share a common ancestor are placed together in superfamilies (Dayhoff et al., 1972; Dayhoff et al., 1978). Based upon extreme structural similarity or a similar peripheral binding site the evolutionary relationships of these superfamilies have been determined (Todd et al., 2001; Yi et al., 2006). Many gene superfamilies have hundreds or thousands of members and hence pose a significant challenge when performing a large-scale phylogenetic analysis (Irving et al., 2004). Some ancestral superfamilies have been more versatile than others during evolution allowing more genetic and functional variation. The differences in genetic versatility between protein families are more attributable to their functional nature rather than the time that they have been evolving (Ranea et al., 2006).

Most recently, phylogenomic conclusion of archaea has been derived on the basis of protein superfamilies (Karthigeyen et al., 2007; Chellapandi et al., 2007a) and of their secondary structures (Chellapandi et al., 2009a). Thus, the main objective of this work was to reconstruct supertree of archaeal domain based on the selected protein superfamilies of key metabolic pathways. The contribution of secondary structures of these protein families have also to be addresssed to know the molecular evolutionary process of methanogens.

METHODS

Sequence Retrieval and Similarity Search

The data set comprising 25 archaeal protein families of different metabolisms were retrieved from NCBI database (http://www.ncbi.nlm.nih.gov/) and the details are summarized in Table 1 and Supplementary. These query sequences were used to retrieve pairwise similarity sequences from complete and incomplete archaeal and bacterial genomes deposited in NCBI database using BLASTp search tool (Altschul et al., 1997) with default parameters. A sequence with low e-value and high identity score was selected and again pairwise aligned with individual query sequence using pairwise similarity program of BioEdit 7.0.4.1 software (Hall, 1999). Multiple sequence alignment was carried out for the sequences showing a reasonable similarity to each query sequence. ClustalX 1.81 software (Thomson et al., 1997) was used for multiple sequence alignment and editing. All the aligned sequences were inspected manually to select the significant sequence alignments for further comparative analyses.

Domain and Motif Similarity Search

The query and best hit of both archaeal and bacterial sequences (high similarity sequences) were subjected to domain and motif predictions. Domain Linker Prediction server (DLP 2.4) (Miyazaki et al., 2002) with default parameters was used to predict the major domain regions in both query and best hit sequences. Motifs region (structural and functional aspects of proteins) of each sequence was predicted and compared with one another by eMotif 3.6 server (Huang and Brutlag, 2001).

Table 1. List of archaeal protein families (data set) used in this study

E.C.	Enzyme	Accession	Organism	Length	Function
3.5.4.27	MH cyclohydrolase	NP_634677	*Methanosarcina mazei* Go1	321	Methane biosynthesis
1.5.99.11	Coenzyme F420	AAB28420	*Methanopyrus kandleri* AV-19	33	Methane biosynthesis
2.8.4.1	Coenzyme M	AB56198	*Methanococcus voltae*	64	Methane biosynthesis
1.2.99.2	CO dehydrogenase	Q8TXX3	*Methnopyrus kandleri*	638	Methane biosynthesis
1.14.99.39	AMO	AAZ38770	*Candidatus Nitrosopumilus maritims*	174	Methane biosynthesis
4.1.1.39	RUBISCO	ZP_00867437	*Methanospirillum hungatei* JF-1	399	Photorespiration
1.6.-.-	NADH oxidase	YP_183712	*Thermococcus kodakarensis* KOD1	442	Photorespiration
1.3.1.75	4- Vinyl reductase	CAB50023	*Pyrococcus abyssi* GE5	339	Photorespiration
4.2.1.-	HN synthase	AAM30567	*Methanosarcina mazei* Go1	349	Photorespiration
2.1.1.133	CoP methyltransferase	O29534	*Archaeoglobus fulgidus*	446	Photorespiration
1.9.3.-	Quinol oxidase I	P98004	*Sulfolobus acidocaldarius*	517	Photorespiration
1.6.5.5	Quinone reductase	NP_615299	*Methanosarcina acetivorans* C2A	213	Energy metabolism
1.8.1.4	DHL dehydrogenase	NP_635303	*Methanosarcina mazei* Go1	487	Energy metabolism
2.3.1.9	3-Oxothiolase	YP_501645	*Methanospirillum hungatei* JF-1	387	Fatty acid metabolism
1.8.98.1	CoHD reductase	P96801	*Methanopyrus kandleri*	656	Vitamin biosynthesis
6.3.3.3	DB synthetase	Q58695	*Methanocaldococcus jannaschii*	214	Vitamin biosynthesis
2.6.1.16	G-6-P synthase	Q8TZ14	*Methanopyrus kandleri*	614	Amino sugar metabolism
3.5.1.54	Allophanate hydrolase	YP_182746	*Thermococcus kodakarensis* KOD1	228	Urea cycle
1.2.1.8	BA dehydrogenase	BAB59541	*Thermoplasma volcanium* GSS1	498	Oslolyte biosynthesis
1.1.99.1	Choline dehydrogenase	AAV45185	*Haloarcula marismortui* ATCC 43049	560	Oslolyte biosynthesis
1.5.3.1	Sarcosine oxidase	NP_126003	*Pyrococcus abyssi* GE5	481	Oslolyte biosynthesis
5.99.1.3	Topoisomerase VIB	NP_614204	*Methanopyrus kandleri* AV19	475	DNA supercoiling
3.1.26.4	Ribonuclease H	YP_445237	*Salinibacter ruber* DSM 13855	307	DNA supercoiling
5.4.99.12	tRNA PU synthase	Q8TZ08	*Methanopyrus kandleri*	338	Translation process
6.1.1.5	Ile-tRNA synthetase	Q8TWY2	*Methanopyrus kandleri*	1080	Translation process

The motifs regions were retrieved from the multiple sequence alignments in the BLOCKS 13.0 and the PRINTS 31.0 databases (Craig et al., 1998). Expectation threshold was set as 1e = -2 with alignment databases. Similarity score, e-value, hit region and conservation plot were computed for every sequence.

Phylogenetic Tree Building and Analysis

Preliminary phylogenetic tree was searched by BioEdit 7.0.4.1 using Neighbor Joining (NJ) (Saitou and Nei, 1987; Gascuel and Steel 2006) and Fitch Margolis (FM) (Fitch and Margoliash, 1967) algorithms. FM tree was constructed on the basis of calculating sum square value of each populations and sequence similarities. The significant cluster of taxa from these trees was selected for supertree building according to biological reliability, orthologous feature, identity score, domain, and motif occurrences. Supertree construction is a phylogenetic approach in which many overlapping source trees, rather than the character data used to derive those trees, are combined to produce a single, larger supertree (Daubin et al., 2001; Bininda-Emonds, 2002; Bininda-Emonds et al., 2004). This approach was used to search phylogenetic tree in our study. The sequences of the selected taxa were manually inspected and clustered with complete deletion of gaps using ClustalX 1.81 and then NJ and FM trees were searched homogeneous patterns among all lineages by MEGA 3.1 software (Kumar et al., 2004) with 1000 bootstraps values (Bradlay et al., 1996) using JTT model (Jones et al., 1992), at uniform rates among sites. Tajma's neutrality value (Tajima, 1989) was computed to each cluster by MEGA 3.1. Tajima's test compares the number of segregating sites per site with the nucleotide diversity. A site is considered a segregating site if there are two or more nucleotides at that site in a comparison of m sequences. Nucleotide diversity is defined as the average number of nucleotide differences per site between two sequences. If all the alleles are selectively neutral, then the product 4Nv (Where N is the effective population size and v is the mutation rate per site) can be estimated in two ways, and the difference in the estimate obtained in these two ways provides indication of non-neutral evolution. Bayes's rule (Huelsenbeck and Ronquist, 2001) can be used to compute a posterior distribution for the unknown quantities. BALSA server was used to calculate the posterior probability of each clade in the supertrees constructed in this study. The BALSA algorithm (Webb et al., 2002) assumes that scoring matrices and gap penalties are viewed as pairs. Branch length and orders of each cluster in phylogenetic trees was computed by MEGA 3.1.

RESULTS

Sequence Similarity Analysis

BLASTp similarity search results have shown that apart from enzymes of methane biosynthesis many protein superfamilies randomly distributed in methanogens and non-methanogens except (Table 2). HN synthase and RUBISCO sequences showed the sequence similarities to bacterial photosynthesis proteins supporting the occurrence of earlier photorespiration in the members of primitive archaea.

Table 2. BLASTp similarity search results of archaea and bacteria obtained for sequences of archaeal proteins families

Protein family	Archaea	Archaea	Score	Accession	Bacteria	Score
CO dehydrogenase	NP_247713	*Methanocaldococcus jannaschii* DSM 2661	501	ZP_00886714	*Caldicellulosiruptor saccharolyticus* DSM	582
MH cyclohydrolase	NP_616637	*Methanosarcina acetivorans* C2A	258	YP_115256	*Methylococcus capsultus* str. Bath	594
Coenzyme F420	NP_613298	*Methanopyrus kandleri* AV 19	62	-	-	-
Coenzyme M	NP_988679	*Methanococcus maripaludis* S2	125	ZP_00657517	*Nocardioides sp.* JS614	130
RUBISCO	NP_142861	*Pyrococcus horikoshii* IOT3	320	YP_422059	*Magnetospirillum magneticum* AMB-1	822
AMA	NP_275420	*Methanothermobacter thermautotrophicus*	148	ZP_00575014	*Moorella thermoacetica* ATCC 39073	470
G- 6-P synthase	YP_448220	*Methanosphaera stadtmanae* DSM 3091	563	NP_346802	*Clostridium acetobutylicum* ATCC 824	574
Quinone reductase	ZP_00540702	*Methanosarcina barkeri str.* Fusaro	143	NP_745853	*Pseudomonas putida* KT2440	303
NADH oxidase	NP_619176	*Methanosarcina acetivorans* C2A	310	ZP_00767983	*Chloroflexus aurantiacus* J-10-fl	2674
Quinol oxidase I	NP_343985	*Sulfolobus solfataricus* P2	630	ZP_00525145	*Solibacter ustitatus* Ellin6076	782
CoP methyltransferase	NP_633019	*Methanosarcina mazei* GO1	223	YP_099801	*Bacteroides fragilis* YCH 46	216
4 - Vinyl reductase	NP_143024	*Pyrococcus horikoshii* OT3	51	YP_123440	*Legionella pneumophila str.* Paris	533
tRNA PU synthase	YP_183922	*Thermococcus kodakarensis* KOD1	142	ZP_00900460	*Pseudomonas putida* F1	417
Ile-tRNA synthetase	NP_276491	*Methanothermobacter thermautotrophicus*	678	ZP_00601058	*Rubrobacter xylanophilus* DSM 9941	1015
Topoisomerase VIB	NP_616520	*Methanosarcina acetivorans* C2A	194	ZP_00400821	*Anaeromyxobacter dehalogenans* 2CP-C	259
Ribonuclease H	YP_331317	*Natronomonas pharaonis* DSM 2160	98	NP_926558	*Gloeobacter violaceus* PCC 7421	51
DB synthetase	NP_988693	*Methanococcus maripaludis* S2	144	NP_952634	*Geobacter sulfurreducens* PCA	234
DHL dehydrogenase	NP_616580	*Methanosarcina acetivorans* C2A	299	NP_712296	*Leptospira interrogans serovar* Lai str. 56601	828
HN synthase	NP_342195	*Sulfolobus solfataricus* P2	385	ZP_00767653	*Chloroflexus aurantiacus* J-10-fl	230
BA dehydrogenase	EAM94461	*Ferroplasma acidaramanus* Fer1	358	YP_146052	*Geobacillus kaustophilus* HTA 426	585
Choline dehydrogenase	NP_279965	*Halobacterium sp.* NRC-1	260	ZP_00946012	*Ralstonia solanacearum* UW 551	498
Allophonate hydrolase	NP_142905	*Pyrococcus horikoshii* OT3	397	YP_074208	*Symbiobacterium thermophilum* IAM 14863	578
Sarcosine oxidase	NP_579524	*Pyrococcus furiosus* DSM 3638	145	NP_563466	*Clostridium perfringens* str. 13	721
CoHD reductase	?	*Methanopyrus kandleri* AV 19	701	YP_359777	*Carboxydothermus hydrogenoformans* Z-2901	1244
3-Oxothiolase	ZP_00562778	*Methanococcoides burtonii* DSM 6242	370	YP_360563	*Carboxydothermus hydrogenoformans* Z-2901	531

Table 3. List of phylosignatures of archaea and bacteria obtained for the sequences of archaeal protein families

Protein family	Phylosignature of archaea	e-Value	Phylosignature of bacteria	e-Value
MH cyclohydrolase	Class II cytochrome C signature I	1.05e+00	ATP synthase alpha and beta subunit, N-terminal	1.06e-01
Coenzyme F_{420}	CDC45-like protein	3.03e-1	Aminotransferases class-II	2.97e+00
RUBISCO	Ribulose bisphosphate carboxylase, large chain	1.16e-10	Ribulose bisphosphate carboxylase, large chain	2.18e-16
AMO	Pyridine nucleotide disulphide reductase class-I signature	1.87e-06	Bacterial-type phytoene dehydrogenase	1.04e-07
G-6-P synthase	Glutamine amidotransferase class-II	3.18e-09	Glutamine amidotransferase class-II	9.70e-09
Quinone reductase	NAD(P)H dehydrogenase (quinone)	8.59e-11	NAD(P)H dehydrogenase (quinone)	1.23e-16
NADH oxidase	Pyridine nucleotide-disulphide oxidoreductase class-II	5.87e-03	Pyridine nucleotide disulphide reductase class-I signature IV	8.40e-05
Quinol oxidase	Cytochrome c oxidase, subunit III	3.77e-06	Pyridine nucleotide disulphide reductase class-I signature IV	9.78e-05
CoP methyltransferase	Precorrin-8X methylmutase CbiC/CobH	1.04e-10	Precorrin-8X methylmutase CbiC/CobH	7.33e-10
4-Vinyl reductase	Herpsevirus integral membrane protein signature V	4.92e-01	Bacterial type II secretion system protein E	4.43e-02
tRNA PU synthase	Pseudouridylate synthase, TruB family, N-terminus	2.56e-15	Pseudouridylate synthase, TruB family, N-terminus	1.51e-17
IIe tRNA synthetase	Cysteinyl-tRNA synthetase signature IV	2.68e-15	Isoleucyl-tRNA synthetase signature II	2.71e-15
Topoisomerase VIB	Type II DNA topoisomerase	5.01e-18	DNA topoisomerase VI subunit A (TOP6A) signature II	1.78e-05
Ribonulease H	Ribonuclease HII	3.08e-10	C_2H_2-type zinc finger signature II	3.54e-01
DB synthetase	Biotin synthase	2.98e-12	NAD/NADP octopine/nopaline dehydrogenase	8.09e-01
DHL dehydrogenase	Pyridine nucleotide disulphide reductase class-I signature II	4.29e-10	Pyridine nucleotide disulphide reductase class-I signature II	3.09e-09
HN synthase	GTP cyclohydrolase II	1.85e-15	Phosphoglycerate kinase family signature III	1.56e+00
BA dehydrogenase	Aldehyde dehydrogenase family	1.32e-08	Aldehyde dehydrogenase family	5.44e-11
Allophanate hydrolase	DUF183	1.76e-45	DUF213	7.59e-19
Sarcosine oxidase	3Fe-4S ferredoxin signature I	9.67e-05	Pyridine nucleotide disulphide reductase class-I signature	1.96e-05
CoHD reductase	3Fe-4S ferredoxin signature I	1.06e-04	Bacterial-type phytoene dehydrogenase	1.10e-06
3-Oxothiolase	7Fe ferredoxin signature III	8.09e-4	Nil	-

Proteins of sulfur metabolism in mesophilic methanogens gave a maximum sequence similarity to different species of thermophilic methanogens (*M. kandleri)* and thermophilic proteobacteria, particularly *Carboxydothermus hydrogenoformans*. Osmoregulatory proteins in *Haloarcula marismortui* and coenzyme F_{420} in *M. kandleri* have a low sequence similarity to closely related species of archaea. Quinone oxidase, NADH oxidase, CO dehydrogenase and Ile-tRNA synthetase provided a maximum sequence similarity to different bacterial groups. The sequence similarity observed to be low for AMO, 4-vinyl reductase, sarcosine oxidase, ribonulease H, DB synthetase, and tRNA PU synthase as compared to bacterial homologs. DHL dehydrogenase of *M. acetivorans* C2A was similar to the same protein family of eubacteria.

Leptospira interrogans serovar Lai str. 56601. Therefore, many of these protein families show to have high sequence similarity to only closly related genera of archaeal domain.

Recombination Frequency Analysis

The posterior probability of G-6-P synthase, Ile-tRNA synthetase and HN synthase with bacteria and of CO dehydrogenase, 4-vinyl reductase, tRNA PU synthase, and BA dehydrogenase provided a strong statistical confidence (~1) for recombination frequency for evolutionary transfer of these families within the members of archaea (Table 4).

Table 4. Gene/protein diversity and segregation sites (recombination frequency) of from archaea and bacteria obtained for archaeal protein families

Protein family	Archaea						Bacteria					
	m	n	s	ps	pi	D	m	n	s	ps	pi	D
CO dehydrogenase	30	25	25	1.00	0.85	8.55	15	596	532	0.89	0.58	4.97
MH cyclohydrolase	17	35	35	1.00	0.55	3.56	5	327	315	0.96	0.70	4.04
RUBISCO	13	387	342	0.88	0.48	3.21	31	442	371	0.83	0.45	4.56
AMO	5	172	164	0.95	0.66	3.40	32	278	261	0.93	0.67	7.25
G-6-P synthase	9	585	386	0.65	0.40	3.53	34	575	466	0.81	0.46	5.09
tRNA PU synthase	24	211	211	1.00	0.77	7.66	22	254	214	0.84	0.38	2.77
Ile- tRNA synthetase	18	375	375	1.00	0.67	5.64	28	800	700	0.87	0.46	4.21
Topoisomerase VIB	16	334	326	0.97	0.63	5.00	11	283	283	1.00	0.84	7.17
Ribonuclease H	8	184	181	0.98	0.71	4.81	22	124	96	0.77	0.45	4.61
DB synthetase	23	169	169	1.00	0.75	7.16	23	183	168	0.91	0.60	5.76
DHL dehydrogenase	19	307	305	0.99	0.70	6.24	18	421	389	0.92	0.63	5.83
HN synthase	16	323	323	1.00	0.61	4.59	16	327	311	0.95	0.53	3.85
Choline dehydrogenase	15	358	337	0.94	0.63	5.33	13	532	363	0.68	0.24	0.56
Allophanate hydrolase	13	222	198	0.89	0.53	3.98	23	199	196	0.98	0.70	6.57
Sarcosine oxidase	10	95	95	1.00	0.83	6.67	40	149	149	1.00	0.73	7.91
3-Oxothiolase	17	352	338	0.96	0.66	5.80	19	319	293	0.91	0.51	3.98
NADH oxidase	3	442	369	0.83	0.55	-	-	-	-	-	-	-
CoHD reductase	13	233	233	1.00	0.74	6.04	-	-	-	-	-	-
BA dehydrogenase	-	-	-	-	-	-	25	447	394	0.88	0.48	4.39
Quinone reductase	-	-	-	-	-	-	18	233	202	0.86	0.43	3.14
4-Vinyl reductase	-	-	-	-	-	-	24	164	164	1.00	0.72	6.83
Coenzyme M	-	-	-	-	-	-	28	234	230	0.98	0.70	7.03
CoP methyltransferase	-	-	-	-	-	-	17	217	195	0.89	0.53	4.41

(m – No. of sequences; n – No. of sites; s – No. of segregating sites; ps - Segregating sites per site; pi - Nucleotide/amino acid diversity; D – Tajima's D statistic).

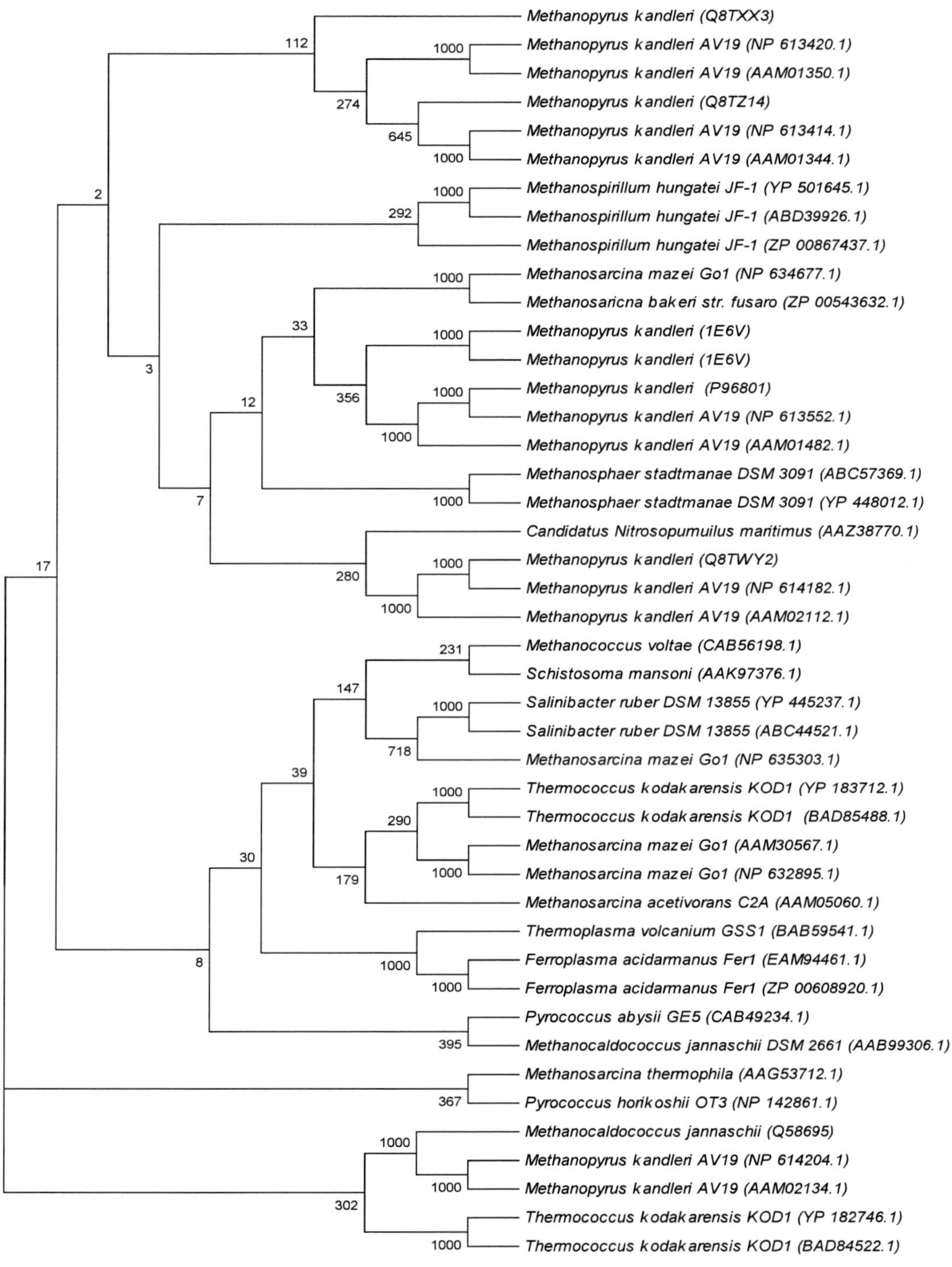

Figure 1. Phylogeny (supertree) based on the sequences of archaeal protein superfamilies constructed by NJ algorithm. Numbers above branches indicate bootstrap support values using 1000 pseudo sequence replicates. Branch length indicates number of inferred amino acid changes per position. Tips of the tree give the names of the organisms. Major clades and sub-clades are indicated by vertical bars, each of which shares a common organization of domains (those in parentheses are variable in their presence or absence within that clade).

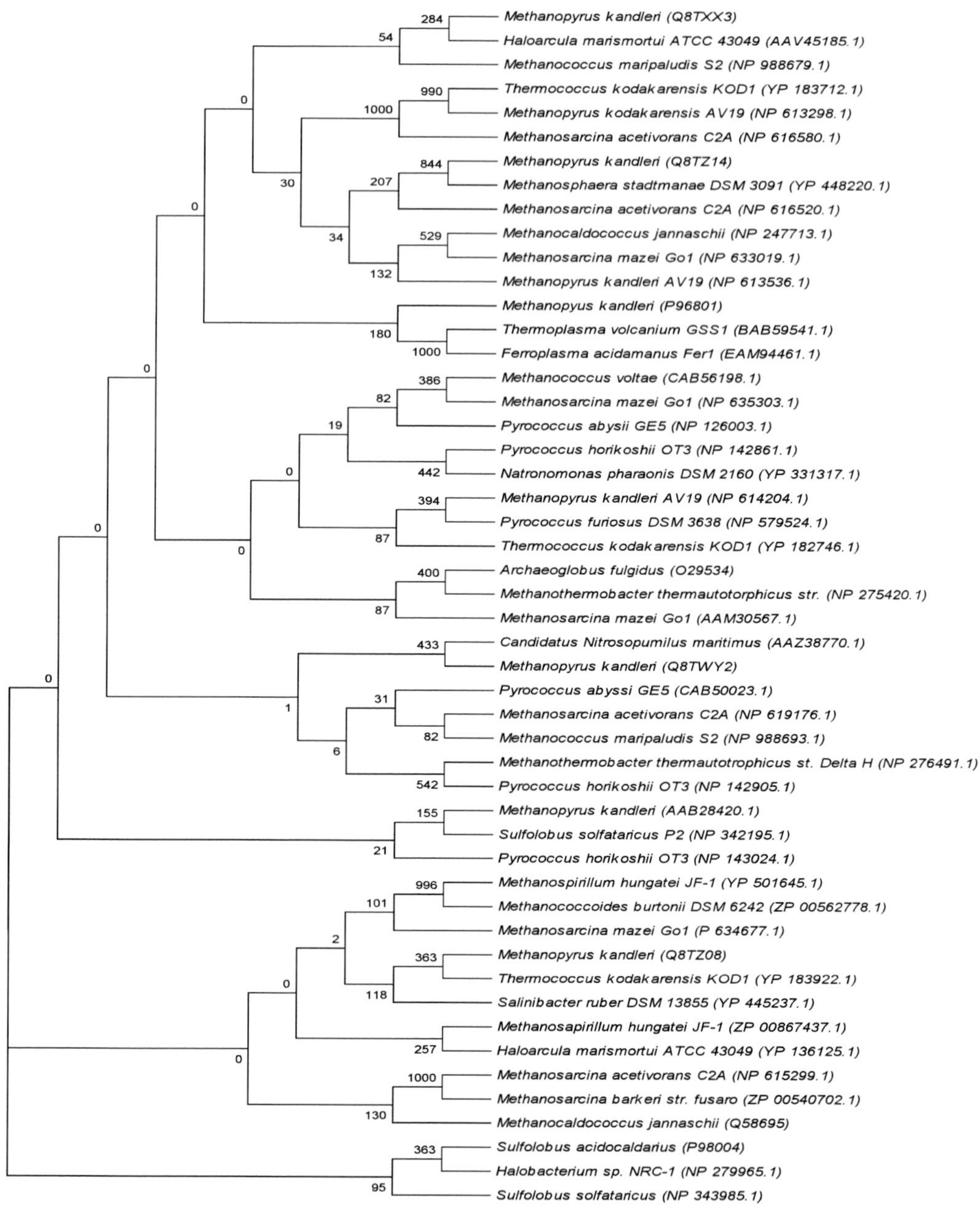

Figure 2. Phylogeny (supertree) based on the sequences of archaeal protein superfamilies constructed by FM algorithm. Numbers above branches indicate bootstrap support values using 1000 pseudo sequence replicates. Branch length indicates number of inferred amino acid changes per position. Tips of the tree give the names of the organisms. Major clades and sub-clades are indicated by vertical bars, each of which shares a common organization of domains (those in parentheses are variable in their presence or absence within that clade).

Table 5. Evolutionary distances (root and pair distances) among archaea and bacteria calculated for the selected protein superfamilies

Query Proteins	With Archaea bacterial species						With bacterial species					
	Root Distance		Pair Distance		Poisson Correction		Root Distance		Pair Distance		Poisson Correction	
	Average	Variation	Average	Variation	Distance	Std. Err.	Average	Variation	Average	Variation	Distance	Std. Err.
CO dehydrogenase	0.457	0.002	0.840	0.008	2.32	0.10	0.335	0.002	0.599	0.010	0.92	0.03
MH cyclohydrolase	0.326	0.016	0.320	0.026	0.99	0.11	0.370	0.075	0.667	0.045	1.55	0.07
Coenzyme M	0.382	0.062	0.641	0.025	1.26	0.05	0.403	0.003	0.748	0.006	1.30	0.05
RUBISCO	0.289	0.010	0.509	0.017	0.73	0.03	0.347	0.038	0.466	0.009	0.71	0.03
AMO	0.456	0.010	0.724	0.055	1.37	0.09	0.414	0.006	0.720	0.006	1.21	0.06
G-6-P synthase	0.355	0.060	0.419	0.021	0.57	0.02	0.251	0.003	0.474	0.002	0.63	0.02
Quinone reductase	0.000	0.000	0.000	0.000	0.84	0.05	0.281	0.032	0.481	0.010	0.62	0.04
NADH oxidase	0.378	0.086	0.605	0.092	1.20	0.07	0.315	0.004	0.591	0.006	1.20	0.03
Quinol oxidase	0.429	0.024	0.859	0.024	1.67	0.09	0.320	0.004	0.591	0.005	1.20	0.07
CoP methyltransferase	0.389	0.000	0.778	0.000	1.01	0.06	0.298	0.004	0.565	0.006	0.79	0.04
tRNA PU synthase	0.588	0.043	0.792	0.013	1.94	0.05	0.327	0.045	0.466	0.011	0.54	0.03
Ile- tRNA synthetase	0.620	0.073	0.715	0.017	1.46	0.04	0.420	0.043	0.523	0.007	0.66	0.02
Topoisomerase VIB	0.542	0.087	0.646	0.024	1.39	0.06	0.457	0.019	0.849	0.016	2.18	0.05
Ribonuclease H	0.431	0.042	0.736	0.037	1.67	0.09	0.258	0.011	0.481	0.006	0.62	0.05
DB synthetase	0.449	0.013	0.759	0.012	1.68	0.08	0.381	0.006	0.650	0.009	0.99	0.06
DHL dehydrogenase	0.396	0.013	0.738	0.009	1.32	0.05	0.438	0.024	0.660	0.009	1.06	0.04
HN synthase	0.468	0.003	0.688	0.020	1.20	0.03	0.305	0.025	0.555	0.015	0.89	0.03
BA dehydrogenase	0.351	0.135	0.479	0.081	0.00	0.00	0.287	0.004	0.521	0.006	0.70	0.03
Choline dehydrogenase	0.600	0.076	0.675	0.018	1.20	0.05	0.140	0.039	0.239	0.016	0.32	0.01
Allophanate hydrolase	0.521	0.091	0.598	0.030	1.01	0.06	0.416	0.023	0.713	0.010	1.47	0.08
Sarcosine oxidase	0.471	0.008	0.832	0.029	2.21	0.10	0.432	0.007	0.745	0.004	1.52	0.07
CoHD reductase	0.550	0.101	0.689	0.038	1.69	0.06	0.363	0.044	0.543	0.011	1.28	0.05
3-Oxothiolase	0.390	0.027	0.669	0.015	1.28	0.05	0.347	0.005	0.555	0.118	0.78	0.04

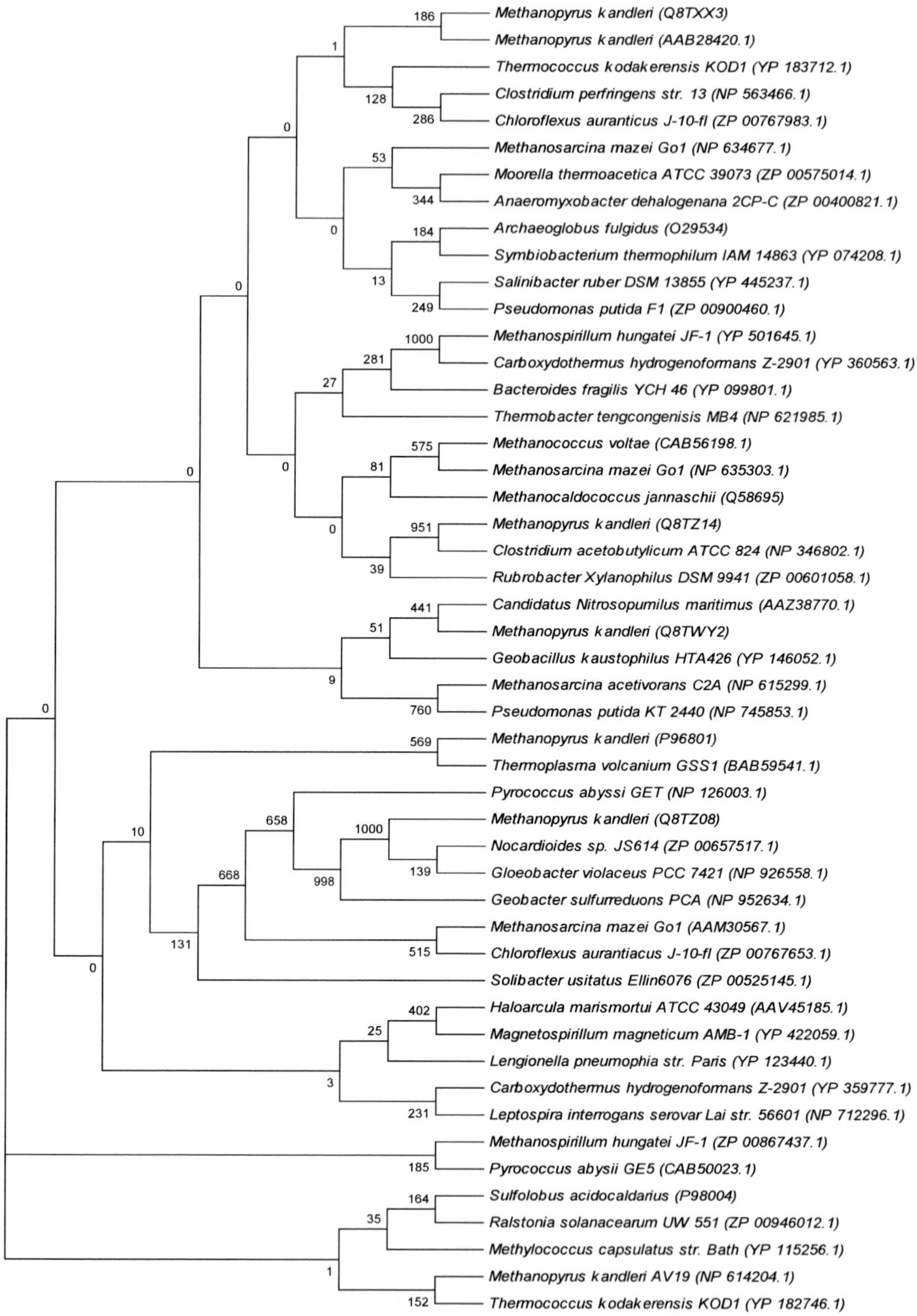

Figure 3. Phylogeny (supertree) based on the sequences of archaeal and bacterial protein superfamilies constructed by FJ algorithm. Numbers above branches indicate bootstrap support values using 1000 pseudo sequence replicates. Branch length indicates number of inferred amino acid changes per position. Tips of the tree give the names of the organisms. Major clades and sub-clades are indicated by vertical bars, each of which shares a common organization of domains (those in parentheses are variable in their presence or absence within that clade).

The results of Tajma's neutrality test revealed that archaeal protein families such as CO dehydrogenase, sarcosine oxidase, tRNA PU synthase, and DB synthetase have not segregated to form new domains or subunits, and susceptible for recombination process with distantly related bacteria. The families of topoisomerase VIB and sarcosine oxidase conserved in bacteria, and have posterior probability observed as low as in archaea. No common segregation sites observed for coenzyme M, NADH oxidase, CoHD reductase and quinol oxidase I. However, some subunits of these enzymes could be relocated to the members of archaea through recombination process.

Phylogenetic Analysis

As shown in Figures 1-2, many of the protein families clustered in a separate cluster in the phylogenetic trees that represented the evolutionary relationship only to closely related organisms. Bacterial choline dehydrogenase and archaeal RUBISCO diverged shortly as the root distance computed to be low among other superfamilies (Table 5 and Figure 3). A divergence rate needed to cluster with topoisomerase VI, DH dehydrogenase and Ile-tRNA synthetase and BA dehydrogenase. CO dehydrogenase, coenzyme M, coenzyme $F_{420,}$ MH cyclohydrolase, topoisomerase VI, ribonulease H and CoHD reductase shared their homologies to the genera of methanogens *Methanopyrus*, *Methanococcus* and *Methanosarcina*. This cluster was again clustered to thermophilic organisms, the genera of *Thermoplasma* and *Ferroplasma* for BA dehydrogenase. The genera of methanogenic archaea, *Methanospirillum*, *Methanosarcina*, *Methanosphaer* and *Candidetus Nitrsopumilus meritimus* grouped to *Methanopyrus* genus for methane biosynthesis, sulfur metabolism and tRNA processing. This cluster joined again for biosynthesis and energy metabolism with two groups of organisms. The first group included the genera of *Thermococcus*, *Thermoplasma*, *Pyrococcus*, *Methanocaldococcus*, *Salinibacter ruber* and second group consisted of the genera of *Methanococcus* and *Schistosoma mensoni*.

The DB synthase, allophanate hydrolase and topoisomerase families showed a phylogenetic correspondence for *Thermococcus* to *Methanopyrus* and *Methanosarcina* to *Thermophila*. This obviously revealed that some metabolic functional modules could be shared between thermophilic archaea and thermophilic methanogens. Protein superfamilies of photorespiration including bacteriochlorophyll and precorin biosynthesis pointed out a phylogenetically resemblance between archaea and the genera of *Geobacter* and *Rubrobacter* (photosynthetic bacteria). RUBISCO of *M. hungatei* provided phylogenetic relatedness to the proteobacteria, *Magnetospirillum magneticum* and *Ralstonia solanacearum*.

Discussion

There are many evolutionary mechanisms including selective pressure, recombination, replication, mutation, domain shuffling, HGT and gene silencing that can result a change in the gene itself, which may or may not have a phenotypic effect (Nelson et al., 1999; Apic et al., 2001; Deppenmeier et al., 2002; Novichkov et al., 2004). Probably, 70% genes from archaea and prokaryotes are shared for common physiological functions. Almost 15% of the proteins encoded by each archaeal genome sequence are members of the signature and therefore unique to archaea (David et al., 2000). In this study, protein superfamilies have a strong functional relationship within archaeal domain, but not to have more resemblance to other organisms except the proteobacteria. The domain similarity of these protein families among prokaryotes suggested the occurrence of domain shuffling in metabolic and stress genes for evolution of a new protein family. Nevertheless, phylosignature analysis of these protein superfamilies revealed a strong phylogenetic proximity occurred between archaea and bacteria for RUBISCO, G-6-P synthase, quinone reductase, CoP methyltransferase, tRNA PU synthase, NAD oxidase, DHL dehydrogenase, BA dehydrogenase and allophanate hydrolase protein families.

The most of the conserved domains gave a significant divergence in many proteins superfamilies among prokaryotes. We significantly noted the sequence similarity relationship between methanogens and methylotrophs (obligate and facultative) for C_1 assimilation pathway. It revealed to support the existence of evolutionary relationship and metabolic shuffling between them. Evolutionary conservation on modulation properties of enzyme structures (methanogenesis) was responsible for reverse methanogenesis of methylotrophic archaea (Hallam et al., 2004). Accordingly, we proposed the hypothesis is that the metabolic genes of methanogenesis and methylotropy (reverse methanogenesis) are clamped together in methylotrophic methanogens. Some common genes of these pathways that crosses the bacterial/archaeal boundaries in accordance to earlier investigations (Hallam et al., 2003; 2004; Deppenmeier, 2004; Kalyuzhnaya et al., 2005; Chistoserdova et al., 2005; Johnson and Mukhopadhyay, 2007). The majority of the genomic proteins in archaea are multidomain proteins, which have evolved through gene duplication, recombination and fusion (Apic et al., 2001). In *Methanosarcina* 1,043 ORFs are homologues in the bacterial domain and 544 of these ORFs reach significant similarity values only in the bacterial domain, particularly proteins involved in gluconeogenesis, proline biosynthesis, transport processes, DNA-repair, environmental sensing, gene regulation, and stress response (Deppenmeier et al., 1996; 1999; Muller et al., 2005; Clementino et al., 2007). A study on domain similarity analysis revealed that more and more number of domains could be introduced the subunits of these enzyme families without changing the conserved region. These domains could be transferred into the members of archaea by domain shuffling. Moreover, the same module could be originated from bacteria into methanogens by subsequent diversification of these enzyme domains.

Dissimilatory sulfite reductase and sulfite reductase of sulfur assimilation pathway in hyperthermophilic archaea were similar as that found in the δ-proteobacterial sulfate-reducing bacteria (Klein et al., 2001; Dhillon et al., 2005). Unlike other kingdoms, since archaea have more uniqueness in many segregation sites of protein superfamily they have a long divergence in protein evolution and gene recombination. Hence, such metabolic pathways (sometimes modules), protein superfamilies and or extreme adaptation could be originated from the host organisms (hyperthermophilic, holophilic or sulfur reducing microbes) to the members of other archaea by gene recombination. Archaea have the deep hyperthermophilic bacterial lineages so that a significant phylogenetic relationship of topoisomerase II found in microorganisms (Wolf et al., 2001; Gadelle et al., 2003). Thus, we have shown that topoisomerase VI (stabilize the genome at high temperature) typically diverged within thermophilic archaea, but the domains of this family could be transferred to mesophilic organisms growing in extreme environments or under starvation. Topoisomerase VI can interact in their respective domain with protein partners that are domain specific, preventing their correct function in a foreign domain (Apic et al., 2001; Gadelle et al., 2003). The results obtained for domain similarity analysis revealed to support the existence of majestic metabolic and protein diversity in archaea, which are not closely corresponded to bacterial metabolism. Phylogenetic analysis of methanogens implied that methanogens have deep evolutionary lineages with haloarchaean, even amino acid composition of halotolerant protein families differed from that of the other archaea (Lake et al., 1985; Cambillau and Claverie, 2000; Muller et al., 2005). Also, osmotic solutes such as (cyclic-2, 3-bisphosphoglycerate, β-mannosylglycerate, i-myo-inositol-1, 1(prm1) (3, 3(prm1))-phosphate, β-mannosylglycerate, di-glycerol-phosphate, α-glutamate, N-δ-acetyl-β-lysine, and glycine betaine have found in *Methanosarcina* genus that was proved its halotolerance (Sowers and Gunsalus, 1995; Martins et al., 1997). Therefore, halotolerance mechanism as such could be acquired from haloarchaean by a long evolutionary process (Nelson et al., 1999; Pagel 1999; Deppenmeier et al., 2002). Overall, cellular machinery genes and stress genes of this domain could be taken a long evolutionary process to diverge into a new protein family.

In this study, NJ phylogenetic trees represented an evolutionary resemblance between methanogens and haloarchaean; methanogens and sulfur utilizing archaean. Methanogens also have strong phylogenetic proximity to thermophilic archaea. It revealed that metabolic modules and or domains of methanogens could be crossed among these groups of organisms thereby methanogens should have a capability to withstand in extreme environmental conditions. When the protein families of archaea bacteria compared with eubacteria, some thermophilic bacteria show to have a phylogenetic resemblance to thermophilic methanogens and provided vertical lineages to present bacteria groups. We have also shown that for energetic metabolisms a phylogenetic correspondence between methanogens and proteobacteria, particularly Clostridium genus. In anaerobic respiration, acidogenic bacteria (Clostridium genus) convert the sugars and amino acids into carbon dioxide, hydrogen, ammonia, and organic acids, and then methanogens convert these products to methane and carbon dioxide (Chellapandi et al., 2007b: 2008; 2009b; Patel and Chellapandi, 2008). Thus, anaerobic respiration and microbial community structures of these groups of organisms may be ascribed for their metabolic propinquity. Supertree construction is currently the only phylogenetic method that can build complete phylogenies of very large clades. Although most conventional phylogenetic support measures are invalid in a supertree context, several supertree-specific measures now exist (Daubin et al., 2001; 2002; Bininda-Emonds et al., 2002; Bininda-Emonds, 2004). Therefore, the present evolutionary hypothesis of archaea can be supported by this supertree representation.

Protein families of photorespiration and of bacteriochlorophyll biosynthesis in archaea have a phylogenetic relationship with primitive photosynthetic bacteria (Brown et al., 1994; Bryant and Frigaard, 2006). In earlier, RUBISCO proved to be responsible to carry out the photorespiration of anoxic archaea, *M. jannaschii* and *A. fulgidus* (Watson et al., 1999). On account of gene degeneration and a long evolutionary process, photosynthetic protein homologs could not be succeeded to the present lineages. Thus, functional metabolic modules of that metabolic capacity could be shared with methanogens, and lasting of them could be degenerated during early evolution process, which is agreed to previous hypotheses (Masinovsky et al., 1992; Andersson and Andersson, 1999; Hallam et al., 2004). Therefore, a complete metabolic pathway for photorespiration could not be existed and established in the present archaeal lineages.

CONCLUSION

A comparative analysis of secondary structures was to reveal more about distant evolutionary relationships than just sequence comparison since structure is usually more conserved than sequence (Chellapandi et al., 2009a). The reconstruction of evolutionary trees has been advanced by considering heterogeneity of evolutionary rates among sequence sites because that heterogeneity rate was strongly associated with structural environment (Goldman et al., 1996; 1998). The distinction between buried and exposed residues is also particularly fruitful for molecular evolutionary studies because it allows consideration of protein regions undergoing different selection pressures. Exposed sites tend to experience ~2X the rate of amino acid replacements experienced by buried sites. A higher rate of replacement for exposed sites is seen for each secondary structure type. The association between accessibility status and replacement rates is a noteworthy feature of protein evolution but has received scant attention in the field of molecular evolution (Goldman et al., 1998). Reconstruction of phylogenies from sequences with known structures involves less uncertainty, and is therefore expected to be more accurate than reconstruction of phylogenies from sequences with unknown structures (Thorne et al., 1996). Thus, the present work pinpoints the strong necessities of sequences and secondary structures of these protein superfamilies for better understanding of the microbial diversity, particularly methanogens and of its evolutionary relationships. Also, this attempt will obviously provide a new view to evolutionary biologists to strengthen the conceptual ideas about diversity of protein superfamilies between archaea and bacteria.

ACKNOWLEDGMENT

The first author is grateful to the University Grants Commission, New Delhi, India for financial assistance (UGC Sanction No. 32-559/2006) to carry out this work.

SUPPLEMENTARY

Table 1. Pairwise sequence alignment (EMBOSE-ALIGN) and sequence similarity of query and best hits retrieved from archaea

Query	Archaea		BLOSUM 62				PAM 250			
Accession	Accession	Enzyme	Length	Identity (%)	Similarity (%)	Score	Length	Identity (%)	Similarity (%)	Score
Q8TXX3	NP_247713	Co dehydrogenase	661	21.6	27.7	641.5	646	20.0	31.4	663.6
NP_634677.1	NP_616637	Methenyl-H_4MPT cyclohydrolase	322	94.4	97.2	159.9	321	94.4	98.4	1545.0
AAB28420	NP_613298	F_{420} – dependent methylene - H_4MPT dehydrogenase	283	11.3	11.3	153.0	283	11.3	11.3	144.0
CAB56198.1	NP_988679	Methyl - CoM reductase	553	11.0	11.2	327.0	554	11.0	11.2	297.0
ZP_00867437	NP_142861	RUBISCO	512	38.1	52.5	1062.4	506	36.4	55.1	1073.5
AAZ38770	NP_275420	Methane monooxygenase	438	20.1	26.5	391.7	436	17.2	28.9	443
Q8TZ14	YP_448220	Glucosamine 6 - phosphate synthetase	687	49.1	64.0	1796.9	676	48.1	68.6	1791.5
NP_615299	ZP_00540702	Quinone reductase	215	78.1	88.8	911.8	215	78.1	90.2	924.8
YP_183712	NP_619176	Quinone oxidase	2066	14.8	18.0	1291.601	2049	13.0	19.3	1325.2
P98004	NP_343985	Heme - Cu oxidase	531	78.5	86.8	2259.8	523	79.0	91.4	2442.3
O29534	NP_633019	Cobalamine adenosyl transferase	482	31.3	41.5	753	478	30.1	45.0	773.4
CAB50023	NP_143024	4 - Vinyl reductase	423	65.5	78.0	1502.1	416	66.3	82.9	1569.1
Q8TZ08	YP_183922	Pseudouridylate synthase	355	62.5	76.1	1189.2	364	59.9	76.4	1197.4
Q8TWY2	NP_276491	Cysteinyl - tRNA synthetase	1230	49.8	61.1	3408.499	1206	49.4	64.8	3640.999
NP_614204	NP_616520	Topoisomerase VI	729	31.6	42.2	1164.0	696	30.9	49.1	1211.4
YP_445237	YP_331317	Ribonuclease H	353	26.9	36.3	479.6	345	26.4	40.0	479.4
Q58695	NP_988693	Dethiobiotin synthetase	227	54.6	69.2	652.7	221	54.8	73.8	662.9
NP_635303	NP_616580	Dihydrolipoamide dehydrogenase	491	82.7	89.4	2147.8	489	82.6	92.0	2126.8
AAM30567	NP_342195	Hydroxy- neurosporene synthase	819	23.2	29.1	834.0	801	20.8	32.2	874.8
BAB59541	EAM94461	Betaine aldehyde dehydrogenase	542	57.6	69.6	1640.2	535	56.3	75.5	1641.0
AAV45185	NP_279965	Choline dehydrogenase	639	48.8	58.1	1631.5	619	48.8	64.3	1563.6
YP_182746	NP_142905	Allophonate hydrolase	393	24.7	36.1	482.4	389	23.9	37.8	550.2

Table 1. (Continued)

Query	Archaea		BLOSUM 62				PAM 250			
Accession	Accession	Enzyme	Length	Identity (%)	Similarity (%)	Score	Length	Identity (%)	Similarity (%)	Score
NP_126003	NP_579524	Sulfur oxidase	488	83.4	90.4	2115.7	488	83.4	91.8	2057.5
P96801	NP_613536	Heterodisulfide reductase	714	61.5	73.7	2402.1	708	60.5	76.1	2434.6
YP_501645	ZP_00562778	Thiolase	412	65.8	78.4	1437.4	410	65.6	82.4	1335.5

Table 2. Pairwise sequence alignment (EMBOSE-ALIGN) and sequence similarity of query and best hits retrieved from bacteria

Query	Archaea		BLOSUM 62				PAM 250			
Accession	Accession	Enzyme	Length	Identity (%)	Similarity (%)	Score	Length	Identity (%)	Similarity (%)	Score
Q8TXX3	ZP_00886714	Co dehydrogenase	720	47.4	62.4	1843.301	740	44.3	62.3	1831.30
NP_634677.1	YP_115256	Methenyl-H_4MPT cyclohydrolase	400	38.0	50.0	804.6	389	36.0	57.3	844.9
CAB56198.1	ZP_00657517	Methyl - CoM reductase	334	12.9	16.5	194.6	334	12.3	17.4	199.7
ZP_00867437	YP_422059	RUBISCO	569	32.2	41.5	902.2	537	30.0	50.7	916.9
AAZ38770	ZP_00575014	Methane monooxygenase	386	20.7	28.8	366.0	383	16.7	30.0	422.7
Q8TZ14	NP_346802	Glucosamine 6 - phosphate synthetase	714	47.8	62.3	1774.101	697	46.6	67.9	1796.90
NP_615299	NP_745853	Quinone reductase	274	36.5	51.5	549.3	263	36.5	59.7	604.0
YP_183712	ZP_00767983	Quinone oxidase	526	43.0	55.7	1134.4	509	42.0	62.7	1162.5
P98004	ZP_00525145	Heme - Cu oxidase	561	21.0	27.5	520.2	563	17.6	28.6	664.3
O29534	YP_099801	Cobalamine adenosyl transferase	578	37.2	48.8	1067.4	565	35.8	54.3	1097.1
CAB50023	YP_123440	4 - Vinyl reductase	672	25.3	36.5	793.4	636	24.5	41.0	929.1
Q8TZ08	ZP_00900460	Pseudouridylate synthase	426	32.2	42.3	664.0	415	31.1	48.4	700.2
Q8TWY2	ZP_00601058	Cysteinyl - tRNA synthetase	1308	40.4	51.5	2816.4	1306	38.6	54.7	3062.4
NP_614204	ZP_00400821	Topoisomerase VI	797	28.5	39.3	114.05	776	26.8	44.6	1220.84
YP_445237	NP_926558	Ribonuclease H	400	19.5	26.0	48.5	371	19.1	35.8	460.9
Q58695	NP_952634	Dethiobiotin synthetase	286	36.7	53.1	542.6	282	35.8	55.3	554.2

Query	Archaea		BLOSUM 62				PAM 250			
Accession	Accession	Enzyme	Length	Identity (%)	Similarity (%)	Score	Length	Identity (%)	Similarity (%)	Score
NP_635303	NP_712296	Dihydrolipoamide dehydrogenase	576	38.2	53.1	1147.7	568	37.7	56.0	1189.3
AAM30567	ZP_00767653	Hydroxy- neurosporene synthase	397	52.1	64.2	1093.6	388	52.1	69.6	1064.8
BAB59541	YP_146052	Betaine aldehyde dehydrogenase	605	39.3	55.5	1265.9	598	37.8	61.2	1323.5
AAV45185	ZP_00946012	Choline dehydrogenase	709	34.1	44.6	1247.3	686	32.7	51.3	1284.8
YP_182746	YP_074208	Allophonate hydrolase	665	20.3	25.3	707.7	661	19.2	26.5	7566
NP_126003	NP_563466	Sulfur oxidase	614	28.7	39.6	852.1	604	25.3	43.7	875.8
P96801	YP_359777	Heterodisulfide reductase	732	53.6	66.1	2100.2	717	53.6	71.1	2170.1
YP_501645	YP_360563	Thiolase	447	49.2	60.2	119.2	458	46.7	61.1	1083.8

Table 3. Description of motif region for archaeal protein sequences

Accession	Enzyme	E- Value	Probability of no. random hit so far (%)	Description	Predicted region
NP_616637	Methenyl-H_4MPT cyclohydrolase	1.05e+00	35.2	PR00608A Class II cytochrome C signature I	2511-AAGIAPIAPVDPDSLKA-2527
NP_613298	F_{420} – dependent methylene - H_4MPT dehydrogenase	3.03e-1	73.9	IPB003874C CDC45-like protein	150-DMLLDERADRED-161
NP_988679	Methyl - CoM reductase	8.09e-4	99.9	PR00354C 7Fe ferredoxin signature III	1391-IDPDHCVGCGLCQLAC-1406
NP_142861	RUBISCO	1.16e-10	99.9	IPB000685F Ribulose bisphosphate carboxylase, large chain	1408-FLLSKWEHIRPVFPVASGGLHPGLMPELIRLFGKDLVIQ-1446

Table 3. (Continued)

Accession	Enzyme	E- value	Probability of no. random hit so far (%)	Description	Predicted region
NP_275420	Methane monooxygenase	1.87e-06	99.9	PR00411A Pyridine nucleotide disulphide reductase class-I signature I	763-DVLIIGAGPAGSLAAREASK-782
YP_448220	Glucosamine 6 - phosphate synthetase	3.18e-09	99.9	IPB000583C Glutamine amidotransferase class-II	315-HNGIIENFMQLREELEDKGHRFDSETDTE-343
ZP_00540702	Quinone reductase	8.59e-11	99.9	IPB003680 NAD(P)H dehydrogenase (quinone)	142-DFLIFQFPIYFTSMPAILKGWIDRVLAPG-170
NP_619176	Quinone oxidase	5.87e-03	99.4	IPB000103E Pyridine nucleotide-disulphide oxidoreductase class-II	315-TSVENVYAAGDVAETKHLIT-334
NP_343985	Heme - Cu oxidase	3.77e-06	99.9	IPB000298D Cytochrome c oxidase, subunit III	1102-YWHLVDIVWIFLFPLLY-1118
NP_633019	Cobalamine adenosyl transferase	1.04e-10	99.9	IPB003722A Precorrin-8X methylmutase CbiC/CobH	516-LRDGCRIIADVHMVRAGL-533
NP_143024	4 - Vinyl reductase	4.92e-01	61.1	PR00333E Herpsevirus integral membrane protein signature V	336-ERSLIEFVYQFLGSKEG-352
YP_183922	Pseudouridylate synthase	2.56e-15	99.9	IPB002501A Pseudouridylate synthase, TruB family, N-terminus	189-TSHEVVAWVKEIFGLSKAGHGGTLDPKVTGVLPIALEKATKIIQTLLPAGKEY-241
NP_276491	Cysteinyl - tRNA synthetase	2.68e-15	99.9	PR00983D Cysteinyl-tRNA synthetase signature IV	5613-DIHVGGQDLVFPHHENEVAQS-5633
NP_616520	Topoisomerase VI	5.01e-18	99.9	IPB002815C Type II DNA topoisomerase	795-ARATRRLLRRLNEELGLPVYVFTDGDPYGWHIYMVIRSGS-834
YP_331317	Ribonuclease H	3.08e-10	99.9	IPB001352A Ribonuclease HII	805-GIDEAGRGPVIGPMV-819

Accession	Enzyme	E- value	Probability of no. random hit so far (%)	Description	Predicted region
NP_988693	Dethiobiotin synthetase	2.98e-12	99.9	IPB002684D Biotin synthase	2505-DKVKVIKEAKKLDLEVCSGGIFGLGESVEERIKMAF-2540
NP_616580	Dihydrolipoamide dehydrogenase	4.29e-10	99.9	PR00411B Pyridine nucleotide disulphide reductase class-I signature II	7843-LGGVCLNVGCIPTKAL-7858
NP_342195	Hydroxy neurosporene synthase	1.85e-15	99.9	IPB000926C GTP cyclohydrolase II	7001-EGRGIGLKEKIKAYELQDKGFDTVEANKAL-7030
EAM94461	Betaine aldehyde dehydrogenase	1.32e-08	99.9	IPB002086E Aldehyde dehydrogenase family	425-YFPPTILDKVSEDMEVAKEEIFGPVLSVMEFDDEADAIDKANMTDYGLAAGVWT-478
NP_142905	Allophonate hydrolase	1.76e-45	99.9	IPB003778B DUF183	1792-GMYGYIAFAGGIECQKILGSCATYSRAKFGRPLKPGDKLKLGY-1834
NP_579524	Sulfur oxidase	9.67e-05	99.9	PR00352A 3Fe-4S ferredoxin signature I	3466-IDEEHCIRCGLC-3477
NP_613536	Heterodisulfide reductase	1.06e-04	99.9	PR00352A 3Fe-4S ferredoxin signature I	352-IDEEHCIRCGLC-363

Table 4. Description of motif region for bacterial protein sequences

Accession	Enzyme	E- value	Probability of no. random hit so far (%)	Description	Predicted region
YP_115256	Methenyl-H4MPT cyclohydrolase	1.06e-01	90	IPB000194A ATP synthase alpha and beta subunit, N-terminal	2749-YGRILDKDGHRID-2761
ZP_00657517	Methyl - CoM reductase	2.97e+00	5.1	IPB001917C Aminotransferases class-II	10962-LVDDMHIIGTAG-10973
YP_422059	RUBISCO	2.18e-16	99.9	IPB000685E Ribulose bisphosphate carboxylase, large chain	2511-HAVIDRQRNHGIHFRVLAKCLRLSGGDHLHSGTVVGK-2547
ZP_00575014	Methane monooxygenase	1.04e-07	99.9	IPB000171A Bacterial-type phytoene dehydrogenase	8326-VVGAGPGGSATAAHLARRGLHVTLLEK-8352
NP_346802	Glucosamine 6 - phosphate synthetase	9.70e-09	99.9	IPB000583C Glutamine amidotransferase class-II	169-HNGIIENYMQLRTWLKSKGYEFKSETDTE-197
NP_745853	Quinone reductase	1.23e-16	99.9	IPB003680 NAD(P)H dehydrogenase (quinone)	1063-DAVILQFPLWWFSMPAIMKGWVERVYAYGFAY-1094
ZP_00767983	Quinone oxidase	8.40e-05	99.9	PR00411D Pyridine nucleotide disulphide reductase class-I signature IV	6902-VTIIGGGAIGLEMAETFVELGKKV-6925
ZP_00525145	Heme - Cu oxidase	9.78e-05	99.9	PR00411D Pyridine nucleotide disulphide reductase class-I signature IV	6902-VTIIGGGAIGLEMAETFVELGKKV-6925
YP_099801	Cobalamine adenosyl transferase	7.33e-10	99.9	IPB003722A Precorrin-8X methylmutase CbiC/CobH	393-LRDGCRIIADVHMVRAGL-410
YP_123440	4 - Vinyl reductase	4.43e-02	95.7	IPB001482B Bacterial type II secretion system protein E	1882-LVSGATGTGKTLMV-1895

Accession	Enzyme	E- value	Probability of no. random hit so far (%)	Description	Predicted region
ZP_00900460	Pseudouridylate synthase	1.51e-17	99.9	IPB002501A Pseudouridylate synthase, TruB family, N-terminus	1868-GMTSHDCVFRLRRLFQTKKVGHTGTLDPEVTGVLPICLGRATKLARFITDEGKRY-1922
ZP_00601058	Cysteinyl - tRNA synthetase	2.71e-15	99.9	PR00984B Isoleucyl-tRNA synthetase signature II	16982-LGWTTTPWTLPANVALAVHPNMEY-17005
ZP_00400821	Topoisomerase VI	1.78e-05	99.9	PR01552B DNA topoisomerase VI subunit A (TOP6A) signature II	2256-FEEQEESDPIIEDLE-2270
NP_926558	Ribonuclease H	3.54e-01	70.2	PR00048B C2H2-type zinc finger signature II	503-LWEHVRGHTG-512
NP_952634	Dethiobiotin synthetase	8.09e-01	44.5	IPB003421A NAD/NADP octopine/nopaline dehydrogenase	1985-VIVEGAGGLMVPLADQL-2001
NP_712296	Dihydrolipoamide dehydrogenase	3.09e-09	99.9	PR00411B Pyridine nucleotide disulphide reductase class-I signature II	5327-MGGDCLNYGCIPSKSL-5342
ZP_00767653	Hydroxyneurospore-ne neurosporene synthase	1.56e+00	20.9	PR00477C Phosphoglycerate kinase family signature III	3851-LSLLENGRLKAGD-3863
YP_146052	Betaine aldehyde dehydrogenase	5.44e-11	99.9	IPB002086E Aldehyde dehydrogenase family	429-FVAPTIFSNVTNDMKIAQEEIFGPVLSIIRYHDVEEAVTMANDTIYGLAAGVWT-482
YP_074208	Allophonate hydrolase	7.59e-19	99.9	IPB003833C DUF213	2633-PRLETPRTEIKAGSVGIAGNQTGIYPLSSPGGWRII-2668
NP_563466	Sulfur oxidase	1.96e-05	99.9	PR00411A Pyridine nucleotide disulphide reductase class-I signature I	13762-DLLVIGSGPAGLMAALTAGR-13781
YP_359777	Heterodisulfide reductase	1.10e-06	99.9	IPB000171A Bacterial-type phytoene dehydrogenase	13850-IIGGGVAGMTSALNLAEQGFNVNLVEK-13876

Table 5. Domain description of query sequences

Enzyme	Region	Sequence
Co dehydrogenase	549-564	IDIPDVPAVASAPQWL
	243-249	ITQTEGG
	586-598	PVPPVTGSELVTK
	39-41	WPV
Methenyl-H_4MPT	252-284	VRKVPSTTSDDFGKPFYQTFKAANFDFFKVDAG
cyclohydrolase	74-78	KWPAI
RUBISCO	310-322	LTEPFFDLKPTFP
	171-176	DQAFCP
	109-110	VP
Glucosamine 6 - phosphate synthetase	193-209	NFLASDIPAILPETNRV
Quinone reductase	48-77	LTEKDFTDRKKPDVFKPFLEAIQASKTGAF
Quinone oxidase	415-426	VFFTDLAYAPPF
	112-140	ASPQVPAIEGIDLPGVFTADLPPDAVAIT
Heme - Cu oxidase	483-497	QPVFSSPSVSSFSMQ
Cobalamine adenosyl	388-394	KHPPRFI
transferase	415-421	PVPSVTT
4 - Vinyl reductase	95-110	SPTNVVGPNVLPNPMD
	344-347	FENF
Pseudouridylate synthase	25-35	TDPRYGCPPEE
Cysteinyl - tRNA synthetase	897-915	EEFPEKVPVAEPRPDKIGP
	207-233	SIFVIFPIEDDSDAEVDLPENSALLIW
	504-516	AVSLPPGEPDLHR
	571-580	GRCDPDDPEA
Topoisomerase VI	142-157	EVKPDPSTNEPVVVRR
	77-90	GIPPERVPKVFGKF
	405-416	NSPFVPFSGPAK
	213-228	WEPLVEELPDPPRVLK
	308-313	PSDVVS
Ribonuclease H	152-166	IPKCITPSLIPSSTP
Dethiobiotin synthetase	51-86	ILNTDDDLDLMNPINLKLPLSPNIAFDVENSPLTLD
Hydroxyneurospo-rene	239-255	PNGKFPSKAAKMLGFTK
synthase	89-97	PTSTIVAQA
Betaine aldehyde	174-189	VVIKPSSFTPLTTIEM
dehydrogenase	359-371	VPKGGLYFPPTIL
	309-313	EKMRP
	35-35	S
Choline dehydrogenase	363-399	TSGGAQSRLPPTSPSRFLDIVRNPTPDALAELATDPL
	150-172	DDNPHGPPRKEPYPLTAFEPSYS
	407-428	ALELPTTSGPPYPLSIRGAGEM
	535-540	TAGANN
	443-445	TDN
	447-447	G
Allophonate hydrolase	188-206	RLFDPSREPPTLLQPGDEV
	142-150	ERIATPRLE
	59-72	LKVTFEDVKKAVEP
Sulfur oxidase	349-355	WRPIPGT
	242-245	PFEN
Heterodisulfide reductase	369-380	GKVVRPSDGKKP
	548-553	DIPSSV
	190-194	TFPTN
	278-284	KPFPQAV
	524-527	HPKL
Thiolase	176-201	DNGAKNPNAQFRKPITLDTVINSTLV
	201-213	VADPLRLFDCSPV

Table 6. Domain description of best hit sequences from archaea

Enzyme	Region	Sequence
Co dehydrogenase	24-37	ETPHTRFPNQFPKC
	597-608	KLIPEPNPYKAA
	538-550	PDLPVAASAPQWL
	123-131	KLGFDANKD
	157-164	LVPKLCPE
Methenyl-H_4MPT	253-284	KKVPSTTSNDFGKPFYQTFKAANFDFFKVDAG
cyclohydrolase	75-78	WPAI
F_{420} – dependent	65-87	PDFIVYGGPNPAAPGPSKAREML
methylene - H_4MPT	170-181	EDEISENDLPKL
dehydrogenase	266-267	HA
	262-264	LRT
Methyl - CoM reductase	17-23	EEDPKEK
	512-522	VKVAFADPMLV
	59-60	LY
	470-472	PLE
RUBISCO	188-197	DENFTSFPFN
	405-413	KAKSSPELK
	154-161	LTATVPKP
Methane monooxygenase	253-259	RNAQPVE
Glucosamine 6 -	499-509	VGILPPGPSYR
phosphate synthetase	543-548	LLFDPE
Quinone reductase	9-81	KASDFPERKKQDVFKPFFEAIQASKTGAFAPDV
	121-129	FGFNPVTNS
Quinone oxidase	362-412	THETPSAPVAGFSATPTSGNAPLTVNFTDQSTNT PTSWLWDFGDNSTDTEQ
	941-975	PEANFSSDKQTGSFPLTVSFKDRSFNANKFFWDF
	1641-1665	GKDSLTSPVANFSADITSGNAPLTVN
	1722-1747	TDYITVSSTPVEPEPVAAFTADVTGG
	1957-1971	LSPLPDQEYAPKDLD
Heme - Cu oxidase	484-495	QGIFSSPSPSTS
4 - Vinyl reductase	99-117	YSPASISGPDVLPNPLDLG
	218-221	PYVK
Pseudouridylat-e synthase	22-38	DEKAETNPKWGFPPEKR
	54-57	PPGP
Cysteinyl - tRNA	854-882	NAKSIKTSTEFPDMKIIARPNPATLGPRL
synthetase	918-931	KKFKLTSEDIVFET
	911-912	FT
Topoisomerase VI	215-235	GNEEVFERATDKIPDPAEEIL
	555-563	AKPEPKVVT
	309-315	IMSPPTD
	432-441	SESKDAIADI
	493-497	ILEKD
	341-345	STRKP
Dethiobiotin synthetase	41-56	LLNPINLKNPLSPNIA
Dihydrolipoamide dehydrogenase	143-151	PSIPPVKGL

Table 6. (Continued)

Enzyme	Region	Sequence
Hydroxyneuros-porene synthase	476-491	SMGSTAPPDVEKPARI
	501-512	QVTRPSIDPYRE
	92-103	KATSISLTPKAN
Betaine aldehyde dehydrogenase	355-381	MQGTPDKGTYFPPTLLTGLTPGQQISD
	312-321	KSGNPMDFET
Choline dehydrogenase	363-399	TSGGAQSRLPPTSPSRFLDIVRNPTPDALAELATD
	150-172	PL
	407-428	DDNPHGPPRKEPYPLTAFEPSYS
	535-540	ALELPTTSGPPYPLSIRGAGEM
	443-445	TAGANN
	447-447	TDN
		G
Allophonate hydrolase	48-59	NPDDSPLLEFTL
	271-274	AQKR
Sulfur oxidase	96-111	NKPVLPPANSNGSKAK
	345-350	NPIPGT

Table 7. Domain description of best hit sequences from bacteria

Enzyme	Region	Sequence
Co dehydrogenase	23-31	DLETSFNRK
	191-196	TSTNLS
	543-543	L
Methenyl-H_4MPT cyclohydrolase	220-235	GTGCAPVPPPHPNFIK
	265-277	ANELPSSVSRDYG
RUBISCO	192-204	NDEPQGNQVFAPF
	125-136	DFYVPPDFLTLF
	228-236	FSANITADD
Methane monooxygenase	72-86	TFFPDGTVTSTPAPG
Quinone reductase	212-219	QLSDDQPL
	53-56	DDFT
Quinone oxidase	422-430	LDLSYAPPF
	113-124	TGGQPGRPPLPG
Pseudouridylate synthase	116-126	RGPISQIPPMY
Cysteinyl - tRNA synthetase	65-80	PVEKPFVFYEGPPTAN
	373-383	VRPDGTFDERV
	513-517	DPLPD
	695-695	E
Topoisomerase VI	34-90	APRRSKPAAPPAPRVPVRKAAGRARPAPAEEPELLPP APEAEPEPAATRSPKRRAT
	455-474	ATKKGRAAKAEAPEVPDAPA
	309-327	PRQAPLSFPRATDELPKEA
Dethiobiotin synthetase	78-78	R
Dihydrolipoamide dehydrogenase	65-90	EKFQISFPKKPEVDFKTLIERISKTV
	140-157	ARPAIPDIPGLAGTPFMT
	90-108	VDDESASILPAYDKNPNIT
	332-344	RTPIEYPPVPHAV

Table 7. (Continued)

Enzyme	Region	Sequence
Hydroxyneurosporene synthase	36-51	SGTPIEAKSVPGPDED
	239-259	PNAKFPQAVAKRLGFTDKQIA
	183-187	TDTPD
	191-191	R
Choline dehydrogenase	157-178	DEDLGSPRSQPYPMAPLPLSFN
	59-82	KMDFMAPYPSTPWAPHPEYGPSPN
	410-432	ILPLPENRIVPSATETDALGIPR
	488-512	PKDSVVDKDCRTFDHPNLFISSSAT
	204-213	SRPYDGRPTC
Allophonate hydrolase	191-210	RLFDPLRDPPTLLQPGDRVR
	305-337	FTGPAAFALTGAPIPAELIPADGGAPRPVPGWT
	471-493	PEGPAEMLSEPIPPGALQVVPSG
	223-240	AAEQRAEPAGPPAFAPGR
	401-419	PLHPPADLAGRRLPADAIP
	429-443	LIPGPQADQFPAASL
	148-150	AAP
	597-597	A
Sulfur oxidase	236-237	IK
Thiolase	198-213	GAPVADPFTVWDCSPT

Table 8. Bayesian phylogenetic analysis of query sequences with archaeal domain

Query	Archaea	BLOSUM	Posterior probability
Q8TXX3	NP_247713	62	1
NP_634677.1	NP_616637	45	0.97
AAB28420	NP_613298	62	0.76
CAB56198.1	NP_988679	62	0.96
ZP_00867437	NP_142861	62	0.99
AAZ38770	NP_275420	45	0.83
Q8TZ14	YP_448220	62	1
NP_615299	ZP_00540702	62	0.99
YP_183712	NP_619176	62	0.99
P98004	NP_343985	62	0
O29534	NP_633019	62	0.99
CAB50023	NP_143024	62	1
Q8TZ08	YP_183922	62	1
Q8TWY2	NP_276491	62	0
NP_614204	NP_616520	62	0.99
YP_445237	YP_331317	50	0.72
Q58695	NP_988693	62	0.99
NP_635303	NP_616580	50	0
AAM30567	NP_342195	45	0.97
BAB59541	EAM94461	62	1
AAV45185	NP_279965	62	1
YP_182746	NP_142905	45	0.13
NP_126003	NP_579524	62	1
P96801	NP_613536	62	0
YP_501645	ZP_00562778	62	1

Table 9. Bayesian phylogenetic analysis of query sequences with bacteria

Query	Bacteria	BLOSUM	Posterior probability
Q8TXX3	ZP_00886714	45	0.99
NP_634677.1	YP_115256	62	0.99
CAB56198.1	ZP_00657517	45	0.42
ZP_00867437	YP_422059	62	0.99
AAZ38770	ZP_00575014	45	0.80
Q8TZ14	NP_346802	62	1
NP_615299	NP_745853	62	0.99
YP_183712	ZP_00767983	62	0.99
P98004	ZP_00525145	45	0.95
O29534	YP_099801	62	0.99
CAB50023	YP_123440	50	0.54
Q8TZ08	ZP_00900460	62	0.99
Q8TWY2	ZP_00601058	62	1
NP_614204	ZP_00400821	62	0.99
YP_445237	NP_926558	62	0.99
Q58695	NP_952634	62	0.99
NP_635303	NP_712296	62	0.99
AAM30567	ZP_00767653	62	1
BAB59541	YP_146052	45	0
AAV45185	ZP_00946012	62	0.99
YP_182746	YP_074208	62	0.99
NP_126003	NP_563466	50	0.65
P96801	YP_359777	62	0
YP_501645	YP_360563	62	1

Table 10. Biochemical reactions of protein superfamily selected in this study

E.C.	Reaction
1.2.99.2	$CO + H_2O + A = CO_2 + AH_2$
3.5.4.27	5,10-methenyl-5,6,7,8-tetrahydromethanopterin + H_2O =5-formyl-5,6,7,8-tetrahydromethanopterin
1.5.99.1	5-methyltetrahydromethanopterin + coenzyme F420 = 5,10-methylenetetrahydromethanopterin + reduced coenzyme F420
2.8.4.1	2-(methylthio)ethanesulfonate (methyl-CoM) + N-(7-mercaptoheptanoyl)threonine 3-O-phosphate (coenzyme B) = CoM-S-S-CoB + methane
4.1.1.39	2 3-phospho-D-glycerate + 2 H+ = D-ribulose 1,5-bisphosphate + CO_2 + H_2O
1.14.99.39	ammonia + AH_2 + O_2 = NH_2OH + A + H_2O
1.6.5.5	NAD(P)H + H+ + a quinone = NAD(P)+ + a hydroquinone
1.6.-.-	Unclassified
1.3.1.75	chlorophyllide a + NADP+ = divinyl chlorophyllide a + NADPH + H+
1.8.1.4	protein N6-(dihydrolipoyl)lysine + NAD+ = protein N6-(lipoyl)lysine + NADH + H+
4.2.1.-	Carotenoid biosynthesi

E.C.	Reaction
1.9.3.-	Oxidative phosphorylation
5.4.99.12	tRNA uridine = tRNA pseudouridine
6.1.1.5	ATP + L-isoleucine + tRNAIle = AMP + diphosphate + L-isoleucyl-tRNAIle
5.99.1.3	ATP-dependent breakage, passage and rejoining of double-stranded DNA
3.1.26.4	Endonucleolytic cleavage to 5'-phosphomonoester
1.2.1.8	betaine aldehyde + NAD+ + H_2O = betaine + NADH + 2 H+
1.1.99.1	choline + acceptor = betaine aldehyde + reduced acceptor
1.5.3.1	sarcosine + H_2O + O2 = glycine + formaldehyde + H_2O_2
1.8.98.1	coenzyme B + coenzyme M + methanophenazine = N-{7-[(2-sulfoethyl)dithio]heptanoyl}-O3-phospho-L-threonine + dihydromethanophenazine
2.3.1.9	2 acetyl-CoA = CoA + acetoacetyl-CoA
2.1.1.133	Cobalt-precorrin 4 + S-Adenosyl-L-methionine <=> Cobalt-precorrin 5A + S-Adenosyl-L-homocysteine
6.3.3.3	ATP + 7,8-diaminononanoate + CO2 = ADP + phosphate + dethiobiotin
2.6.1.6	L-glutamine + D-fructose 6-phosphate = L-glutamate + D-glucosamine 6-phosphate
3.5.1.54	urea-1-carboxylate + H_2O = 2 CO_2 + 2 NH_3

REFERENCES

Alqueres SMC, Almeida RV, Clementino MM, Vieira RP, Almeida WI, Cardoso AM and Martins OB (2007) Exploring the biotechnological applications in the archaeal domain. *Brazilian Journal of Microbiology* 38: 398–405.

Altschul SF, Madden TL, Schaffer AA, Zhang J, Zhang Z, Miller W and Lipman DJ (1997) Gapped BLAST and PSI-BLAST, a new generation of protein database search program. *Nucleic Acids Research* 25: 3389–3402.

Andersson JO and Andersson SGE (1999) Insights into the evolutionary process of genome degradation. *Current Opinion in Genetics and Development* 9: 664–671.

Apic G, Gough J and Teichmann SA (2001) Domain combinations in archaeal eubacterial and eukaryotic proteomes. *Journal of Molecular Biology* 310: 311–325.

Bininda-Emonds ORB (2004) The evolution of supertrees. *Trends in Ecology and Evolution* 19: 315–322.

Bininda-Emonds ORB, Gittleman JL and Steel MA (2002) The (super) tree of life: procedures problems and prospects. *Annual Review of Ecology and Systemics* 33: 265–289.

Bradley E, Elizabeth H and Susan H (1996) Bootstrap confidence levels for phylogenetic trees. *Proceedings of National Academy of Science*, USA 93:13429–13429.

Brown JR and Doolittle WF (1997) Archaea and the prokaryote-to-eukaryote transition. *Microbiology and Molecular Biology Reviews* 61: 456–502.

Brown JR, Masuchi Y, Robb FT and Doolittle WF (1994) Evolutionary relationships of bacterial and archaeal glutamine synthetase genes. *Journal of Molecular Evolution* 38: 566–576.

Bryant DA and Frigaard NU (2006) Prokaryotic photosynthesis and phototrophy illuminated. *Trends in Microbiology* 14: 488–496.

Cambillau C and Claverie JM (2000) Structural and genomic correlates of hyperthermostability. *Genome Research* 275: 32383–32386.

Chellapandi P, Karthigeyan C and Sivaramakrishnan S (2007a) Phylogenomic analysis of archaeal domain: An overview. *Research and Review Journal of Bioscience* 1: 160–168

Chellapandi P, Karthigeyen C and Sivaramakrishnan S (2009a) Evolutionary implication of protein secondary structure among archaea and bacteria. *Internet Journal of Genomics and Proteomics* 4.

Chellapandi P, Prabaharan D and Uma L (2008) A preliminary study on co-digestion of ossein factory waste for methane production. *EurAsian Journal of Bioscience* 2: 110–114.

Chellapandi P, Prabaharan D and Uma L (2009b) Evaluation of methanogenic activity of biogas plant slurry for monitoring anaerobic co-digestion of ossein factory waste and cyanobacterial biomass. *Applied Biochemistry and Biotechnology* 162: 524–535.

Chellapandi P, Suman LS and Sivaramakrishnan S (2007b) Biomethanation of fungal predigested caster seed cake in acclimatized seed. *Biotechnology, An Indian Journal* 1: 19–24.

Chistoserdova L, Vorholt JA and Lidstrom ME (2005) A genomic view of methane oxidation by aerobic bacteria and anaerobic archaea. *Genome Biology* 6: 208.

Ciaramella M, Cannio R, Moracci M, Pisani FM and Rossi M (1995) Molecular biology of extremophiles. *World Journal of Microbiology and Biotechnology* 11: 71–84.

Clementino MM, Fernandes CC, Vieira RP, Cardoso AM, Polycarpo CR and Martins OB (2007) Archaeal diversity in naturally occurring and impacted environments from a tropical region. *Journal of Applied Microbiology* 103: 141–151.

Craig G, Nevill-Manning, Thomas DW and Douglas LB (1998) Highly specific protein sequence motifs for genome analysis. *Proceedings of the National Academy of Sciences, USA* 95: 5865–5871.

Daubin V, Gouy M and Perriere G (2001) Bacterial molecular phylogeny using supertree approach. *Genome Informatics* 12: 155–164.

Daubin V, Gouy M and Perriere G (2002) A phylogenic approach to bacterial phylogeny: Evidence of core genes sharing a common history. *Genome Research* 12: 1080–1090.

David EG, Ross O, Olsen GJ and Woese CR (2000) An archaeal genomic signature. *Proceedings of the National Academy of Sciences*, USA 97: 3304–3308.

Dayhoff MO, Eck RV and Park CM (1972) A model of protein evolutionary change in proteins. In: Dayhoff MO (ed.) *Atlas of protein sequence and structure*. National Biomedical Research Foundation, Washington DC. Volume 5: 89–99.

Dayhoff MO, Schwartz RM and Orcutt BC (1978) A model of evolutionary change in proteins. Dayhoff MO (ed.) *Atlas of protein sequence and structure*. National Biomedical Research Foundation, Washington DC. Volume 5: 345–352.

Deppenmeier U (2004) The membrane-bound electron transport system of *Methanosarcina species. Journal of Bioenergetics and Biomembrane* 36: 55–64.

Deppenmeier U, Johann A, Hartsch T, Rainer R, Schmitz RA, Martinez-Arias R, Henne A, Wiezer A, Bäumer S, Jacobi C, Brüggemann H, Lienard T, Christmann A, Bömeke M, Steckel S, Bhattacharyya A, Lykidis A, Overbeek R, Klenk H-P, Gunsalus RP, Fritz H-J and Gottschalk G (2002) The genome of *Methanosarcina mazei*: evidence for lateral gene transfer between bacteria and archaea. *Journal of Molecular Microbiology and Biotechnology* 4: 453–461.

Deppenmeier U, Lienard T and Gottschalk G (1999) Novel reactions involved in energy conservation by methanogenic archaea. *FEBS Letters* 457: 291–297.

Deppenmeier U, Muller V and Gottschalk G (1996) Pathways of energy conservation in methanogenic Archaea. *Archives of Microbiology* 165: 149–163.

Dhillon A, Goswami S, Riley M, Teske A and Sogin M (2005) Domain evolution and functional diversification of sulfite reductases. *Astrobiology* 5: 18–29.

Ebenhoh O, Handorf T and Heinrich R (2005) A cross species comparison of metabolic network functions. *Genome Informatics* 16: 203–213.

Fitch WM and Margoliash E (1967) Construction of phylogenetic trees. *Science* 155: 279–284.

Gadelle D, File′e J, Buhler C and Forterre P (2003) Phylogenomics of type II DNA topoisomerases. *BioEssays* 25: 232–242.

Gascuel O and Steel M (2006) Neighbor-joining revealed. *Molecular Biology and Evolution* 23: 1997–2000.

Goldman N, Thorne JL and Jones DT (1996) Using evolutionary trees in protein secondary structure prediction and other comparative sequence analyses. *Journal of Molecular Biology* 263: 196–208.

Goldman N, Thorne JL and Jones DT (1998) Assessing the impact of secondary structure and solvent accessibility on protein evolution. *Genetics* 149: 445–458.

Hall TA (1999) BioEdit: A user-friendly biological sequence alignment editor and analysis program for Window 95/98/NT. *Nucleic Acids Symposium Series* 41: 95–98.

Hallam SJ, Girguis PR, Preston CM, Richardson PM and DeLong EF (2003) Identification of methyl coenzyme M reductase A (*mcr*A) genes associated with methane oxidizing archaea. *Applied and Environmental Microbiology* 69: 5483–5491.

Hallam SJ, Putnam N, Preston CM, Detter JC, Rokhsar D, Richardson PM and DeLong EF (2004) Reverse methanogenesis: Testing the hypothesis with environmental genomics. *Science* 305: 1457–1462.

Hong SH, Kim JS, Lee SY, In: Choi SS, Rih JK, Kim KS, Jeong H, Hur CG and Kim JJ (2004b) The genome sequence of the capnophilic rumen bacterium *Mannheimia succiniciproducens*. *Nature Biotechnology* 22: 1275–1281.

Hong SH, Kim TY and Lee SY (2004a) Phylogenetic analysis based on genome-scale metabolic pathway reaction content. *Applied Microbiology and Biotechnology* 65: 203-210.

Huang JY and Brutlag DL (2001) The eMotif Database. *Nucleic Acids Research* 29: 202–204.

Huelsenbeck JP and Ronquist F (2001) MRBAYES: Bayesian inference of Phylogenetic trees. *Bioinformatics* 17: 754-755.

Irving JA, Askew DJ and Whisstock JC (2004) Computational analysis of evolution and conservation in a protein superfamily. *Methods* 32: 73–92.

Johnson EF and Mukhopadhyay B (2007) A novel coenzyme F_{420} dependent sulfite reductase and a small sulfite reductase in methanogenic archaea. In: *Microbial sulfur metabolism*. Editied by Dahl C and Friedrich CG, Springer, pp.202–216.

Kalyuzhnaya MG, Bowerman S, Nercessian O, Lidstrom ME and Chistoserdova L (2005) Highly divergent genes for methanopterin-linked C_1 transfer reactions in Lake Washington assessed via metagenomic analysis and mRNA detection. *Applied and Environmental Microbiology* 71: 8846–8854.

Karthigeyan C, Sivaramakrishnan S and Chellapandi P (2007) Phylogenomic analysis of archaeal domain. *Bioinformatics Trends* 2: 37–55.

Klein M, Friedrich M, Roger AJ, Hugenholtz P, Fishbain S, Abicht H, Blackall LL, Stahl DA and Wagner M (2001) Multiple lateral transfers of dissimilatory sulfite reductase genes between major lineages of sulfate-reducing prokaryotes. *Journal of Bacteriology* 183: 6028–6035.

Kumar S, Tamura K and Nei M (2004) MEGA 3: Integrated software for Molecular Evolutionary Genetics Analysis and Sequence Alignment. *Briefings in Bioinformatics* 5: 150–163.

Lake JA, Clark MW, Henderson E, Fay SP, Oakes M, Scheinman A, Thornber JP and Mah RA (1985) Eubacteria halobacteria and the origin of photosynthesis: The photocytes. *Proceedings of the National Academy of Sciences*, USA 82: 3716–3720.

Masinovsky Z, Lozovaya GI and Sivash AA (1992) Some aspects of the early evolution of photosynthesis. *Advance Space Research* 12: 199–205.

Mattanovich D and Borth N (2006) Applications of cell sorting in biotechnology. *Microbial Cell Factories* 5: 12.

Miyazaki S, Kuroda Y and Yokoyama S (2002) Characterization and prediction of linker sequences of multi-domain proteins by neural network. *Journal of Structural and Functional Genomics* 2: 37–51.

Muller V, Spanheimer R and Santos H (2005) Stress response by solute accumulation in archaea. *Current Opinion in Microbiology* 8: 729–736.

Naylor GJP and Brown WM (1997) Structural biology and phylogeny estimation. *Nature* 388: 527–528

Nelson KE, Clayton RA, Gill SR, Gwinn ML, Dodson RJ, Haft DH, Hickey EK, Peterson JD, Nelson WC, Ketchum KA, McDonald L, Utterback TR, Malek JA, Linher KD, Garrett MM, Stewart AM, Cotton MD, Pratt MS, Phillips CA, Richardson D, Heidelberg J, Sutton GG, Fleischmann RD, Eisen JA, White O, Salzberg SL, Smith HO, Venter JC and Fraser CM (1999) Evidence for lateral gene transfer between Archaea and bacteria from genome sequence of *Thermotoga maritima*. *Nature* 399: 323–329.

Novichkov PS, Omelchenko MV, Gelfand MS, Mironov AA, Wolf YI and Koonin EV (2004) Genome-wide molecular clock and horizontal gene transfer in bacterial evolution. *Journal of Bacteriology* 186: 6575–6585.

Pagel M (1999) Inferring the historical patterns of biological evolution. *Nature* 401: 877–884.

Patel CN and Chellapandi P (2008) Anaerobic digestion of cotton seed cake using developed mixed consortia. *Electronic Journal of Environment, Agriculture and Food Chemistry* 7: 3035–3046.

Ranea JAG, Sillero A, Thornton JM and Orengo CA (2006) Protein Superfamily Evolution and the Last Universal Common Ancestor (LUCA). *Journal of Molecular Evolution* 513: 513–525.

Saitou N and Nei M (1987) The neighbor-joining method: A new method for reconstructing phylogenetic trees. *Molecular Biology and Evolution* 4: 406–425.

Schiraldi C, Giuliano M and De Rosa M (2002) Perspectives on biotechnological applications of archaea. *Archaea* 1:75–86.

Sowers KR and Gunsalus RP (1995) Halotolerance in *Methanosarcina* sp.: Role of N(sup(epsilon))-acetyl-(beta)-lysine, (alpha)-glutamate, glycine betaine, and K(sup+) as compatible solutes for osmotic adaptation. *Applied and Environmental Microbiology* 61: 4382–4388.

Tajima F (1989) Statistical methods to test for nucleotide mutation hypothesis by DNA polymorphism. *Genetics* 123: 585–595.

Thompson JD, Gibson TJ, Plewniak F, Jeanmougin F and Higgins DG (1997) The ClustalX windows interface: Flexible strategies for multiple sequence alignment aided by quality analysis tools. *Nucleic Acids Research* 24: 4876–4882.

Thorne JL, Goldman N and Jones DT (1996) Combining protein evolution and secondary structure. *Molecular Biology of Evolution* 13: 666–673.

Todd AE, Orengo CE and Thornton JM (2001) Evolution of function in protein superfamilies, from a structural perspective. *Journal of Molecular Biology* 307: 1113–1143.

Watson GMF, Yu JP and Tabita F R (1999) Unusual ribulose 1,5-bisphosphate carboxylase/oxygenase of anoxic archaea. *Journal of Bacteriology* 181: 1569–1575.

Webb BM, Liu JS and Lawrence CE (2002) BALSA: Bayesian Algorithm for Local Sequence Alignment. *Nucleic Acids Research* 30: 1268–1277.

Wolf YI, Rogozin IB, Grishin NV, Tatusov RL and Koonin EV (2001) Genome trees constructed using five different approaches suggest new major bacterial clades. *BMC Evolutionary Biology* 1: 8–29.

Yi Z, Vitek O, Qasim MA, Lu SM, Lu W, Ranjbar M, Li J, Laskowski MC, Bailey-Kellogg C and Laskowski M Jr (2006) Functional evolution within a protein superfamily. *Proteins* 63: 697–708.

In: Biochemistry Research Updates
Editor: Simon J. Baginski

ISBN 978-1-61209-700-8

Chapter 7

APTAMERS: THE NEW BIORECOGNITION ELEMENT FOR PROTEOMIC BIOSENSING

Mònica Mir*

Nanobioengineering Laboratory, Institute for Bioengineering of Catalonia (IBEC), Barcelona Science Park, Centro de Investigación Biomédica en Red de Bioingeniería, Biomateriales y Nanomedicina, (CIBER-BBN), Spain

ABSTRACT

Aptamers are single stranded artificial nucleic acid ligands that can be generated against almost any kind of target, such as ions, metabolites aminoacids, drugs, toxins, proteins or whole cells. They are isolated from combinatorial libraries of synthetic nucleic acids by an iterative process of adsorption, recovery and amplification, know as SELEX (Systematic Evolution of Ligands by EXponential enrichment) process.

Aptamers, the nucleic acid equivalent to antibodies, are easy to synthesise, is not required the use of animals for its synthesis, for this reason it can be developed again toxins and small molecules that do not produce immune response in animals and can be tuned for affinity in closer to assay conditions permitting recognition out of the physiological state. So, aptamers posses numerous advantages that make them preferred candidates as biorecognition elements.

In view of the advantages and simple structure of aptamers, they have been used in a wide range of applications such as therapeutics, diagnosis, chromatography, environmental detection, among other.

Keywords: Aptamer, biosensor, aptasensor, SELEX, proteomic, electrochemistry, beacon, aptazyme.

* Corresponding author; Baldiri Reixac, 10-12, 08028 Barcelona, Spain, Phone; +34934037178, Fax; +34934037181. E-mail: mmir@ibec.pcb.ub.es.

1. Aptamers

The term aptamer defines small single stranded synthetic oligonucleotides (40-100 bases) isolated from random combinatorial libraries by exponential enrichment, coined SELEX (Selective Expansion of Ligands by Exponential enrichment), which provides a unique three-dimensional structure for the high specific binding of their target. Aptamers can specifically recognise and bind to virtually any kind of target including, ions (Smironov and Shafer, 2000), whole cell (Famulok et al., 2000), drugs (Gallego and Varani, 2001), toxins (Jayasena, 1999), low molecular weight ligands (Famulok, 1999), peptides (Wilson and Szostak, 1999) and proteins (McCauley et al., 2003).

The affinity of aptamers depends on the kind of target. Aptamers against small molecules have affinities in the micromolar range, as in the case of dopamine, 2.8 μM (Lorsch and Szostak, 1994) or ATP, 6 μM (Kiga et al., 1998). Affinities in the nanomolar and subnanomolar range are found against some proteins, as Vascular endothelial growth factor (VEGF) with an affinity of 100 pM (Ruckman et al., 1998), and keratinocyte growth factor, 1pM (Pagratis et al., 1997).

High affinity does not always mean high specificity but in the case of aptamers it does. Using aptamers can be discriminated targets on the basis of subtle structural differences such as the presence or absence of a hydroxyl group (Sassanfar and Szostak, 1993) or a methyl (Haller and Sarnow, 1997). Furthermore, enzymes with similar activity can be distinguished, it is demonstrated with the detection of α-thrombin and γ-thrombin (Paborsky et al., 1993). More impressive is the ability of aptamers to distinguish between enantiomers, as was reported by Geiger (Geiger et al., 1996), when L and D-arginine were discriminated. Aptamers are known for their remarkable specificity. The high degree of specificity often seen in aptamers, sometimes even better than antibodies (Jenison et al., 1994), is a result of the selective process in the SELEX.

The aptamer generation emerged from the experimentation of three independent groups, all of whom published their work in 1990. First, the Joyce group (Robertson and Joyce, 1990) was looking for a new enzymatic activity of RNA, They use in vitro mutation, selection and amplification in order to isolate the enzymatic RNA, which became the basis for the current in vitro selection of aptamers.

Then, the Gold group was trying to identify in vitro the sequences of T4 DNA polymerase (Tuerk and Gold, 1990). The library was based on the natural structure but with the eight-loop randomised, they baptised the process of in vitro selection as SELEX, this process was able to identify the natural target of the enzyme as the predominant. The authors patented their process. A month later, Szostak and Ellington reported the use of in vitro selection to isolate molecules with specific ligand binding activities (Ellington and Szostak, 1990). They create a library with 100 nucleotides of randomised sequences, related to any known oligonucleotide sequence. The targets had no previously identified nucleic acids ligand. This group coined the term aptamer, which comes from the Latin *aptus* meaning to fit.

2. Types of Aptamers

Aptamers had been initially developed with single stranded DNA or RNA molecules for a direct interaction with the target. However, lately appeared other kind of aptamer conformations or chemical modifications of the DNA structure in order to improve the features of these bioreceptor molecules, enlarging aptamers family.

2.1. Spiegelmers

Spiegelmers comes from the German word spiegel, which means mirror in English language. Due to this kind of molecules are the mirror image of aptamers. In nature, nucleic acids are composed of D-nucleotides, and nucleases, the nucleic acid degrading enzyme, are L-amino acids. Consequently, the structure of nucleases is also inherently chiral, thus resulting in stereospecific substrate recognition. Hence, these enzymes only accept substrate molecules in the correct chiral configuration. Therefore L- oligonucleotide (spiegelmers) escapes from enzymatic recognition and subsequent degradation.

However, due to this same principle nature developed no enzymatic activity to amplify such mirror image, thus it cannot be directly obtained by employing the SELEX process.

The first functional spiegelmers were designed for binding to the small molecules arginine and adenosine (Klussmann et al., 1996, Nolte et al., 1996).

2.2. Beacons

Beacons are a single strand oligonucleotide designed to reveals the ligand interaction through a change in the 3-D structure of this strand. The structure used form a stem-loop structure with a fluorophore on one end of the stem and a quencher on the other end, so in the absence of the complementary target, the fluorophore is quenched by energy transfer to the quencher in close proximity. The loop sequence is complementary to the target nucleic acid. In the presence of the target, a duplex is formed between the molecular beacon and the target When this occurs, the stem is broken, separating the fluorophore from the quencher. Consequently, the fluorophore is no longer quenched and fluorescence is restored.

Aptamer beacons are based on the same hypothesis. An additional sequence is added to the aptamer forming a stable stem-loop and destabilises the binding structure. Then the aptamer beacon exists in a quenched stem-loop structure in the absence of a target molecule. Formation of the aptamer beacon-target complex would alter the equilibrium between quenched, non-native and unquenched, native structures generating a change in observed fluorescence intensity. The first aptamer beacon had been developed by Yamamoto (Yamamoto et al., 2000) to detect Tat protein of HIV-1.

Electrochemical aptamer beacon was also developed based on the ferrocene labelled DNA beacon demonstrated by Plaxco (Fan et al., 2003). The aptamer beacon 3-D change of the structure when interact with the target is transduced by observing the significant change in the distance-dependent voltamperometric behaviour of the redox label (Mir and Katakis, 2007) (Figure 1).

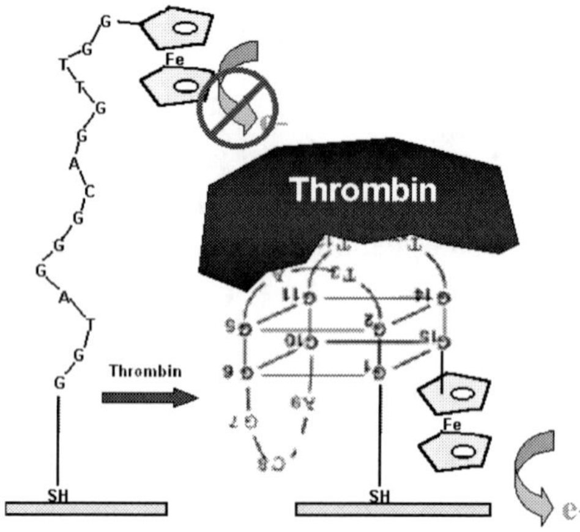

Mir and Katakis, 2007.

Figure 1. Schematic drawing of signalling mechanism of electrochemical thrombin aptamer beacon.

2.3. Circular DNA Aptamers

Circular DNA aptamers or captamers combines aptameric recognition with multitasking nucleic acid functions through the use of modular engineering principles derived from DNA nanotechnology.

DNA captamers have been demonstrated in 2-, 3- and 4-headed versions, focussing upon the pleiotropic activity that can be obtained by combining different functions together (Di Giusto and King, 2004). These scaffolds provide a framework for spatial orientation, allowing multiple binding activities to be combined into single molecules, their circularity imparting both enhanced thermal stability and exonuclease resistance (Figure 2).

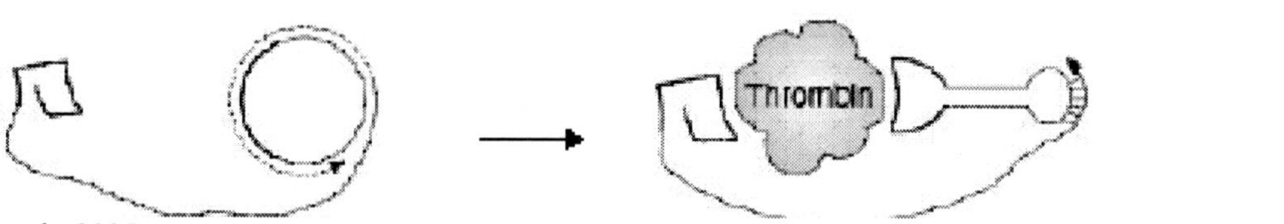

Di Giusto et al., 2005.

Figure 2. Captamers for the detection of human thrombin by proximity extension.

2.4. Aptazymes

The word ribozyme or deoxyribozyme comes from ribonucleic or desoxiribonucleic acid and enzyme, these are also called RNA/DNA enzyme or catalytic RNA/DNA. The ribozyme or desoxiribozyme is an RNA or DNA molecule that catalyses a chemical reaction. Many natural ribozymes catalyse either their own cleavage or the cleavage of other RNAs, but synthetic ribozymes have also been found that function as polymerases, dephosphorylases and restriction endoribonucleases (Cech et al., 1997). The Nobel Prize in Chemistry in 1989 was awarded to Thomas R. Cech and Sidney Altman for their discovery of catalytic properties of RNA.

The ribozyme/deoxyribozyme linked to an aptamer is called allosteric ribozyme/ deoxyribozyme or aptazyme. This system undergoes alteration of the catalytic activity in the presence of a molecule binding to a receptor site, the aptamer. In this way, aptazymes directly transduce molecular recognition into a quantifiable catalytic event. Aptazymes have been adapted to chip array that discriminated between a varieties of metabolites (Seetharaman et al., 2001). Furthermore, this kind of aptamers were used to construct catalytic molecular beacons (Stojanovic et al., 2000), which amplify the recognition event through cleavage of a doubly-labelled fluorescent substrate.

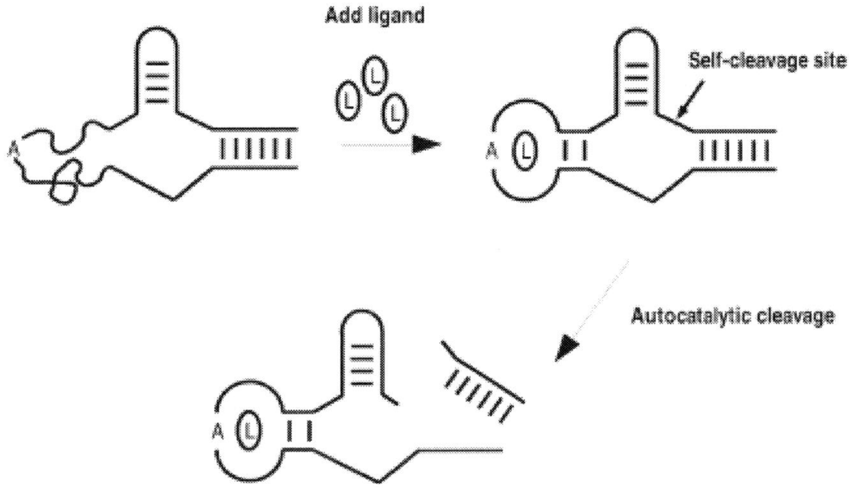

Figure 3. This figure (Koizumi et al., 1999), illustrates a simple self-cleaving aptazyme. The structure consists of a selfcleaving hammerhead ribozyme with an internal stem replaced by an aptamer sequence ("A"). Upon binding the ligand ("L"), the aptamer becomes well-ordered, reconstituting the disrupted hammerhead stem. Once the stem forms, the ribozyme can adopt an active conformation and undergo self-cleavage.

2.5. Chemical Modifications of DNA Chain

Usually chemical modifications of DNA molecules have been carried out in order to increase the stability of these molecules.

Peptide nucleic acids (PNA) are synthetic molecules similar to DNA, in which the natural nucleobases are appended to a polyamide backbone (Figure 4). The lack of a negative charge along the PNA chain contributes to very high affinity for the complementary DNA strand. PNA have been widely applied in DNA biosensors (Egholm et al., 2002) and additionally have been adopted for aptamer sensors (Datta and Armitage, 2001).

Another chemical alternative to the ribose sugar backbone is the (3',2')-α-L-threose nucleic acid (TNA), with a four-carbon sugar backbone instead of two-carbon sugar backbone in DNA (Figure 4).

TNA is capable of antiparallel, Watson-Crick base-pairing with complementary DNA, RNA, and TNA oligonucleotides (Schoning et al., 2000). This property is remarkable, given that the TNA repeat unit is one atom shorter than that of DNA or RNA. Furthermore protein polymerases can incorporate TNA monomers by primer extension on DNA templates (Chaput and Szostak, 2003), and can also use TNA to template DNA. A system for selecting functional TNA structures from libraries of random molecules was recently devised (Ichida et al., 2005).

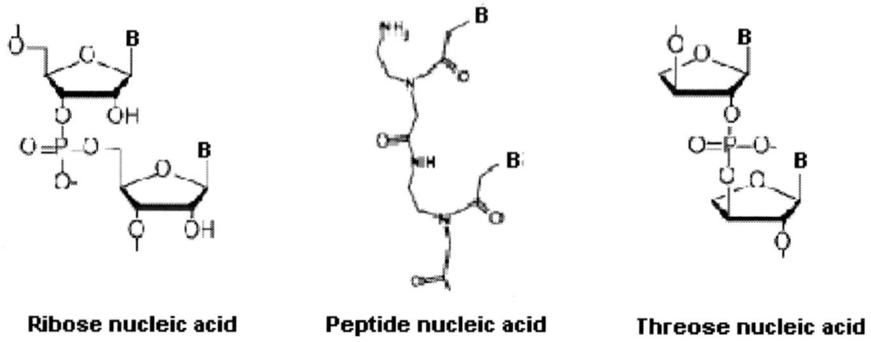

Figure 4. Chemical structure of PNA, TNA and RNA.

Another way to modify oligonucleotide structure is to introduce sulphur in the heterocycle by using phosphorothioate oligonucleotides (figure 5). Incorporation of a thioether into the 2'-substitutent of ssDNA has resulted in improved binding to human serum albumin (Prakash et al., 2002). Kawakami and Sugimoto (Kawakami et al., 2000), carried out an in vitro selection of aptamers from an RNA library with phosphorothioate linkages instead of normal phosphodiesters.

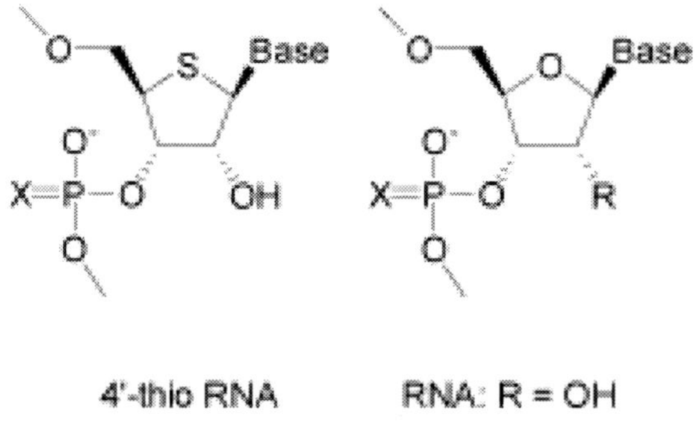

Dande et al., 2006.

Figure 5. Chemical structure of thio-RNA and RNA.

2.6. Peptides Aptamers

Peptide aptamers (paptamers, thioredoxin-insert proteins (TIPs), perturbagen) are proteins that are designed to interfere with other protein inside cells. These are also coined aptamers because they were developed through a combinatorial library to bind with proteins.

They consist of a variable peptide loop attached at both ends to a protein scaffold. The variable loop length is typically comprised of 10 to 20 amino acids, and the scaffold may be any protein, which has good solubility and compactly folding. Currently, the bacterial protein Thioredoxin-A is the most used scaffold protein, the variable loop being inserted within the reducing active site, which is a -Cys-Gly-Pro-Cys- (Cysteine-Glycine-Prolamin-Cysteine) loop in the wild protein. Selection of Ligand Regulated Peptide Aptamers (LiRPAs) has been demonstrated by displaying 7 amino acid peptides from a novel scaffold protein based on the trimeric FKBP-rapamycin-FRB structure (Binkowski et al., 2005).

2.7. Anticalins

Anticalins are ligand–binding proteins constructed on the basis of the lipocalins proteins as scaffolds. The principal element of this protein architecture is the β-barrel structure of eight antiparallel strands, which support four loops at its open end. These loops form the natural binding site of the lipocalins and can be reshaped in vitro by extensive amino acid replacement, creating novel binding specificities. The bilin-binding protein (BBP)-was employed as a model system for the preparation of a random library with 16 selectively mutagenised residues. Several anticalins have been selected from this library, exhibiting binding activity for compounds like fluorescein or digoxigenin (Schlehuber and Skerra, 2005).

3. SELEX

SELEX (Systematic Evolution of Ligands by EXponential enrichment) is the technique for selecting and isolating functional synthetic nucleic acids specific for a target interaction. This process is an in vitro screening of large random libraries of oligonucleotides by an iterative process of adsorption, recovery and amplification of the oligonucleotide sequences (Figure 6) (Kopylov and Spiridonova, 2000). A public database (HTPSELEX) provides access to primary and derived data from high-throughput SELEX experiments, this data base has free access in ftp://ftp.isrec.isb-sib.ch/pub/databases/htpselex/ and http://www.isrec.isb-sib.ch/htpselex.

The starting round of a typical SELEX requires the synthesis around 10^{14}-10^{15} of random single stranded individual oligonucleotide sequences. The ssDNA chemically synthesised is amplified by PCR in order to generate the corresponding double-stranded DNA (dsDNA). Depending on the kind of aptamer that is wanted, the library is converted into ssDNA by strand separation or into RNA by in vitro transcription. The ss nucleic acids are incubated with the target of interest. The nucleic acids that adopt a conformation that allow them to bind with the target are separated by filtration through a nitrocellulose column or in the case of small molecules by affinity column. The aptamer-target duplex is eluted and amplified by RT-PCR, in the case of RNA aptamers or PCR in the case of DNA aptamers, to have a lower and enriched pool of molecules with which to repeat the cycle. The process requires several iterations of the cycle, usually around 8-15 cycles, carried out under increasingly stringent conditions to achieve an aptamer of high affinity and specificity. SELEX takes a long time, around one month, nevertheless this process has been automated (Cox et al., 2002) and miniaturised in a microfluidic SELEX prototype by Hybarger (Hybarger et al., 2006), reducing the time required.

Innovation around this process has improved the binding efficiency of aptamers. One of the examples is PhotoSELEX, this technique substitutes one of the bases in the nucleic acid library with a photoactivatable nucleotide. In the presence of light this modified nucleotide is activated forming a photoinduced covalent bond to the target molecule, which provides an enhancement in the aptamer affinity (Golden et al., 2000).

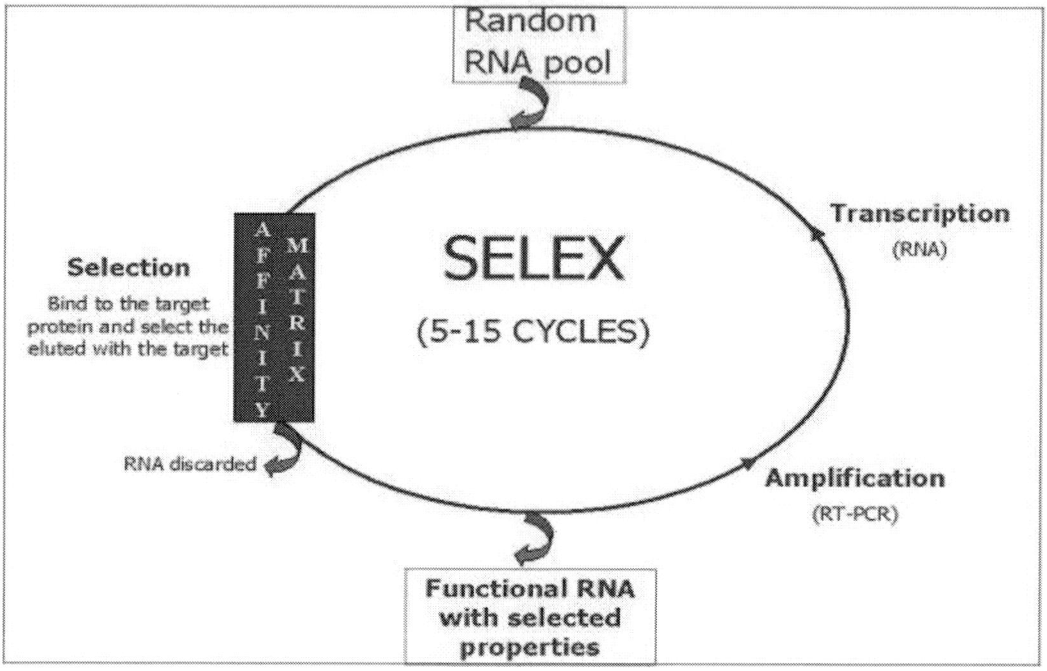

Figure 6. Scheme of SELEX process.

Kinetic homogeneous affinity method termed non-equilibrium capillary electrophoresis of equilibrium mixtures (NECEEM), demonstrated an efficiency, which is at least two orders of magnitude higher than SELEX (Berezovski et al., 2003). The high efficiency of NECEEM allows with just one SELEX round the selection of the appropriate aptamer sequence. The concept of NECEEM is incubate the DNA library with the target protein in order to form the equilibrium mixture that consists of free DNA, DNA-protein complexes and free protein. This mixture is introduced into the capillary and a high voltage is applied. The equilibrium fraction of DNA·Protein is separated from the equilibrium fraction of DNA and protein free by gel-free capillary electrophoresis under non-equilibrium conditions.

Counter-SELEX is a highly specific SELEX method that discards ligands that have ability to bind the target as well as closely related structural analogous to the target (Jenison et al., 1994). During selection, the population of aptamers bound to the target is subjected to affinity elution with structural analogous and the sequences eluted are discarded. Thus, the counter-SELEX strategy could be a valuable tool in identifying aptamers aimed at a unique target in a complex mixture. Furthermore, in some diagnostic applications of small molecule targets, it may be important to measure the analyte as well certain structural variants of the analyte. In that case, aptamers retained on the target could be specifically eluted with the structural analogous to select the species that do not discriminate them.

FluMag-SELEX, incorporates fluorescent labels for DNA quantification and use of magnetic beads for target immobilisation in the SELEX process. The use of magnetic beads enables easy handling, use of very small amounts of target for the aptamer selection and a rapid and efficient separation of bound and unbound molecules (Stoltenburg et al., 2005).

Another example is the post-SELEX process in which a modification of the ligands is carried out after the traditional SELEX process for the enhanced of the binding activity and other properties such as improved pharmacokinetics, tissue distribution, and nuclease resistance (Green et al., 1995).

4. Interactions between Aptamer-Target

The principal interactions between protein and aptamers are van der Waals, hydrogen bonds and water mediated contacts (Mandelgutfreund et al., 1995). About 30% of all interactions are non-specific bonds, mediated through van der Waals, however they are extremely important for the stability of the complex. The majorities of the specific interactions involve hydrogen bonds (Schwabe, 1997) and demonstrate some amino acid-base preferences, nevertheless is not generalised. Cowan (Cowan et al., 2000), demonstrated that 80% of the binding energy was contributed by hydrogen bounds. The relative contributions of this and other forces, such as interaction with aromatic rings and van der Waals forces has been also reported (Hermann and Pate, 2000). Water molecules also facilitate the contact by filling the empty spaces at the protein-aptamer interface and have and important role in complex formation (Tuerk and Gold, 1990).

5. Aptamers Versus Antibodies

Microarray technology provides a tool to analyse a large number of molecules simultaneously. Antibody-based microarrays are being developed for the analysis of a large number of proteins (Dove, 1999). Randox laboratories launched the first commercial protein array in 2003, nevertheless numerous advantages makes aptamers a preferable candidate as biorecognition element in proteomics arrays.

One of the antibodies inconvenient is its identification process depending on an animal host. The expression process is further complicated when the targets are toxins or low molecular weight molecules, which trigger minimal immunogenic response. The antibody's performance is lot specific, and although monoclonal preparations have reduced the variability, the preparation still requires a lengthy purification protocol. Furthermore, antibodies are large and complex molecules that are sensitive to acid and basic exposure and temperature, which highly limit their use in reusable sensors.

Conversely, aptamers are designed in vitro. Identification of aptamers utilises automated oligonucleotide synthesis and screening systems. The binding specificity, affinity and stability of aptamers can be improved by molecular evolution techniques or by rational design (Bacher and Ellington, 1998). Additionally the identification process can be changed to obtain an aptamer that interacts with specific regions of the target or under differing binding condition. Once identified aptamers are readily produced with extremely high reproducibility and purity using established chemical synthesis processing. Aptamers can also be easily labelled with reporter molecule without affecting their affinity. An aptamer sensor can support operation in a wide variety of sample matrixes including non-physiological buffers and temperature conditions that would denature typical antibody formulations. Kinetics parameters in aptamer interaction can be changed on demand. Aptamers can undergo repeated cycles of denaturation and renaturation without damaging their structure. As well, they can be transported at ambient temperature, are stables to long-term storage and can be subjected to numerous freeze and thaw cycles.

6. Applications

The first application of aptamers was therapeutics. Their advantages of small size, quick elimination, low production cost, biocompatibility, biodegradability and no cross reactivity with antibody binding receptors makes them a good candidates for therapeutic applications. However, because blood is rich in nucleases, an appropriate modification of aptamers should be done in order to avoid nuclease attack. Aptamers can block the receptors and inhibit protein activities with high affinity and specificity and it can validate targets and screen drugs.

Anti-thrombin aptamer has been widely used to inhibit physiologic blood coagulation events (Bock et al., 1992, Tasset et al., 1997). Other examples in the literature are an aptamer against C5 developed by Biesecker (Biesecker et al., 1999) to inhibit human serum hemolytic activity and anti-tenascin C to inhibit the activity of trypsin like protease of non-structured protein (NS3) for the control of polyprotein secretin in HCV (hepatitis C virus) (Hicke et al., 2001).

Aptamers has also go into the separation chemistry field. Affinity chromatography is one of the most powerful techniques among the purification methods for rapid selection and separation of biomolecules from complex mixtures. Antibodies are useful receptor molecules in stationary phases due to their high affinity and selectivity. Nevertheless a number of constraints explained above limit their application in chromatography. Their large molecular size limits the surface loading and antibodies can be denatured and degraded. This is further compound by the difficulty in obtaining antibodies for small molecules or toxins. As a result of antibody limitations, aptamers have become a promising tool for affinity chromatography. A high affinity aptamer has been used by Romig (Romig et al., 1999) for purification of L-selectin. Also adenosine and analogues were successfully purified, using aptamers supported on a stationary phase (Deng et al., 2001). Furthermore aptamers have been used with the same aim in the separation of amino acids and polycyclic aromatic hydrocarbons (Kotia et al., 2000).

In view of the advantages and simple structure of aptamers, they have been used as biorecognition elements in the biosensor development (more details in next section). Due to the wide range of application of biosensors it has opened the possibility to apply aptamers in a good deal of fields such as environmental (Stratis-Cullum et al., 2005), industrial (Mann et al., 2005), defence (Paddle, 2003, Kiel et al., 2004), or diagnosis (Dill and McShea, 2005, Gander and Brody, 2005).

7. APTASENSORS

DNA biosensors are the subject of intense research and development, which has yielded several commercial platforms. The results of the human genome project increase further the possible applications of these biosensors. However, the DNA sequence information provides a static picture of the possibilities in which the cell might use its proteins and the cell life is a dynamic process. Therefore, an important value of the genetic information lies in its ability to predict the outcome, the phenotype. Nevertheless, the translation of RNA leads to a differential protein expression and is present a posttranslational modification of translated proteins that determine functional forms, so at the DNA level is difficult to asses the functionality of gene product. Due to the discrepancy between the genotype and the phenotype it would be wealthier detect directly proteins to better understand the cellular function and for early detection of diseases and monitoring the progression of existing diseases or the effect produced by a drug in the organism

Aptasensor is the word to define biosensors using aptamers as the biorecognition element. The first examples on aptasensors date back to 1996, when Davis and co-workers, investigated the feasibility of using aptamers for this purpose (Davis et al., 1996). They described a model system consisting of human neutrophil elastase (HNE) coated on beads that interact with fluorescent tagged aptamers. Eight months later two new examples of aptasensors appeared, a radiolabelled aptamer was used for the detection of different protein kinase C isozymes (Ellington and Szostak, 1990, Conrad and Ellington, 1996) and Drolet, described an enzyme-linked sandwich assay that uses a SELEX-derived oligonucleotide fluorescently labelled (Romig et al., 1999).

Afterwards, hundred of aptasensors have been reported using different kind of transductors (Table 1). Fluorescence labels have been widely used, L-adenosine detection was transduced by evanescent wave-induced fluorescence (Kleinjung et al., 1998). Thrombin detection was performed with aptamer-fluorescence-quenching pairs (Hamaguchi et al., 2001), or by a competitive assay of fluorescently-labelled thrombin with thrombin on an aptamer modified optical fibre (Kleinjung et al., 1998). Optical analysis has been also used to detect the catalytic activation of thrombin through the dissociation of a thrombin-aptamer complex (Pavlov et al., 2005).

For detection of non-labelled aptamers, quartz crystal microbalance (QCM) (Furtado et al., 1999, Liss et al., 2002, Pavlov et al., 2004) and surface plasmon resonance (SPR) (Fukusaki et al., 2001, Rhodes et al., 2001, Okumoto et al., 2002) are two widely described techniques. A label-free protein detection strategy using a microfabricated cantilever-based sensor that was functionalised with aptamers in order to act as receptor of Taq DNA polymerase was described (Savran et al., 2004). Atomic force microscopy (AFM) has been used to measure the specific interaction between a protein immunoglobulin E (IgE) and an aptamer (Jiang et al., 2003). As well, a low-wave biosensor array was designed by coupling aptamers to detect thrombin and HIV-1 Rev peptide (Schlensog et al., 2004).

Table 1. Resume of high sensitive transducers in aptamer based sensors

Transducer	Detection Limit	Detection Technique	Analyte	Label	Reference
Optical	100 nM	Surface plasmon resonance	Oligoadenylate sintetase	Label-free	(Tombelli, 2005)
	5 pM	Evanescent wave induced fluorescence anisotropy	Thrombin	Fluorophore	(Potyrailo, 1998)
	40 fM	Fluorescent proximity ligation assay	Cytokine platelet derived growth factor	Fluorophore	(Fredriksson, 2002)
Piezoelectric	500 pM	Quartz crystal microbalance	IgE	Label-free	(Liss et al., 2002)
Electrochemistry	500 pM	Differential pulse voltammetry	Thrombin	Ferrocene	(Radi et al., 2006)
	100 pM	Impedance spectroscopy	IgE	Label-free	(Xu et al., 2001)
	500 fM	Square wave stripping voltammetry	Thrombin	Semiconductor nanocrystal	(Hansen, 2006)

More recently examples in electrochemical aptasensors have been reported. Ikebukuro, developed a biosensor using a two aptamer sandwich configuration (Ikebukuro et al., 2005). Impedance spectroscopy was used to detect interaction of an aptamer with lysozyme and thrombin (Rodriguez et al., 2005, Radi et al., 2006).

Mir reported different aptasensor configurations to detect thrombin electrochemically (Mir et al., 2006). Applications of an electrochemical beacon have also been published. The first example was reported by Bang, where methylene blue was intercalated into the beacon sequence as an electrochemical label, which is released when the protein interacts with the aptamer (Bang et al., 2005). A switch on ferrocene labelled aptabeacon for real detection of thrombin was published (Mir and Katakis, 2007). An electronic aptamer-based beacon was developed for cocaine detection (Baker et al., 2006).

Conclusion

This review provides a snapshot on the properties of aptamers, the advantages that these bioreceptors offer and the applicability of this kind of molecules. The versatility demonstrated by these single stranded DNA to bind strongly and specifically to targets other than the obvious complementary DNA sequence allows its promising application in different fields such as therapeutics, separation techniques, and biosensors.

References

Bacher JM, Ellington AD. Nucleic acid selection as a tool for drug discovery. *Drug Discovery Today,* 1998;3(6),265-273.

Baker BR, Lai RY, Wood MS, Doctor EH, Heeger AJ, Plaxco KW. An Electronic, Aptamer-Based Small-Molecule Sensor for the Rapid, Label-Free Detection of Cocaine in Adulterated Samples and Biological Fluids. *J. Am. Chem. Soc*, 2006;128(10),3138-3139.

Bang GS, Cho S, Kim BG. A novel electrochemical detection method for aptamer biosensors. *Biosensors and Bioelectronics,* 2005;21(6),863-870.

Berezovski M, Nutiu R, Li Y, Krylov SN. Affinity Analysis of a Protein-Aptamer Complex Using Nonequilibrium Capillary Electrophoresis of Equilibrium Mixtures. *Anal. Chem,* 2003;75(6),1382-1386.

Biesecker G, Dihel L, Enney K, Bendele RA. Derivation of RNA aptamer inhibitors of human complement C5. *Immunopharmacology,* 1999;42,219-324.

Binkowski BF, Miller RA, Belshaw PJ. Ligand-Regulated Peptides: A General Approach for Modulating Protein-Peptide Interactions with Small Molecules. *Chemistry and Biology*, 2005;12,847-855.

Bock LC, Griffin LC, Lathan JA, Vermaas EH, Toole JJ. Selection of single-stranded DNA molecules that bind and inhibit human thrombin. *Nature*, 1992;355,564-568.

Cech TR, Arthur ZJ, Michael B. RNA ribozyme polymerases, dephosphorylases, restriction endoribonucleases and methods. *Biotechnology Advances*, 1997;15(3-4),666-669.

Chaput JC, Ichida JK, Szostak JW. DNA Polymerase-Mediated DNA Synthesis on a TNA Template. *J. Am. Chem. Soc,* 2003;125(4),856-857.

Chaput JC, Szostak JW. TNA Synthesis by DNA Polymerases. *J. Am. Chem. Soc,* 2003; 125,9274-9275.

Conrad R, Ellington AD. Detecting Immobilized Protein Kinase C Isozymes with RNA Aptamers. *Anal. Biochem*, 1996;242,261-265.

Cowan JA, Ohyama T, D. W, Natarajan K. Recognition of a cognate RNA aptamer by neomycin B: quantitative evaluation of hydrogen bonding and electrostatic interactions. *Nucleic acid Res*, 2000;28,2935-2942.

Cox JC, Rajendran M, Riedel T, Davidson EA, Sooter LJ, ′Bayer TS, et al. Automated Acquisition of Aptamer Sequences. Combinatorial Chemistry and High Throughput Screening, 2002;5,289-299.

Dande P, Prakash TP, Sioufi N, Gaus H, Jarres R, Berdeja A, et al. Improving RNA interference in mammalian cells by 4 '-thio-modified small interfering RNA (siRNA): Effect on siRNA activity and nuclease stability when used in combination with 2 '-O-alkyl modifications. *J. Med. Chem*, 2006;49,1624-1634.

Datta B, Armitage BA. Hybridization of PNA to Structured DNA Targets: Quadruplex Invasion and the Overhang Effect. *J. Am. Chem. Soc,* 2001;123,9612-9619.

Davis KA, Abrams B, Y. L, Jayasena SD. Use of a high affinity DNA ligand in flow cytometry. *Nucleic acid research*, 1996;24,702-706.

Deng Q, German I, Buchaman D, Kenedy RT. Retention and Separation of Adenosine and Analogues by Affinity Chromatography with an Aptamer Stationary Phase. *Anal. Chem,* 2001;73,5415-5420.

Di Giusto DA, Wlassoff WA, Gooding JJ, Messerle BA, King GC. Proximity extension of circular DNA aptamers with real-time protein detection. *Nucleic Acids Res*, 2005;33,e64.

Di Giusto DG, King GC. Construction, Stability, and Activity of Multivalent Circular Anticoagulant Aptamers. *J. Biol. Chem,* 2004;279,46483-46489.

Dill K, McShea A. Recent advances in microarrays. *Drug Discovery Today: Technologies,* 2005; 2(3),261-266.

Dove A. Millennium mergers bring pipeline. *Nat. Biotechnol*, 1999;17,1151-1152.

Egholm M, Buchardt O, Nielsen PE, Berg RH. Peptide nucleic acids (PNA). Oligonucleotide analogs with an achiral peptide backbone. *J. Am. Chem. Soc,* 2002;114,1895-1897.

Ellington AD, Szostak JW. Invitro selection of RNA molecules that bind specific ligands. *NATURE,* 1990; 346(6287),818-822.

Famulok M. Oligonucleotide aptamers that recognize small molecules. *Current Opinion in Structural Biology,* 1999;9,324-329.

Famulok M, Mayer G, Blind M. Nucleic acid aptamers - From selection in vitro to applications in vivo. *Acc. Chem. Res,* 2000;33,591-599.

Fan C, Plaxco KW, Heeger AJ. Electrochemical interrogation of conformational changes as a reagentless method for the sequence-specific detection of DNA. *Proc. Natl. Assoc.* Sci., 2003;100,9134-9137.

Fredriksson S. Protein Detection Using Proximity-Dependent DNA Ligation Assays. *Nature Biotechnology*, 2002;20,473–477.

Fukusaki EI, Hasunuma T, Kajiyama SI, Okazawa A, Itoh TJ, Kobayashi A. SELEX for tubulin affords specific T-rich DNA aptamers. *Bioorganic and Medicinal Chemistry Letters,* 2001;1(11),2927-2930.

Furtado LM, Su H, Thomson M. Interactions of HIV-1 TAR RNA with Tat-Derived Peptides Discriminated by On-Line Acoustic Wave Detector. *Anal. Chem*, 1999;71,1167-1170.

Gallego J, Varani G. Targeting RNA with small-molecule drugs: Therapeutic promise and chemical challenges. *Acc. Chem. Res.*, 2001;34(10),836-843.

Gander TR, Brody EN. Photoaptamer chips for clinical diagnostics. *Expert Review of Molecular Diagnostics,* 2005; 5(1),1-3.

Geiger A, Burgstaller P, von der Eltz H, Roeder A, Famulok M. RNA aptamers that bind L-arginine with sub-micromolar dissociation constants and high enantioselectivity. *Nucleic Acids Res,* 1996;24,1029-1036.

Golden MC, Collins BD, Willis MC, Koch TH. Diagnostic potential of PhotoSELEX-evolved ssDNA aptamers. *J. Biotechnol.*, 2000;81(2-3),167-178.

Green LS, Jellinek D, Bell C, Beebe LA, Feistner BD, Gill SC, et al. Nuclease-resistant nucleic acid ligands to vascular permeability factor vascular endothelian growth factor. *Chem. Biol*, 1995;2(10), 683-695.

Haller AA, Sarnow P. In vitro selection of a 7-methyl-guanosine binding RNA that inhibits translation of capped mRNA molecules. *Proc. Natl. Acad. Sci. USA,* 1997; 94(16),8521-8527.

Hamaguchi N, Ellington A, Stanton M. Aptamer Beacons for the Direct Detection of Proteins. *Anal. Biochem*, 2001;294,126-131.

Hansen JA. Quantum-Dot/Aptamer-Based Ultrasensitive Multi-Analyte Electrochemical biosensor. *Journal of the American Chemical Society*, 2006;128,2228–2229.

Hermann T, Pate lDJ. Adaptive Recognition by Nucleic Acid Aptamers. *Science,* 2000;287,820-825.

Hicke BJ, Marion C, Chang YF, Gould T, Lynott CK, Parma D, et al. Tenascin-C Aptamers Are Generated Using Tumor Cells and Purified Protein. *J. Biol. Chem,* 2001;276(52), 48644-48654.

Hybarger G, Bynum J, Williams RF, Valdes JJ, Chambers JP. A microfluidic SELEX prototype. *Anal. Bioanal. Chem*, 2006;384(1),191-198.

Ichida JK, Zou K, Horhota A, Yu B, McLaughlin LW, Szostak JW. An in Vitro Selection System for TNA. *J. Am. Chem. Soc*, 2005;127,2802-2803.

Ikebukuro K, Kiyohara C, Sode K. Novel electrochemical sensor system for protein using the aptamers in sandwich manner. *Biosensors and Bioelectronics*, 2005;20,2168-2172.

Jayasena SD. Aptamers: An emerging class of molecules that rival antibodies in diagnostics. *Clin. Chem,* 1999;45(9),1628-1650.

Jenison RD, Gill SC, Pardi A, Polisky B. High-resolution molecular discrimination by RNA. *Science*, 1994;263,1425-1434.

Jiang Y, Zhu C, Ling L, Wan L, Fang X, Bai C. Specific Aptamer-Protein Interaction Studied by Atomic Force Microscopy. *Anal. Chem*, 2003;75,2112-2116.

Kawakami J, Tokitoh H, Tanabe Y, Sugimoto N. Specific recognition of the basic N peptide by an RNA. *Nucleic Acids Symposium Series,* 2000;49,353-354.

Kiel JL, Holwitt EA, Parker JE, Vivekananda J, Franz V. Nanoparticle labeled DNA capture elements for detection and identification of biological agents. P*roceedings of SPIE - The International Society for Optical Engineering*, 2004;5617,382-387.

Kiga D, Futamura Y, Sakamoto K, Yokoyama S. An RNA aptamer to the xanthine/guanine base with a distinctive mode of purine recognition. *Nucleic Acids Res*, 1998;26,1755-1760.

Kleinjung F, Klussmann S, Erdmann VA, Scheller FW, Bier FF. High-Affinity RNA as a Recognition Element in a Biosensor. *Anal. Chem,* 1998;70,328-331.

Klussmann S, Nolte A, Bald R, Erdmann VA, Furste JP. Mirror-image RNA that binds D-adenosine. *Nat. Biotechnol.*, 1996;14,1112-1115.

Koizumi M, Soukup GA, Kerr JN, Breaker RR. Allosteric selection of ribozymes that respond to the second messengers cGMP and cAMP. *Nat. Struct. Biol*,, 1999;6,1062-1071.

Kopylov AM, Spiridonova VA. Combinatorial chemistry of nucleic acids: SELEX. *Molecular Biology,* 2000;34(6),940-954.

Kotia RB, Li L, McGown LB. Separation of Nontarget Compounds by DNA Aptamers. *Anal. Chem*, 2000;72,827-832.

Liss M, Petersen B, Wolf H, Prohaska E. An Aptamer-Based Quartz Crystal Protein Biosensor. *Anal. Chem*, 2002;74,4488-4495.

Lorsch JR, Szostak JW. In vitro selection of RNA aptamers specific for cyanocobalamin. *Biochemistry*, 1994;33(4),973-982.

Mandelgutfreund MY, Schueler O, Margalit H. Comprehensive analysis of hydrogen bonds in regulatory protein DNA complex in search of common principles. *J. Mol. Biol,* 1995; 253(2),370-382.

Mann D, Reinemann C, Stoltenburg R, Strehlitz B. In vitro selection of DNA aptamers binding ethanolamine. *Biochemical and Biophysical Research Communications,* 2005; 338(4), 1928-1934.

McCauley TG, Hamaguchi N, Stanton M. Aptamer-based biosensor arrays for detection and quantification of biological macromolecules. *Anal. biochem,* 2003;319(2),244-250.

Mir M, Katakis I. Aptamers as elements of bioelectronic devices. *Molecular BioSystems,* 2007;3,620-622.

Mir M, Vreeke M, Katakis I. Different Strategies to Develop an Electrochemical Thrombin Aptasensor. *Electrochem. Comm,* 2006;8,505-511.

Nolte A, Klussmann S, Bald R, Erdmann VA, Furste JP. Mirror-design of L-oligonucleotide ligands binding to L-arginine. *Nat. Biotechnol,* 1996;14(9),1116-1119.

Okumoto Y, Ohmichi T, Sugimoto N. Immobilized Small Deoxyribozyme to Distinguish RNA Secondary Structures. *Biochemistry,* 2002;41,2769-2775.

Paborsky LR, McCurdy SN, Griffin LC, Toole JJ, Leung LL. The Single-Stranded DNA Aptamer-Binding Site of Human Thrombin. *J. Biol. Chem*, 1993;268(28),808-811.

Paddle BM. Therapy and prophylaxis of inhaled biological toxins. *Journal of Applied Toxicology*, 2003;23(3),139-170.

Pagratis NC, Fitzwater T, Jellinek D, Dang C. Potent 2'-amino-, and 2'-fluoro-2'-deoxyribonucleotide RNA inhibitors of keratinocyte growth factor. *Nature Biotechnol,* 1997;15(1),68-73.

Pavlov V, Shlyahovsky B, Willner I. Fluorescence Detection of DNA by the Catalytic Activation of an Aptamer/Thrombin Complex. *J. Am. Chem. Soc*, 2005;127,6522-6523.

Pavlov V, Xiao Y, Shlyahovsky B, Willner I. Aptamer-Functionalized Au Nanoparticles for the Amplified Optical Detection of Thrombin. *J. Am. Chem. Soc,* 2004;126,11768-11769.

Potyrailo RA. Adapting Selected Nucleic Acid Ligands (Aptamers) to Biosensors. *Analytical Chemistry*, 1998;70,3419-3425.

Prakash TP, Manoharan M, Kawasaki AM, Fraser AS, Lesnik EA, Sioufi N, et al. 2'-O-[2-(Methylthio)ethyl]-Modified Oligonucleotide: An Analogue of 2'-O-[2-(Methoxy)-ethyl]-Modified Oligonucleotide with Improved Protein Binding Properties and High Binding Affinity to Target RNA. *Biochemistry*, 2002;41,11642-11648.

Radi AE, Acero JL, Baldrich E, O'Sullivan CK. Reagentless, Reusable, Ultrasensitive Electrochemical Molecular Beacon Aptasensor. *J. Am. Chem. Soc,* 2006;128,117-124.

Ray A, Norden B. Peptide nucleic acid (PNA): its medical and biotechnical applications and promise for the future. *FASEB Journal,* 2000;14(9),1041-1060.

Rhodes A, Smither N, Chapman T, Parson S, Rees S. The generation and characterisation of antagonist RNA. aptamers to MCP-1. *FEBS. Lett,* 2001;506,85-93.

Robertson DL, Joyce GF. Selection in vitro of an RNA enzyme that specifically cleaves single-stranded DNA. *Nature*, 1990;344, 467-468.

Rodriguez MC, Kawde A, Wang J. Aptamer biosensor for label-free impedance spectroscopy detection of proteins based on recognition-induced switching of the surface charge. *Chem. Commun,* 2005;34,4267-4269.

Romig TS, Bell C, Drolet DW. Aptamer affinity chromatography: combinatorial chemistry applied to protein purification. *J. Chromatogr* B, 1999;731(2),275-282.

Ruckman J, Green LS, Beeson J, Waugh S, Gillette WL, Henninger A. DD.2'-Fluoropyrimidine RNA-based aptamers to the 165-amino acid form of vascular endothelial growth factor (VEGF165). *J. Biol. Chem*, 1998;273(32),556-567.

Sassanfar M, Szostak JW. An RNA motif that binds ATP. *Nature*, 1993;364,550-553.

Savran CA, Knudsen SM, Ellington AD, Manalis SR. Micromechanical Detection of Proteins Using Aptamer-Based Receptor Molecules. *Anal. Chem*, 2004;76,3194-3198.

Schlehuber S, Skerra A. Anticalins as an alternative to antibody technology. *Expert Opinion on Biological Therap*, 2005;5(11),1453-1462.

Schlensog MD, Gronewold TMA, Tewes M, Famulok M, Quandt E. A Love-Wave Biosensor Using Nucleic Acids as Ligando. *Sensors and Actuators, B: Chemical,*, 2004;101(3),308-315.

Schoning KU, Scholz P, Guntha S, Wu X, Krishnamurthy R, Eschenmoser A. Chemical etiology of nucleic acid structure: The alpha-threofuranosyl-(3 '-> 2 ') oligonucleotide system. *Science,* 2000;290,1347-1351.

Schwabe JWR. The role of water in protein DNA interactions. *Curr. Opin. Struct. Biol,* 1997;7(1),126-34.

Seetharaman S, Zivarts M, Sudarsan N, Breaker RR. Immobilized RNA switches for the analysis of complex chemical and biological mixtures. *Nature Biotechnol,* 2001; 19(4),336-340.

Smironov I, Shafer RH. Effect of Loop Sequence and Size on DNA Aptamer Stability. *Biochemical and Biophysical Research Communications,* 2000;39,1462-168.

Stojanovic MN, de Prada P, Landry DW. Homogeneous assays based on deoxyribozyme catalysis. *Nucleic Acids Res,* 2000;28(15),2915-2919.

Stoltenburg R, Reinemann EC, Strehlitz EB. FluMag-SELEX as an advantageous method for DNA aptamer selection. *Anal. Bioanal. Chem*, 2005;383(1),83-91.

Stratis-Cullum DN, Wade KL, Pellegrino PM. Development of a rapid method for the detection of biological threats in water. *Proceedings of SPIE - The International Society for Optical Engineering,* 2005,5994-5999.

Tasset DM, Kubik MF, Steiner W. Oligonucleotide inhibitors of human thrombin that bind distinct epitopes. *J. Mol. Biol.*, 1997;272(5),688-698.

Tombelli S. Aptamer-Based Biosensors for the Detection of HIV-1 Tat Protein. *Bioelectrochemistry,* 2005;67,135–141.

Tuerk C, Gold L. Systematic evolution of ligands by exponential enrichment: RNA ligands to bacteriophage T4 DNA polymerase. *Science,* 1990;249,505-510.

Wilson DS, Szostak JW. In Vitro selection of functional nucleic acids. *Annu. Rev. Biochem,* 1999; 68,611-647.

Xu S, Cai X, Tan X, Zhu Y, Lu B. A fiber-optic biosensor using aptamer as receptors. *Proceedings of SPIE - The International Society for Optical Engineering,* 2001; 4414,35-37.

Yamamoto R, Baba T, Kumar PK. Molecular beacon aptamer fluoresces in the presence of Tat protein of HIV-1. *Genes Cells*, 2000;5,389 - 396.

In: Biochemistry Research Updates
Editor: Simon J. Baginski
ISBN 978-1-61209-700-8

Chapter 8

EFFECT ON HEAVY OILS BY BACTERIA

Ruixia Hao, Guanyu Wang and Anhuai Lu
School of Earth and Space Sciences,
Peking University, Beijing, China

ABSTRACT

The objective of this chapter is to demonstrate the basic characteristics of halophilic strain TM-1 and Bacillus SP-5 which were isolated from a reservoir of the Shengli oil field in East China and evaluate the effects of strains TM-1 and SP-5 on different heavy oils. Strain TM-1 is a halophilic gram-positive coccus and strain SP-5 is gram-positive rod with spores. The cells of strains TM-1 and SP-5 were grown at temperatures up to 58°C in the neutral to alkaline pH range. They could grow steadily, and produce various metabolites, have various surface features and use different organic substrates (acetate, D-glucose, fructose, glycerol, maltose, pyruvate, starch, sucrose, and xylose). Laboratory studies have demonstrated that the strains TM-1 and SP-5 affected different heavy oils; converted and degraded various components and changed the physical and chemical properties of heavy oils. The bioconversion of heavy oils leads to an enrichment in lighter hydrocarbons and an overall redistribution of these hydrocarbons. The interactions of microorganisms with heavy oils are variable and depend on the microbial species and the chemical compositions of heavy oils.

Keywords: Bacterium, Metabolite, Heavy oil, Degradation, Conversion.

INTRODUCTION

Interactions between microorganisms and crude oils under laboratory and reservoir conditions have been studied for many decades. A wide range of studies has dealt with biodegradation, the role of sulfate-reducing bacteria in reservoir fouling, corrosion and microbial enhanced oil recovery (MEOR). Results from these studies have been extensively reviewed as provided by Tissot and Welte (1978), Yen (1990), and Premuzic and Woodhead (1993).

Petroleum reservoirs constitute ecological niches from which various physiological types of microorganisms have been isolated or detected. Physiological types isolated from these biotopes include sulfate reducers (Beeder et al., 1994; Stetter et al., 1993; Tardy-Jacquenod et al., 1996), sulfur reducers (Stetter et al., 1993), methanogens (Ng et al., 1989), fermentative bacteria (Davey et al., 1993; Grassia et al., 1996), manganese and iron reducers (Greene et al., 1997) and dibenzothiophene-degrading bacteria (Bahrami et al., 2001). Although many bacteria have already been isolated from the reservoirs, the ones useful to MEOR and bioremediation are fewer.

Microbial enhanced oil recovery involves the use of microorganisms and their metabolic products to increase oil recovery. However, it is necessary that the microorganisms can grow and metabolize in the reservoir. The use of microbes in MEOR involves knowledge of the many biochemical changes they impart to oil. This study examines the bioconversion and biodegradation in various heavy oils.

A microbial culture's ability to metabolize a chosen substrate can be determined by monitoring (a) growth on the substrate, (b) the loss of the substrate from the culture medium, or (c) the production of a specific metabolic end product (Fedorak et al., 1983). In studying the conversion and degradation of different components in heavy oils, all three approaches have been used.

Heavy oil has a high viscosity and density and contains many resins and asphaltenes. Currently, physical methods are used to recover heavy oil; however, such methods are expensive and result in pollution of the ambient environment. As a result, microbial systems have been used in heavy oil reservoirs to overcome such problems, thereby improving oil recovery. Conversely, the level of environmental pollution becomes more severe as the quantity of heavy oil produced increases. Therefore, it is important to develop methods that enable the remediation of heavy oil-contaminated environments by microbes to address this critical problem (Hao and Lu, 2009).

Halophiles are usually found in salted products, salt fields, salt lakes, brines, saline soils, cold saline habitats, alkaline saline habitats and unusual habitats (desert plants and desert animals) (Ventosa et al., 1998). Halophiles exhibit optimum growth at a salinity of approximately 15–20% but are often able to grow in saturated brine with salinities as high as 32%. The extreme halophiles that have been isolated to date belong to the genus *Halobacterium* (Ling and Dong, 1999).

Bacterial metabolites are of great value as enhancers of oil recovery due to their surfactant activity and bioemulsifying properties (Banat, 1995). Because the conditions in oil deposits are often saline, the use of salt-resistant metabolites may be advantageous to the recovery of oil. Furthermore, hypersaline water and soil are often contaminated with crude oils, heavy metals or other toxic compounds from anthropogenic sources. However, conventional microbiological treatment processes do not function at high salt concentrations; therefore, the use of moderately halophilic bacteria should be considered.

Many studies have been conducted to evaluate halophilic bacteria; however, halophilic bacteria isolated from oilfields have been relatively less studied, and the interaction between halophilic bacteria and heavy oils has not been evaluated until now (Hao and Lu, 2009).

There are many kinds of *Bacillus,* which are distributed widely and have a large proportion in bacteria (Li et al., 2006). However, the numbers of *Bacillus* which affected crude oils are smaller; the reported ones now are *Bacillus licheniformis* and *Bacillus subtilis* (Das and Mukherjee, 2007; Jenneman et al., 1983; Lin et al., 1990, 1993, 1998;

Pornsunthorntawee et al., 2008). The properties of *Bacillus subtilis* have been studied and reported in a lot of literature (Carballido-López and Formstone, 2007; Copper et al., 1981; Earl et al., 2008; Harwood, 1992; Lin and Jiang, 1997; Mulligan et al., 1989; Zheng and Slavik, 1999), but *Bacillus subtilis* isolated from oil reservoirs and its metabolism on crude oil have been reported in a very few cases (Fox et al., 1993; Jenneman et al., 1983).

This chapter was conducted to describe the basic characteristics of halophilic strain TM-1 and *Bacillus* SP-5, which were isolated from the Shengli oil field in East China. In addition, in this chapter, we evaluated the effect of strains TM-1 and SP-5 on various types of heavy oil.

Materials and Methods

Organism

The halophilic bacterium TM-1 and *Bacillus* SP-5, isolated from the Shengli oil field, were used for this study.

Collection of Samples

The samples were collected at the wellhead of the producing wells in sterile glass bottles, which were then tightly sealed with rubber stoppers. The samples were transported to the laboratory and stored during transport and afterward at ambient temperatures.

Medium and Growth Conditions

Culture Media

The experimental medium (EM) used for culture, stock maintenance and experimental studies contained the following compounds (per liter):

For strain TM-1: $NaNO_3$, 0.5 g; $MgSO_4 \cdot 7H_2O$, 0.01 g; NaCl, 180.0 g; K_2HPO_4, 2.7 g; NH_4Cl, 0.1 g; yeast extract, 0.5 g; and glucose, 3.0 g.

For strain SP-5: $NaNO_3$, 0.5 g; $MgSO_4 \cdot 7H_2O$, 0.01 g; NaCl, 5.0 g; K_2HPO_4, 2.7 g; NH_4Cl, 0.1 g; peptone, 0.5 g; and glucose, 5.0 g.

In addition, medium Y was used to evaluate the capacity of the strain to use organic substrates. Medium Y for TM-1 contained the following compounds (per liter): $NaNO_3$, 0.5 g; $MgSO_4 \cdot 7H_2O$, 0.01 g; NaCl, 180.0 g; K_2HPO_4, 2.7 g; NH_4Cl, 0.1 g; and yeast extract, 0.2 g. Medium Y for strain SP-5 contains per liter: $NaNO_3$, 0.5 g; $MgSO_4 \cdot 7H_2O$, 0.01 g; NaCl, 5.0 g; K_2HPO_4, 2.7 g; and NH_4Cl, 0.1 g.

Finally, medium YG was used for the metabolites assay. For strain TM-1, medium YG was comprised of the following components (per liter): $NaNO_3$, 0.5 g; $MgSO_4 \cdot 7H_2O$, 0.01 g; NaCl, 180.0 g; K_2HPO_4, 2.7 g; NH_4Cl, 0.1 g; yeast extract, 0.5 g, and glucose, 5.0 g. Medium YG for strain SP-5 consists of the following components (in g/l): 5.0 glucose, 0.5 $NaNO_3$, 0.01 $MgSO_4 \cdot 7H_2O$, 5.0 NaCl, 2.7 K_2HPO_4, and 0.1 NH_4Cl.

Strains TM-1 and SP-5 were respectively cultured in 100 ml of medium EM at 37℃ and atmospheric pressure for 24 h, after which they were respectively grown on agar EM plates at 37°C for 24 h.

Growth Characteristics

Strains TM-1 and SP-5 were characterized using biochemical tests including the hydrolyzation of arginine, starch and glycerol, the reduction of nitrite, the indole test, the oxidase test and the catalase test. All tests were conducted following the methods described by Dong and Cai (2001).

To evaluate the ability of strains TM-1 and SP-5 to use organic substrates, the low concentration complex, medium Y, was supplemented with different substrates (citrate, D-glucose, D-mannitol, D-xylose, fructose, glycerol, L-arabinose, lactose, L-rhamnose, maltose, starch and sucrose), each at a concentration of 2 g/l.

Analysis of Metabolic Products

After 72 hours of growth in the medium YG at 37℃, cells were removed by centrifugation (8000×g, 30 min, 4℃). The supernatant was extracted by mixed chloroform-methanol (2:1), and then distilled. The final supernatant was collected with sterile bottles and analyzed by gas chromatography-mass spectrometry (Shimadzu QP5050A). The samples were chromatographed with a gradient temperature program. The initial temperature of 80℃ was held for 2 min, and then the temperature rose to 250℃ at 10℃/min and held at 250℃ for 10 min. The injection temperature was 280℃, and the transport line was maintained at 260℃. The helium flow was 0.5 ml/min, and the injection volumes were 2 μl.

Surface Tensions Assay

The strains were incubated in EM broth at 37°C for 3 days and sampled every 12 hours; the broths were centrifuged and then the surface tensions of the supernatant were measured. The surface tensions of the supernatant were determined using an interfacial tension meter (CSC-NOS.70545).

Surface tensions (air-liquid) were measured by the Dunouy method. The pure water was used to calibrate the system and measurements were taken in replicates of 7. The reported values are expressed as the means and standard deviations (the standard deviations range from 0.1 to 0.2). All surface tension measurements were made at room temperature.

Bacterial Influence on Crude Oil

All tests were conducted under aerobic conditions, and all heavy oils evaluated in this study were collected from the Shengli oil reservoir. The selected oils were divided into three types according to viscosity: heavy oil A, which had a density of 0.9803 g/cm3 and a viscosity of 6.031 Pa•s, heavy oil B, which had a density of 0.9538 g/cm3 and a viscosity of 1.283 Pa•s, and heavy oil C, which had a density of 0.9668 g/cm3 and a viscosity of 1.657 Pa•s. The viscosities of the various heavy oils were measured using a DVIII Brook Viscometer at 50 rpm and 50 °C. To acclimatize strains TM-1 and SP-5 to the different crude oils, they were pre-cultured in 100 ml of EM with 5 ml of the three different oils individually at 37 °C for 24 h prior to use in the experiment. All oils used for the experiment were autoclaved. The acclimated cultures and 65 ml of the three different sterilized oils were then added to 500 ml conical flasks that contained 200 ml of EM. In addition, 300 ml of EM and 70 ml of each of the three various sterilized oils were also added to 500 ml conical flasks as controls for comparison. The flasks were then sealed with foam stoppers and incubated in a shaker at 150 rpm and 37 °C for 5 days, at which time the crude oils were sampled and analyzed. The oil components were analyzed according to the industrial analytical standards of the China National Petroleum Corporation (CNPC). The group components (saturated hydrocarbons, aromatic hydrocarbons, resins, and asphaltenes) were extracted by chloroform (standard code: SY/T 5118-1995) and then measured using an IATROSCAN MK-5 thin layer chromatography/flame ionization detector (TCL/FID, standard code: SY/T6338-1997), which has an analytical error of 0.1%. The components of the saturated and aromatic hydrocarbons were analyzed using a gas chromatograph mass spectrometer (HP5890II GC/5970B MSD), which has an analytical error of 0.1%. The samples were chromatographed using a gradient temperature program, which was as follows: an initial temperature of 110°C, which was held for 1 min and then increased to 320 °C at 4 °C/min, where it was held for 22 min for saturated hydrocarbons, or to 335 °C at 3 °C/min, where it was held for 11.68 min for aromatic hydrocarbons. The column used to analyze the saturated hydrocarbons was a DB5-MS (60 m×0.25 mm; 0.25μm film), and a BPX5 (30 m × 0.22 mm; 0.25μm film) was used to analyze the aromatic hydrocarbons. The injection temperature was 320 °C, and the transport line was maintained at 320 °C. The helium flow rate was 25 cm/s, and the injection volumes were 2μl. The contents of carbon, hydrogen, nitrogen and sulfur were analyzed using an Elemental Analyzer (Germany Elementar vario EL), which has an analytical error of 0.1% for carbon and hydrogen, and of 0.01% for nitrogen and sulfur.

RESULTS

Growth Characteristics

Strain TM-1

Cells of strain TM-1 stained as gram-positive and were found to be cocci without spores (Figure 1). The diameter of the cells ranged from 0.5 μm to 1.0μm. After incubation in liquid EM at 37 °C for 24 h, the cells were distributed in a tetrad formation. Strain TM-1 was grown on EM amended with 15 g/l agar. After incubation at 37 °C for 1 day, round colonies with diameters of approximately 0.1–0.3 mm that were milk-white and smooth had developed.

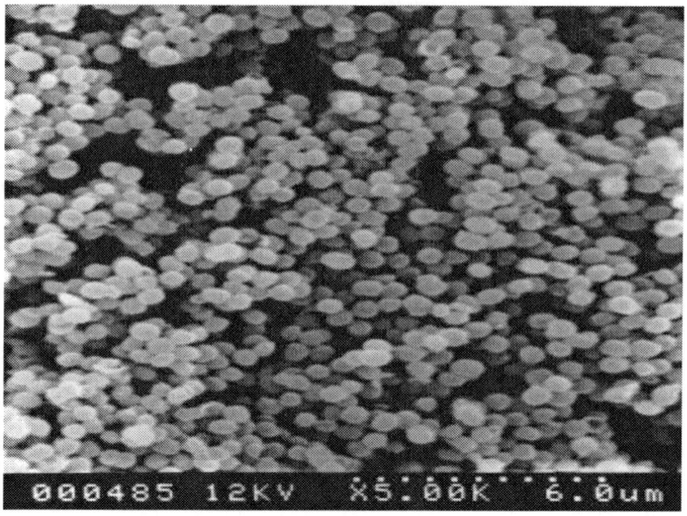

Figure 1. Electron micrograph of strain TM-1 at 37℃ for 24 h.

Next, the ranges of temperature, pH and salinity that strain TM-1 was capable of growth in were determined. The results of the experiments indicated that the temperature growth range for strain TM-1 was 32–58 °C, with the maximal growth rate occurring at 37 °C. In addition, the strain was capable of growth at a pH of between 4.5 and 8.5, with the optimal growth occurring at approximately pH 6.5–7.0. Strain TM-1 required NaCl for growth, and could grow in medium containing up to 18% NaCl.

Strain SP-5

Cells of strain SP-5 were gram-positive and rods with central and subterminal elliptic spores which cytocysts did not dilate, the average cell length being 2.0 μ m to 4.0 μ m, and the average cell width 0.5 μ m to 0.6 μ m (Figure 2). There were no poly- β -hydroxybutyrate (PHB) particles in cells with lateral flagella. The strain was incubated in liquid medium EM at 37°C for 24 hours, cells were distributed in single or pairing at the exponential growth phase and occurred in chain or massed at the stationary growth phase.

The strain SP-5 was put onto Petri dishes within the EM medium containing 15g/l agar. After 24 hours of incubation at 37°C, round lacteous colonies with diameters of about 1-4 mm had developed. The colonies were desiccative, rough, opaque, and with irregular borders.

Temperature growth range for strain SP-5 was 20°C to 58°C with maximal growth rate between 32 and 48°C. The strain was able to grow at pH values comprised between pH5.5 and 8.5, and thrives optimally around pH 7.0-7.5. It can grow in 7% NaCl solution.

Packed cells showed gas production after a 3% (vol./vol.) H_2O_2 solution was added, indicating the presence of catalase activity.

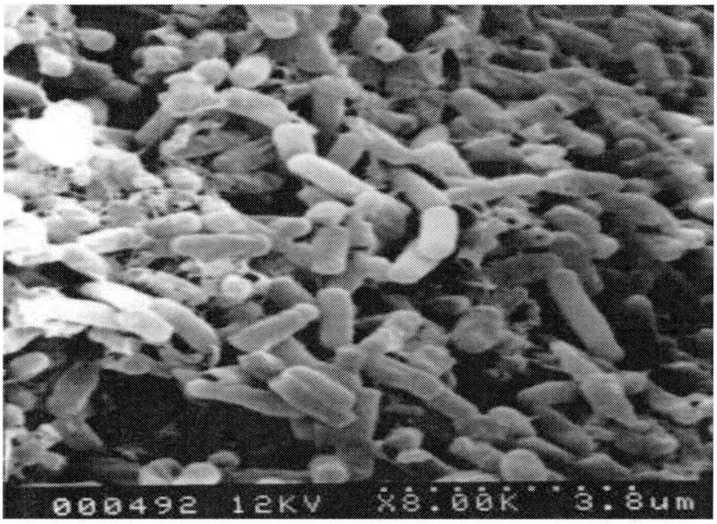

Figure 2. Electron micrograph of strain SP-5 at 37℃ for 24 h.

Metabolism and Metabolic Products

Strain TM-1

Strain TM-1 was a facultative aerobe that was capable of growth under anaerobic conditions. In addition, strain TM-1 was a homofermentative lactic bacterium that fermented glucose and produced lactic acids. The strain could not hydrolyze arginine, starch or glycerol, and could not reduce nitrite or produce indole. Strain TM-1 was found to be oxidase and catalase negative.

Strain TM-1 could utilize various organic substrates, with good growth being observed on D-glucose, fructose, maltose and D-mannitol. No growth occurred on L-arabinose, citrate, D-xylose, L-rhamnose, lactose, sucrose, starch or glycerol.

When grown in YG medium, strain TM-1 produced butylated hydroxytoluene ($C_{15}H_{24}O$); 1,2-benzenedicarboxylic acid-bis ester ($C_{16}H_{22}O_4$); and dibutyl phthalate($C_{16}H_{22}O_4$). The ester, which is a better organic solvent and also a low-molecular-weight surfactant with lower surface tension, was by far the most abundant metabolite, accounting for approximately 63.3% of the total products.

Strain SP-5

The strain SP-5 was grown chemoorganotrophically and facultatively under aerobic culture condition with oxygen as the electron acceptor. It could ferment D-glucose, sucrose, D-fructose, xylose, arabinose, maltose and mannitol. The oxidase and Voges-Proskauer test were positive, but Indole test negative. The litmus milk showed alkaligenous reduction reaction. The strain could reduce nitrate, liquefy gelatin and hydrolysis starch.

The analytical results of Shimadzu chromatograph and mass spectrograph indicate that the strain SP-5 also produces 4-hydroxy-4-methyl-2-pentanone ($C_6H_{12}O_2$), 3-chloro-2-methyl-2-pentanol ($C_6H_{13}ClO$), butylated hydroxytoluene ($C_{15}H_{24}O$) and 1, 2-benzenedicarboxylic acid bis (2-methylpropyl) ester ($C_{16}H_{22}O_4$). The ester proportion among metabolites is larger, that is 42.01%, and the second is ketone (37.38%). The molecular weights of these metabolic products are above 100, and the highest one is 278.

These metabolites produced by the strain SP-5 play different roles in oil reservoirs. The alcohol and ketone can be as the co-surfactants, which are used in the micro-emulsion to stabilize and lower interfacial tension and improve emulsification process. The butylated hydro-xytoluene and 1,2-benzenedicarboxylic acid bis (2-methyprophl) ester are the better organic solvents. Moreover, the butylated hydroxytoluene has the amphoteric group and shows surface activity, and plays the role of surfactant. The ester produced by the strain is also small-molecular surfactant with lower surface tension, which is formed by esterification with fatty acids.

Surface Characters

The close attention to oil-degrading bacterium is whether it can produce surfactants. Strains TM-1 and SP-5 could produce surfactants, but the period and amount of producing surfactants for various strains were different.

For strain TM-1, the surface tension decreased with cell growth, and the surface tension curve was W shaped (Figure 3). In the early growth of bacteria was relatively less surfactants, along with the growth of bacteria, surfactants continuously increasing; and the surface tension was minimum in 84 hours, for 30.25 mN/m.

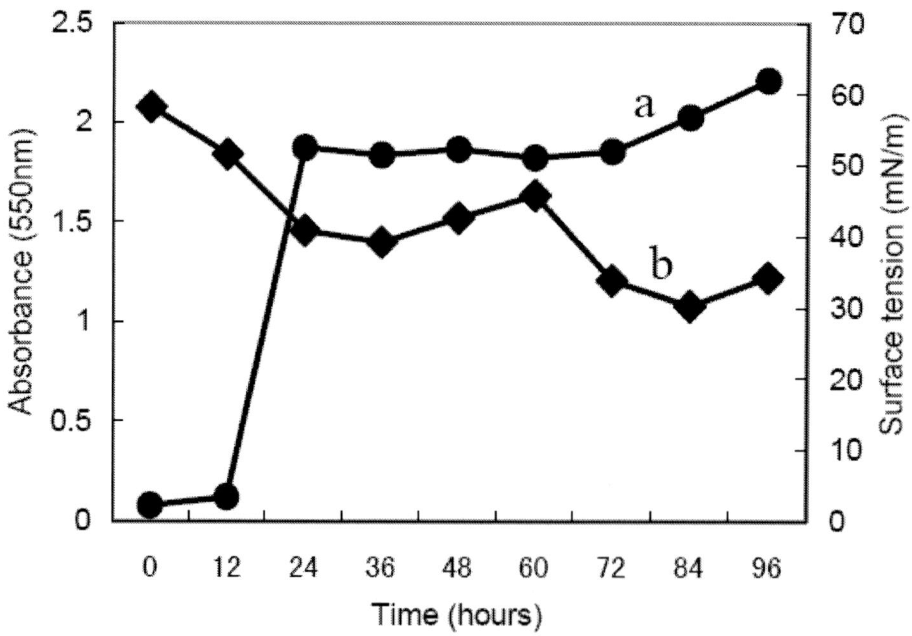

Figure 3. Changes (b) of surface tension with TM-1 cell growth (a) at 37℃.

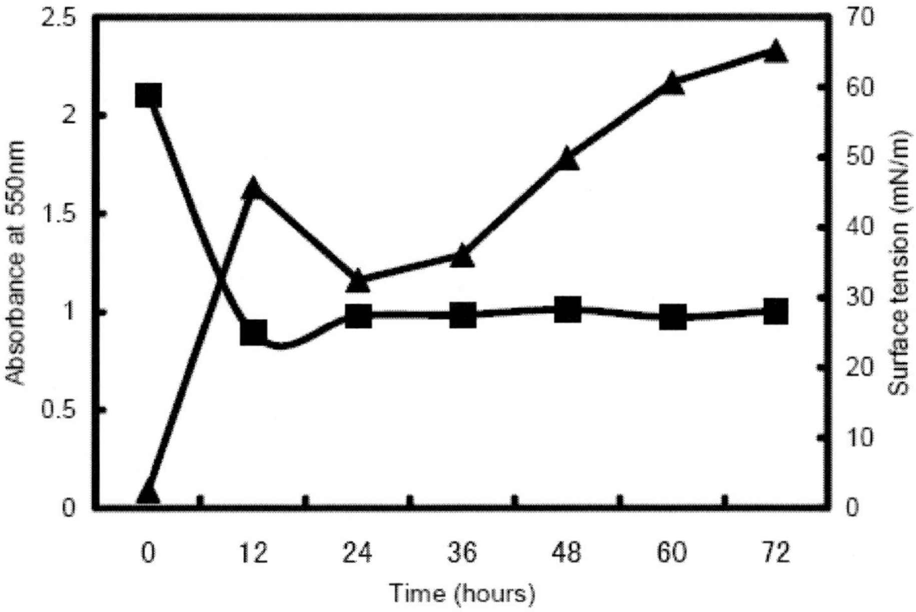

Figure 4. Surface tension (■) and strain SP-5 cell growth as optical density (▲) over time. Culture reactors were maintained at 37℃.

The strain SP-5 produced some surfactants at the short growth period. Figure 4 shows the absorbance (growth) and surface tension (surfactant production) plotted versus time for strain SP-5. Surfactant production was directly proportional to cell growth; the cell density increased with decreasing surface tension. The greatest decrease in surface tension was obtained during log phase (within 12 hours). Surface tension dropped from 58.86 mN/m to 25.04 mN/m within the first 12 hours and then stabilized over the next 72 hours to the values between 25.04 mN/m and 28.16 mN/m.

Effect of Strains on Crude Oils

Bioconversion often brings about some changes in contents of saturated hydrocarbons, aromatic hydrocarbons, resins and asphaltenes, and a reduction in contents of sulfur and nitrogen. Biodegradation is the complete transformation of organic molecules into inorganic substances. Strains TM-1 and SP-5 had different effects on various heavy oils and could convert or degrade different components of crude oils.

Strain TM-1

The effects of induced bioconversion of various heavy oils on four major fractions are shown in Table 1. When compared with sterile controls (SC), the relative percentages of saturated hydrocarbons, aromatic hydrocarbons, resin and asphaltene vary. However, in oil A and oil C, each of these factors decreased, with the exception of saturated hydrocarbons. In addition, the contents of carbon, hydrogen, nitrogen and sulfur in various oils were lower following biotreatment when compared with the sterile controls (Table 2).

Table 1. Distribution of saturated hydrocarbons, aromatic hydrocarbons, resins and asphaltenes in various heavy oils before and after biotreatment with strain TM-1

Crude	Saturated hydrocarbon (% w/w)		Aromatic hydrocarbon (% w/w)		Resin (% w/w)		Asphaltene (% w/w)	
Oil	SC	BT	SC	BT	SC	BT	SC	BT
A	38.44	41.85	25.08	21.54	28.66	28.62	6.19	4.00
B	38.70	37.72	24.86	20.76	27.40	23.68	5.93	4.76
C	38.00	39.83	23.99	21.75	27.08	26.55	5.46	3.95

SC: sterile control.
BT: biotreatment.
Hao and Lu, 2009.

Table 2. Variations in the concentrations of carbon, hydrogen, nitrogen and sulfur in various heavy oils before and after biotreatment with strain TM-1

Crude	Carbon (% w/w)		Hydrogen (% w/w)		Nitrogen (% w/w)		Sulfur (% w/w)		H/C (atom/atom)	
oil	SC	BT	SC	BT	SC	BT	SC	BT	SC	BT
A	86.6	80.9	13.1	12.0	0.6	0.5	0.2	0.1	1.82	1.78
B	86.7	86.3	12.9	12.2	0.6	0.5	0.2	0.1	1.79	1.70
C	86.6	84.8	12.5	11.5	0.7	0.5	0.2	0.1	1.73	1.63

SC: sterile control.
BT: biotreatment.
Hao and Lu, 2009.

Table 3. Variation of parameters of different heavy oils before and after biotreatment with strain TM-1

	$Pr/n\text{-}C_{17}$		$Ph/n\text{-}C_{18}$		Pr/Ph		Carbon number of main peak	
Crude oil	SC	BT	SC	BT	SC	BT	SC	BT
A	2.73	0.80	3.08	1.02	0.63	0.61	25	22
B	2.50	1.92	3.17	2.53	0.85	0.69	15	15
C	1.83	1.28	1.94	1.60	0.61	0.70	14	14

SC: sterile control.
BT: biotreatment.
Hao and Lu, 2009.

Table 4. Changes of paraffin content of various heavy oils before and after biotreatment with strain TM-1

	Content of paraffin (%)		
Crude oil	SC	BT	Percent decreased (%)
A	2.71	2.60	4.06
B	6.16	4.43	28.08
C	6.34	2.85	55.05

Precision on paraffin content is 1%. SC: sterile control.
BT: biotreatment. Hao and Lu, 2009.

Halophilic TM-1 was capable of degrading various components of different heavy oils. Consistent with the aforementioned analyses, biodegradation of these oils leads to enrichment with lighter hydrocarbons and an overall redistribution of these hydrocarbons, as shown by the peak clusters at retention times of 20–30, 30–40 and 40–50 min. The analytical results of the gas chromatograph analysis of saturated hydrocarbons in various heavy oils state that the strain degraded different fractions of crude oils (Figure 5). When added to oil A, strain TM-1 effectively transformed alkanes C_{12}–C_{15}, C_{23}, C_{25}, C_{26} and C_{28} and completely degraded C_{32} (Figure 5a). When strain TM-1 was tested against oil B, it effectively transformed C_{12} and C_{19} and completely degraded alkanes longer than C_{19} (Figure 5b). When strain TM-1 was tested against oil C, it completely degraded alkanes longer than C_{18} (Figure 5c). For all oils, the mass fraction of undecane increased. These results indicate that the differences in the degradation of different crude oils by strain TM-1 primarily depend on the physicochemical properties of crude oils and the growth potential of strain TM-1.

The analytical results of GC of the saturated hydrocarbons in various heavy oils demonstrate that the values of parameters of heavy oils changed following biotreatment (Table 3). The ratios of Pr/n-C_{17} and Ph/n-C_{18} of all oils decreased. There are two possible reasons for this decrease in the ratios. One possibility is that strain TM-1 may exert a greater effect on pristane and phytane than on C_{17} and C_{18}. However, it is also possible that most of the heavy fractions were converted to light fractions, which resulted in increased concentrations of C_{17} and C_{18}. As shown in Figure 5, the fraction of light saturated hydrocarbons in the heavy oils increased following biotreatment. In addition, with the exception of oil C, the ratios of Pr/Ph were reduced. Taken together, these findings indicate that strain TM-1 degraded pristane before phytane (Hao and Lu, 2009).

As shown in Table 4, the paraffin contents of the heavy oils were 4.06–55.05% lower in the treated oils than in the sterile controls.

Strain SP-5

The strain SP-5 had obvious effects on four major fractions of various heavy oils (Table 5). While compared to the sterile controls (abbr. SC) the relative percentages of saturated hydrocarbon, aromatic hydrocarbon, resin and asphaltene vary; there is a decrease in all cases except for the saturated hydrocarbons in oil A and C. The relative percentages of aromatic hydrocarbons were respectively reduced by 17.82% (oil A), 11.34% (oil B) and 13.92% (oil C); the ones of resins were also decreased in 15.67% (oil A), 9.74% (oil B) and 5.13% (oil C).

Table 5. Distribution of saturated hydrocarbons, aromatic hydrocarbons, resins and asphaltenes in various heavy oils before and after biotreatment with strain SP-5

Heavy	Saturated hydrocarbon (% w/w)		Aromatic hydrocarbon (% w/w)		Resin (% w/w)		Asphaltene (% w/w)	
Oil	SC	BT	SC	BT	SC	BT	SC	BT
A	38.44	39.19	25.08	20.61	28.66	24.17	6.19	4.58
B	38.70	37.90	24.86	22.04	27.40	24.73	5.93	4.57
C	38.00	40.05	23.99	20.65	27.08	25.69	5.46	5.29

SC: sterile control.

BT: biotreatment.

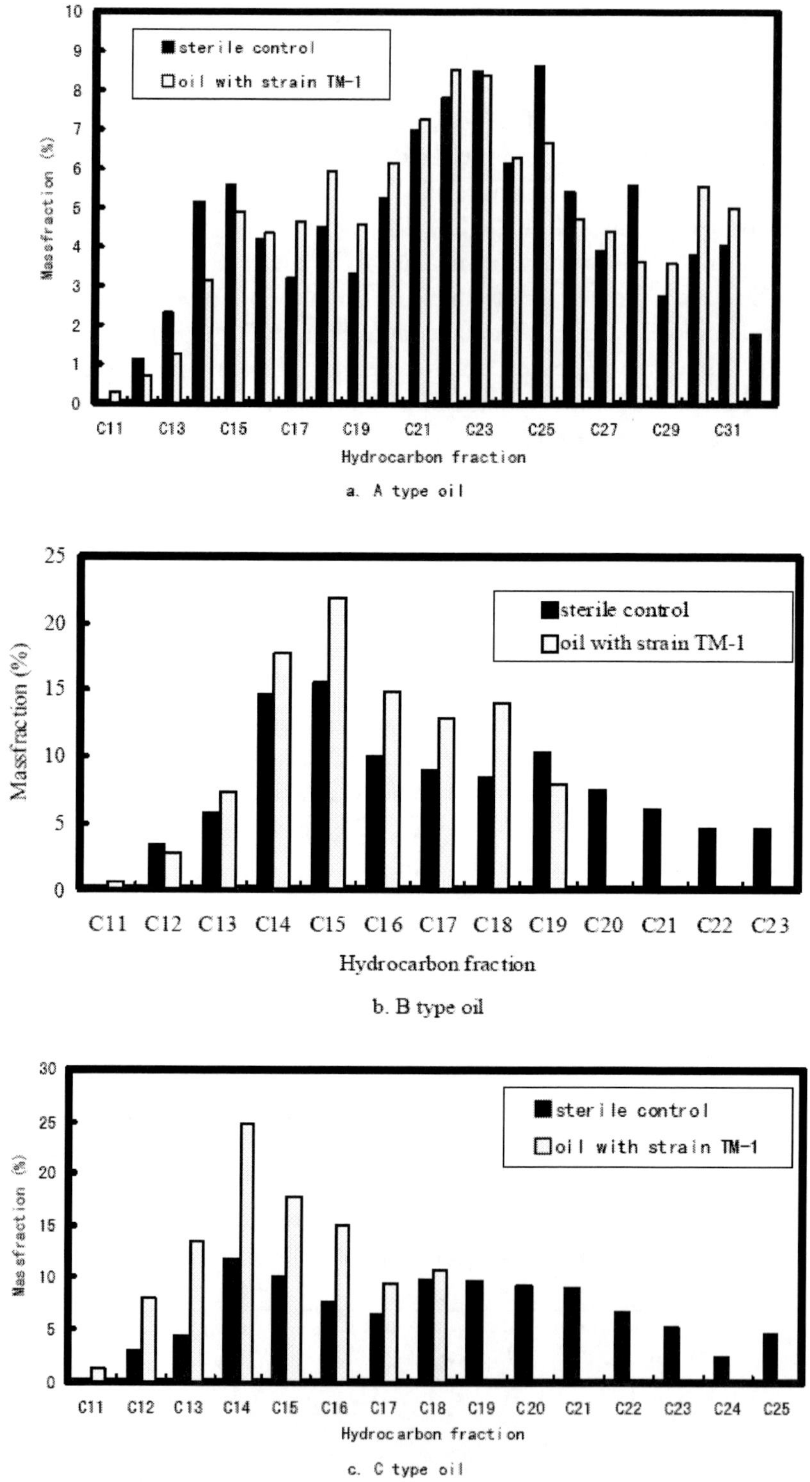

Hao and Lu, 2009.

Figure 5. Distribution of different alkanes in various heavy oils before and after biotreatment with strain TM-1.

The relative percentages of asphaltenes of oil A and oil B were largely reduced in 26.01% (oil A) and 22.93% (oil B). The contents of carbon, hydrogen and sulfur of various heavy oils, compared to the sterile controls, were decreased after biotreatment (Table 6). The nitrogen contents of oil A and B were not changed, but that of oil C was decreased (Table 6).

Table 6. Variations in the concentrations of carbon, hydrogen, nitrogen and sulfur in various heavy oils before and after biotreatment with strain SP-5

Heavy	Carbon (% w/w)		Hydrogen (% w/w)		Nitrogen (% w/w)		Sulfur (% w/w)		H/C (atom/atom)	
Oil	SC	BT	SC	BT	SC	BT	SC	BT	SC	BT
A	86.6	85.8	13.1	12.5	0.6	0.6	0.2	0.1	1.82	1.75
B	86.7	86.3	12.9	12.0	0.6	0.6	0.2	0.1	1.79	1.67
C	86.6	85.5	12.5	11.7	0.7	0.6	0.2	0.1	1.73	1.64

SC: sterile control.
BT: biotreatment.

Table 7. Variation of parameters of different heavy oils before and after biotreatment with strain SP-5

Heavy oil	$Pr/n\text{-}C_{17}$		$Ph/n\text{-}C_{18}$		Pr/Ph		Carbon number of main peak	
	SC	BT	SC	BT	SC	BT	SC	BT
A	2.73	7.73	3.08	4.14	0.63	0.69	25	14
B	2.50	2.36	3.17	2.89	0.85	0.66	15	14
C	1.83	1.38	1.94	1.60	0.61	0.53	14	14

SC: sterile control.
BT: biotreatment.

Strain SP-5 could convert and/or degrade various components of different heavy oils. To oil A, strain SP-5 effectively converted nonadecane-C_{19} and alkanes before nonadecane, and completely degraded alkanes larger than nonadecane (Figure 6a). To oil B, the strain effectively transformed alkanes before nonadecane and completely degraded alkanes larger than octadecane (Figure 6b). To oil C, the strain effectively concerted C_{12}-C_{19} and completely degraded alkanes C_{20}-C_{25} (Figure 6c). Differences in conversion and degradation of various heavy oils by strain SP-5 depend mainly on the physicochemical conditions and the growth potential of strain SP-5. On the whole, the trends of conversion and degradation toward different heavy oils by strain SP-5 are the same, the heavy fractions of oils decreased and the light fractions of oils increased.

The analytical results of GC of saturated hydrocarbons in various heavy oils demonstrate that the values of parameters of heavy oils were changed with biotreatment (Table 7). Except for oil A, the ratios of $Pr/n\text{-}C_{17}$, the ratios of $Ph/n\text{-}C_{18}$ and the ratios of Pr/Ph of oil B and oil C were decreased. The carbon numbers of main peak of GC profiles of saturated hydrocarbons in oil A was changed from 25 to 14, and the one of oil B was moved from 15 to 14, but that of oil C was not changed.

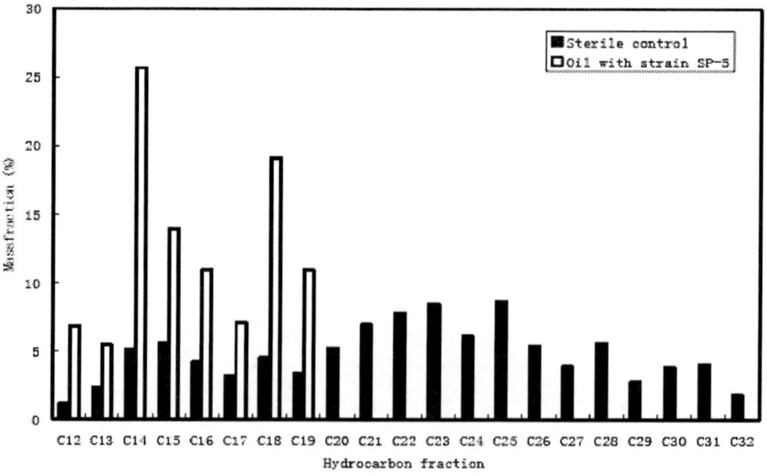

a. A type oil

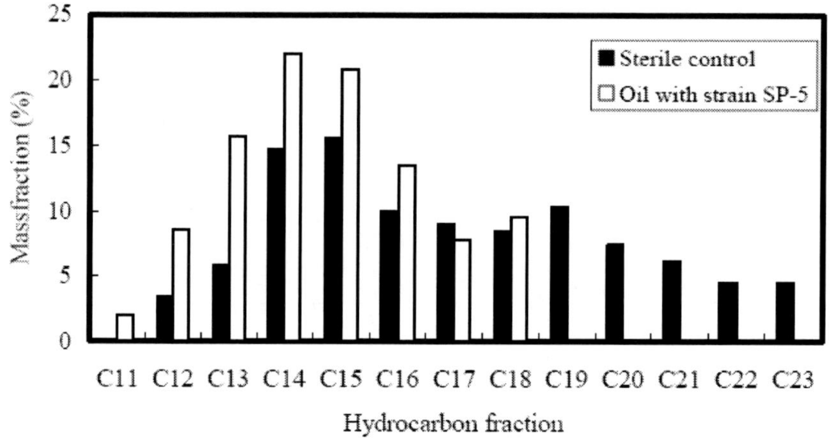

b. B type oil

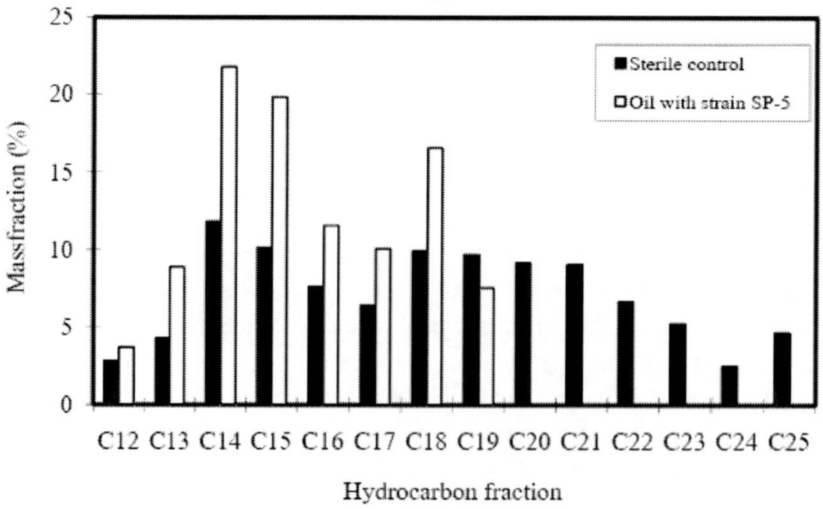

c. C type oil

Figure 6. Distribution of different alkanes in various heavy oils before and after biotreatment with strain SP-5.

The physical and chemical properties of heavy oils were changed with biotreatment. The experimental results demonstrate that the viscosity of oil A was reduced from 6.031Pa.s to 1.192Pa.s, the decreased percentage was 80.24%; the paraffin contents of oil B and oil C, with compared to the sterile controls, and were decreased respectively by 64.12% and 43.69%.

DISCUSSION

The interaction between bacteria and crude oils is a complex biochemical process. In general, bacteria can have direct and indirect effects on the properties of crude oils that include (1) inducing changes in the physical and chemical features of oils and (2) biodegradation of crude oils. These functions occur as a result of microbial growth, bacterial enzymes, and metabolites. Microbial degradation occurs in the entire petroleum range from gas to residuum (Premuzic et al., 1996).

The halophilic coccus TM-1 was able to grow in high salinity environments. In addition, strain TM-1 was capable of metabolizing several low-molecular-weight organic substrates which other halophilic bacteria cannot. As a result, treatment with TM-1 led to transformation of the components of the different heavy oils evaluated in this study.

Pfiffner et al. (1986) isolated halotolerant bacterial strains capable of producing extracellular polysaccharides from oil wells and oil well-associated environments. Additionally, moderate halophiles isolated from the Great Salt Lake (Ward and Brock, 1978) and Antarctic saline lakes (McMeekin et al., 1993) have been shown to degrade hydrocarbons. Furthermore, Rosenberg (1983) isolated *Pseudomonas halodurans* capable of degrading benzoate and other aromatic compounds. Finally, moderate halophiles belonging to the family Halomonadaceae, which were isolated from highly saline sites, were found to utilize chloroaromatic compounds as sources of carbon and energy (Maltseva et al., 1996).

Strain TM-1 differs from other halophiles in that it primarily affects heavy oils. Specifically, strain TM-1 and its metabolites changed the chemical properties of heavy oils by decreasing the concentration of high-molecular alkanes in the oils, which resulted in an increase in the light fractions. In addition, the contents of aromatic hydrocarbons, resins and asphaltenes in the various oils evaluated in this study were reduced by strain TM-1. As a result, the paraffin contents of heavy oils A, B and C were reduced by 4.06%, 28.08% and 55.05%, respectively.

The *Bacillus* SP-5 is able to grow at extensive temperature range, and contribute different metabolic products. Evidence from surface features also shows that the strain produces surfactants. It is that in the metabolic process these products act as bridging between the bacteria and heavy oils and make them contact each other. Indeed, it has been reported that bacterial adherence to hydrocarbons facilitates the reproduction of microbes in crude oils during the biodegradation of petroleum (Boochan et al., 2000; Jain et al., 1992; Kirscner et al., 1980; Rosenberg et al., 1983).

The strain SP-5 and its metabolites changed the physical and chemical properties of heavy oils. The heavy components of oil samples were decreased and the light fractions increased. Also, the contents of aromatic hydrocarbons, resins and asphaltenes in various heavy oils were reduced. As a result, the viscosity and paraffin content of heavy oils were decreased, for example, the viscosity of oil A was decreased by 80.24%, and the paraffin contents of oil B and C were reduced respectively by 64.12% and 43.69%. Moreover, the physical properties of heavy oils including the flow ability were improved.

The interactions between microorganisms and different heavy oils occur through complex biochemical and chemical reactions. These reactions depend on multiple variables within and at the interface of a multicomponent system comprised of organic and inorganic components. The biochemical action on bacteria and heavy oil occurred rapidly within the first 7 days, after which time it gradually slowed down (Aldrett et al., 1997). In this study, the interactions between strains and different heavy oils lasted for 5 days; however, if sufficient nutrients were available, especially nitrogen and phosphorus, the biochemical action could be sustained for as long as 28 days. Such an extended length of biochemical action would lead to the components of the heavy oils being further degraded.

Some chemical markers can be used to diagnose the biochemical interactions between bacteria and heavy oils. They include the distribution of saturated hydrocarbons, aromatic hydrocarbons, resins and asphaltenes, and the contents of sulfur and nitrogen. The use of chemical markers in the monitoring of the interaction between different bacteria and various heavy oils allows us to determine the efficiency of the biochemical conversion and/or degradation of the petroleum.

In this study, the concentrations of each group of compounds in heavy oils A, B and C were changed (Table 1 and 5). Based on the data described in Figure 5 and 6, strains TM-1 and SP-5 biodegraded heavy hydrocarbons into light compounds, which implies that the hydrogen contents (and H/C ratio) of those oils increased as the carbon number decreased. However, the H and C contents, as well as the H/C ratios of the oils, were reduced (Table 2 and 6) due to the partial degradation of organic compounds into inorganic compounds.

The experimental data of this study and several reports (Fedorak et al. 1983; Forrester et al. 1983; Premuzic et al. 1996; Rabus et al. 1999; Solano-serena et al. 1999) indicated the biotreatments of microorganisms on crude oils have resulted in following changes: (1) emulsification, (2) acidification, (3) qualitative and quantitative transformations in the light and heavy fractions of crude oils, (4) decrease in organic sulfur content and chemical changes in oil fractions containing sulfur compounds, and (5) reduction in organic nitrogen content with concurrent biochemical conversion of polar nitrogen containing compounds.

CONCLUSION

The strains TM-1 and SP-5 isolated from the oil field are the mesophile, growing over a wide temperature range and producing different metabolites, and they can convert and degrade saturated hydrocarbons, aromatic hydrocarbons, resins, asphaltenes, and polar organic sulfur compounds and organic nitrogen compounds in heavy oils. The strains can also utilize the part components of heavy oils as the carbon source and energy source for growth.

The optimum condition of biochemical reaction between bacterium and crude oil depends on both the bacterial species used and the chemical compositions of the heavy oils. The strains TM-1 and SP-5, which can metabolize various heavy oils, will be used widely in oil fields and environmental pollution control to enhance oil recovery and remedy the oil-contaminated soils and waters.

ACKNOWLEDGMENTS

This work was supported by the National Basic Research Program of China (2007CB815601) and CNPC Innovation Funds. In addition, we thank the Shengli Petroleum Administrative Bureau for providing samples and information regarding the oil reservoirs.

REFERENCES

Aldrett, S., Bonner, J. S., Mills, M. A., Autenrieth, R. L. and Stephens, F. L. (1997). Microbial degradation of crude oil in marine environments tested in a flask experiment. *Water Res.* 31, 2840-2848.

Bahrami, A., Shojaosadati, S.A., and Mohebali, G. (2001). Biodegradation of dibenzothiophene by thermophilic bacteria. *Biotechnol. Lett.* 23, 899–901.

Banat, I. M. (1995). Biosurfactants production and possible uses in microbial enhanced oil recovery and oil pollution remediation: a review. *Biores. Tech.* 51, 1–12.

Beeder, J., Nilsen, R. K., Torsvik, T., and Lien, T. (1994). Archaeoglobus fulgidus isolated from hot North Sea Oilfield waters. *Appl. Environ. Microbiol.* 60, 1227–1231.

Boochan, S., Britz, M.L. and Stanlty, G. A. (2000). Degradation and mineralization of high-molecular- weight polycyclic aromatic hydrocarbons by defined fungal-bacteria cocultures. *Appl. Environ. Microbiol.* 66, 1007-1019.

Carballido-López, R. and Formstone, A. (2007). Shape determination in *Bacillus subtilis*. *Curr. Opin. Microbio.* 10, 611-616

Copper, D. G., Macdonald, C. R., Duff, S. J. B., and Kosaric, N. (1981). Enhanced production of surfactin from *Bacillus subtilis* by continuous product removal and metal cation additions. *Appl. Environ. Microbiol.* 42, 408–412.

Das, K. and Mukherjee, A. K. (2007). Crude petroleum-oil biodegradation efficiency of *Bacillus subtilis* and *Pseudomonas aeruginosa* strains isolated from a petroleum-oil contaminated soil from North-East India. *Biores. Tech.* 98, 1339–1345

Davey, M. E., Wood, W. A., Key, R., Nakamura, K., and Stahl, D. A. (1993). Isolation of three species of *Geotoga* and *Petrotoga*: Two new genera, representing a new lineage in the bacterial line of descent distantly related to the '*Thermotogales*'. *Syst. Appl. Microbiol.* 16, 191–200.

Dong, X. and Cai, M. (2001). Manual of systematically determinative bacteria. Beijing: Science Press (in Chinese).

Earl, A., M., Losick, R., and Kolter, R. (2008). Ecology and genomics of *Bacillus subtilis*. *Trends Microbiol.* 16, 269-275

Fedorak, P. M., Foght, J. M. and Westlake, D. W. S. (1983). Comparative studies on microbial degradation of aromatics and saturates in crude oil. In J. E. Zajic, D. G. Cooper, T. R. Jack and N. Kosaric (Eds.) Microbial Enhanced Oil Recovery. Tusa: Pennwell Publishing Company, 162-172.

Forrester, P. I., Pearson, D. and Rotw, J. (1983). Microbial degradation of Athabasca bitumen. In J. E. Zajic, D. G. Cooper, T. R. Jack and N. Kosaric (Eds.) Microbial Enhanced Oil Recovery. Tusa: Pennwell Publishing Company, 156-161.

Fox, S. L., Brehm, M. A., Robertson, E. P., Jackson, J. D., Thomas, C. P., and Bala, G. A. (1993). Comparative analysis of microbially mediated oil recovery by surfactants produced by *Bacillus licheniformis* and *Bacillus subtilis*. In E. T. Premuzic and A. Woodhead (eds.), *Microbial Enhancement of Oil Recovery-Recent Advances*. Amsterdam: Elsevier Science Publishers, 143–150.

Grassia, G. S., Mclean, K. M., Glenat, P., Bauld, J., and Sheehy, A. J. (1996). A systematic survey for thermophilic fermentative bacteria and archaea in high temperature petroleum reservoirs. *FEMS Microbiol. Ecol.* 21, 47–58.

Greene, A. D., Patel, B. K. C., and Sheehy, A. J. (1997). *Deferribacter thermophilus* gen. nov., sp. nov., a novel thermophilic manganese- and ion-reducing bacterium isolated from a petroleum reservoir. *Int. J. Syst. Bacteriol.* 47, 505–509.

Hao, R. and Lu, A. (2009). Biodegradation of heavy oils by halophilic bacterium. *Prog. Nat. Sci.* 19, 997–1001.

Harwood, C. R. (1992). *Bacillus subtilis* and its relatives: Molecular biological and industrial workhorses. *Trends Biotechnol.* 10, 247–255.

Jain, D. K., Lee, H. and Trevors, J. T. (1992). Effect of addition of *Pseudomonas aeruginosa* UG2 inocula or biosurfacants on biodegradation of selected hydrocarbons in soil. *J. Ind. Microbiol.* 10, 87-93.

Jenneman, G. E., Mcinerney, M. J., Knapp, R. M., Clark, J. B., Ferro, J. M., Revus, D. E., and Menzie, D. E. (1983). A halotolerant, biosurfactants producing *Bacillus* species potentially useful for enhanced oil recovery. *Dev. Ind. Microbiol.* 24, 485–492.

Kirscner, Z., Rosenberg, I. Z. and Gutnick, D. (1980). Incorporation of 32P and growth of *Pseudomonas* UP-2 on n-tetracosane. *Appl. Environ. Microbiol.* 40, 1086-1093.

Li, J., Lian, B., Hao, J., Zhao, J., and Zu, L. (2006). Non-parallelism between the effect of microbial flocculants on sewerage disposal and the flocculation rate. *Chinese J. Geochem.* 25, 139–142.

Lin, S-C, Lin, K-G, Lo, C-C, and Lin, Y-M (1998). Enhanced biosurfactant production by a *Bacillus licheniformis* mutant. *Enz. Microbial Technol.* 23, 267–273.

Lin, S-C and Jiang, H-J (1997). Recovery and purification of the lipopeptide biosurfactant of *Bacillus subtilis* by ultrafiltration. *Biotechnol. Tech.* 11, 413–416.

Lin, S-C, Sharma, M. M., and Georgiou, G. (1993). Production and deactivation of biosurfactants by *Bacillus licheniformis* JF-2. *Biotechnol. Prog.* 9, 138–145.

Lin, S-C, Goursaud, J-C, Kramer, P. J., Georgiou, G., and Sharma, M. M. (1990). Production of biosurfactants by *Bacillus licheniformis* strain JF-2. In E. C. Donaldson (ed.), *Microbial Enhancement of Oil Recovery-Recent Advances*. Amsterdam: Elsevier Science Publishers, 219–226.

Ling, D., and Dong, X. (1999). Classification and identification methods of lactic acid bacteria. Beijing: Chinese Light Industry Press (in Chinese).

Maltseva, O., McGowan, C., Fulthorpe, R. and Oriel, P. (1996). Degradation of 2,4-dichlorophenoxyacetic acid by haloalkaliphilic bacteria. *Microbiol.* 142, 1115-1122.

McMeekin, T. A., Nichols, P. D., Nichols, S. D., Juhasz, A. and Franzmann, P. D. (1993). Biology and biotechnological potential of halotolerant bacteria from Antarctic saline lakes. *Experientia.* 49, 1042-1046.

Mulligan, C. N., Chow, T. Y. K., and Gibbs, B. F. (1989). Enhanced biosurfactants production by a mutant *Bacillus subtilis* strain. *Appl. Microbiol Biotechnol.* 31, 486–489.

Ng, T. K., Weimer, P. J., and Gawel, L. J. (1989). Possible nonanthropogenic origin of two methanogenic isolates from oil-producing wells in the San Miguelito field, Ventura County, California. *Geomicrobiol. J.* 7, 185–192.

Pfiffiner, S. M., McInerney, M. J., Jenneman, G. E, and Knapp, R. M. (1986). Isolation of halotolerant, thermotolerant, facultative polymer-producing bacteria and characterization of the exopolymer. *Appl. Environ. Microbiol.* 51, 1224–1229.

Pornsunthorntawee, O., Arttaweeporn, N., Paisanjit, S., Somboonthanate, P., Abe M., Rujiravanit, R., and Chavadej, S. (2008). Isolation and comparison of biosurfactants produced by *Bacillus subtilis* PT2 and *Pseudomonas aeruginosa* SP4 for microbial surfactant-enhanced oil recovery. *Biochem. Eng. J.* 42, 172-179.

Premuzic, E. T., Lin, M. S., Lian, H, Zhou, W. M., and Yablon, J. (1996). Microbial interactions in crude oils: possible impact on biochemical versatility on the choice of microbial candidates. In:Proceedings of the 1996 international conference on microbial enhanced oil recovery, 551–569.

Premuzic E.T. and Woodhead A. (1993). *Microbial Enhancement of Oil Recovery-Recent Advances*. New York: Elsevier.

Rabus, R., Wilkes, H., Schramm, A., Harms, G., Behrends, A., Amann, R. and Widdel, F. (1999). Anaerobic utilization of alkylbenzenes and n-alkanes from crude oil in an enrichment culture of denitrifying bacteria affiliating with the β-subclass of Proteobacteria. *Environ. Microbiol.* 2, 145-157.

Rosenberg, A. (1983). Pseudomonas halodurans sp. nov., a halotolerant bacterium. *Arch. Microbiol.* 136, 117–123.

Rosenberg, M., Gutnick, D. L. and Rosenberg, E. (1983). Bacterial adherence to hydrocarbons. In J. E. Zajic, D. G. Cooper, T. R. Jack and N. Kosaric (Eds.) Microbial Enhanced Oil Recovery. Tusa: Pennwell Publishing Company, 114-123.

Solano-serena, F., Marchal, R., Ropars, M., Lebeault, J. M. and Vandecasteele, J. P. (1999). Biodegradation of gasoline: kinetics, mass balance and fate of individual hydrocarbons. *J. Appl. Microbiol.* 86, 1008-1016.

Stetter, K. O., Huber, R., Blochl, E., Kurr, M., Eden, R. D., Fielder, M., Cash, H., and Vance, I. (1993). Hyperthermophilic archaea are thriving in deep North Sea and Alaskan reservoirs. *Nature*. 365, 743–745.

Tardy-Jacquenod, C., Caumette, P., Matheron, R., Lanau, C., Arnauld, O., and Magot, M. (1996). Characterization of sulfate-reducing bacteria isolated from oil-field waters. *Can. J. Microbiol.* 42, 259–266.

Tissot, B. P. and Welte, D. H. (1978). *Petroleum Occurrence and Formation*. New York: Springer-Verlag.

Ventosa, A., Nieto, J. J., and Oren, A. (1998). Biology of moderately halophilic aerobic bacteria. *Microbiol Mol. Biol. Rev.* 62, 504–44.

Ward, D. M. and Brock, T. D. (1978). Hydrocarbon biodegradation in Hypersaline environments. *Appl. Environ. Microbiol.* 35, 353–359.

Yen, T. F. (1990). *Microbial Enhanced Oil Recovery: Principle and Practice*. Florida: CRC Press.

Zheng, G. and Slavik, M. F. (1999). Isolation, partial purification and characterization of a bacteriocin produced by a newly isolated *Bacillus subtilis* strain. *Lett. Appl. Microbiol.* 28, 363–368.

In: Biochemistry Research Updates
Editor: Simon J. Baginski

ISBN 978-1-61209-700-8

Chapter 9

EFFECT OF FOREST ENVIRONMENTS ON HUMAN URINARY ADRENALINE

Qing Li* and Tomoyuki Kawada
Department of Hygiene and Public Health,
Nippon Medical School, Tokyo, Japan

ABSTRACT

Humans have enjoyed forest environments for ages because of the quiet atmosphere, beautiful scenery, mild climate, and fresh, clean air. Empirically, forest environments may reduce stress and have relaxing effect on human. Adrenaline is released from the adrenal medulla, and adrenaline levels increase under circumstances of novelty, anticipation, unpredictability, and general emotional arousal. Measurement of free adrenaline in urine provides a reliable measure of the circulating concentration of adrenaline in the bloodstream and thus is a measure of sympathoadrenal medulla activity. To explore the relaxing effect of forest environments on humans, we investigated the effect of trips to forest parks on urinary adrenaline and noradrenaline in both male and female subjects. We found that three-day/two-night trips to forest parks significantly reduce the concentration of urinary adrenaline and/or noradrenaline. Moreover, a day trip to a forest park also has a similar effect on urinary adrenaline. The Profile of Mood States (POMS) test showed that forest environments significantly increased the score for vigor and decreased the scores for anxiety, depression, anger, confusion, and fatigue suggesting that the subjects were under conditions of lower stress during the forest trips, which support the findings on urinary adrenaline. Other studies have reported that forest environments reduce the concentration of cortisol in saliva, reduce prefrontal cerebral activity, reduce blood pressure, and stabilize autonomic nervous activity in humans, which also support our findings on urinary adrenaline. These findings indicate that forest environments may reduce stress and have relaxing effect on human by reducing adrenaline level.

Keywords: Adrenaline, Cortisol, Forest environment, Noradrenaline, POMS, Stress.

* Corresponding author: Qing Li, MD, Ph.D, Department of Hygiene and Public Health, Nippon Medical School, 1-1-5 Sendagi, Bunkyo-ku, Tokyo 113-8602, Japan. Tel: ++81-3-3822-2131, Fax: ++81-3-5685-3065. e-mail: qing-li@nms.ac.jp. URL: http://forest-medicine.

INTRODUCTION

Forest environments have been enjoyed by humans for ages because of the quiet atmosphere, beautiful scenery, mild climate, and clean fresh air found in forests. A forest bathing trip is a short, leisurely trip to visit a forest park, called "Shinrinyoku" in Japanese, which is similar to natural aromatherapy [1-6]. A forest bathing trip involves visiting a forest park for relaxation and recreation while breathing in volatile organic substances, called phytoncides derived from trees, such as α-pinene and limonene [1-8]. Incorporating forest bathing trips into a good lifestyle was first proposed in 1982 by the Forest Agency of Japan [5, 9]. It has now become a well-recognized relaxation and/or stress management activity in Japan [1-6, 10]. Forest bathing trips significantly increased the score for vigor and decreased the scores for anxiety, depression, anger, confusion, and fatigue as investigated by the Profile of Mood States (POMS) test [1,3-4]. Moreover, forest bathing trips can reduce blood pressure, reduce the concentration of cortisol in saliva, reduce prefrontal cerebral activity, and stabilize autonomic nervous activity in humans [11-12]. Because forests occupy 67% of the land in Japan, forest bathing is easily accessible [13]. Moreover, forest bathing is possible in similar forest environments throughout the world.

EFFECT OF FOREST ENVIRONMENTS ON HUMAN URINARY ADRENALINE IN MALE SUBJECTS

Adrenaline is released from the adrenal medulla, and adrenaline levels increase under circumstances of novelty, anticipation, unpredictability, and general emotional arousal. Noradrenaline is the predominant neurotransmitter released by the sympathetic system, and some of this enters the blood; the level of noradrenaline increases during increased physical activity [14]. Measurement of free adrenaline and noradrenaline in urine provides a reliable measure of the circulating concentration of adrenaline and noradrenaline in the bloodstream and thus is a measure of sympathoadrenal medulla activity [15]. The concentrations of adrenaline and noradrenaline in urine have been used to evaluate work related stress in nurses [16] and lorry drivers [17], in which the subjects showed decreases in adrenaline and/or noradrenaline in urine with the lower stress. Empirically, forest environments may reduce stress and have relaxing effect on human. Thus, we speculate that walk in forest parks may reduce urinary adrenaline level. To investigate the effect of forest environments on urinary adrenaline level, twelve healthy male subjects, aged 35-56 years (mean 45.1±6.7), were selected from four large companies in Tokyo, Japan for the present study [2]. Information gathered from a self-administered questionnaire, including age and lifestyle habits that asked about cigarette smoking, alcohol consumption, eating breakfast, sleeping hours, working hours, physical exercise, nutritional balance, and mental stress, have been reported previously [2, 18]. Written informed consent was obtained from all subjects after a full explanation of the study procedures. The Ethics Committee of the Nippon Medical School approved this study. The subjects experienced a three-day/two-night trip to three different forest fields at Agematsu town in Nagano prefecture located in northwest Japan in early September, 2006. On the first day, the subjects walked for two hours in the afternoon in a forest field, and then stayed at a nearby hotel within the forest. On the second day, the subjects walked for 2 hours in the morning and afternoon, respectively, in two different forest fields. On day 3, the subjects finished the trip and returned to Tokyo after collecting urine and completing the questionnaire. In contrast, in the city tourist visit, eleven subjects experienced a three-day/two-night trip to Nagoya city in mid-May, 2006. Nagoya is located in Aichi prefecture roughly in the center of Japan, and is an important crossroads for transportation in Japan. On the first day, the subjects walked for two hours in the afternoon along a tourist route, which is through an old style district in Nagoya, and then stayed at a hotel in Nagoya. On the second day, the subjects walked for 2 hours around Nagoya baseball Dome in the morning and 2 hours around/in Nagoya airport nearby Nagoya city in the afternoon. There are some areas of trees in Nagoya city, but there are almost no trees in the areas visited. The class of hotels was the same and the hotel experience was the same for the city and the forest trips. The distance between Tokyo and the forest fields at Agematsu town in Nagano prefecture is almost the same as the distance between Tokyo and Nagoya. Each course in both trips was 2.5 km, closely resembling normal physical activity for the subjects on an average of normal working days. Daily physical activity of the subjects was monitored with a pedometer, and the level of background walking steps of the subjects on normal working days were monitored for a week, and the averaged walking distance of all the subjects was about 5.0 km/day. Then, we set the walking distances in both trips based on the result. The duration of sleep was measured with a piezo-electric accelerometer, Actiwatch(R) (Mini Mitter Co. Inc., Sunriver), worn on the wrist of the non-dominant arm. The validation study was previously reported [19]. Urine was sampled on the second and third days during the trips, and three days prior to the trips as a control. Adrenaline and noradrenaline concentrations in urine were determined by HPLC. It was found that forest bathing trips significantly decreased adrenaline concentration in urine in male subjects (Figure 1A) [2], while a city tourist visit had no effect (Figure 1B) [2], suggesting that the subjects were under conditions of lower stress during the forest bathing trips [1-2]. Other studies have reported that forest bathing trips reduce the concentration of cortisol in saliva, reduce prefrontal cerebral activity, reduce blood pressure, and stabilize autonomic nervous activity in humans [7, 10-11]. In addition, forest environments significantly increased the score for vigor and decreased the scores for anxiety, depression, and anger in the POMS test, suggesting that the subjects were physiologically relaxed during the forest bathing trips [1, 3-4]. Moreover, a forest bathing trip significantly increased the proportions of lymphocytes and monocytes and decreased the proportions of granulocytes in WBC [1]. It has been reported that dominance by the parasympathetic nervous system causes an increase in circulating lymphocytes and decrease in granulocytes in peripheral blood [20]. This suggests indirectly that the parasympathetic nervous system of the subjects was dominant, which is associated with relaxation and decreased stress, during the forest bathing trips. In addition, phytoncides, such as α-pinene and β-pinene were detected in forest air [1-2].

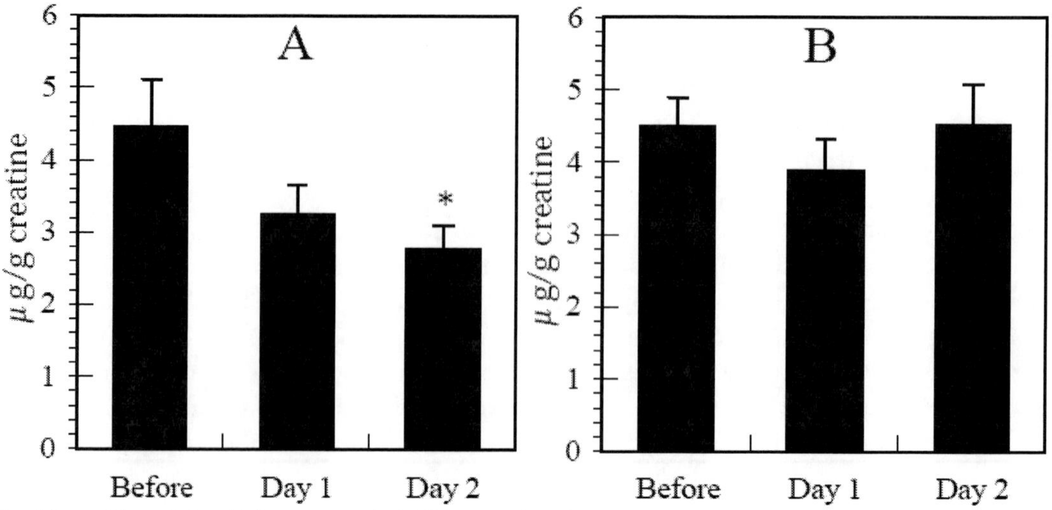

Cited from Li et al. Int. J. Immunopathol. Pharmacol. 2008;21:117-128.

Figure 1. Effect of the forest bathing trip (A) and city tourist visit (B) on urinary adrenaline concentration in male subjects. Data are presented as the mean+SE (n=12 in A and n=11 in B). Two-way ANOVA with no-repeated measures indicated that the forest bathing trip significantly affected the adrenaline concentration in urine ($p<0.05$). *: $p<0.05$, significantly different from before the trip by paired t-test.

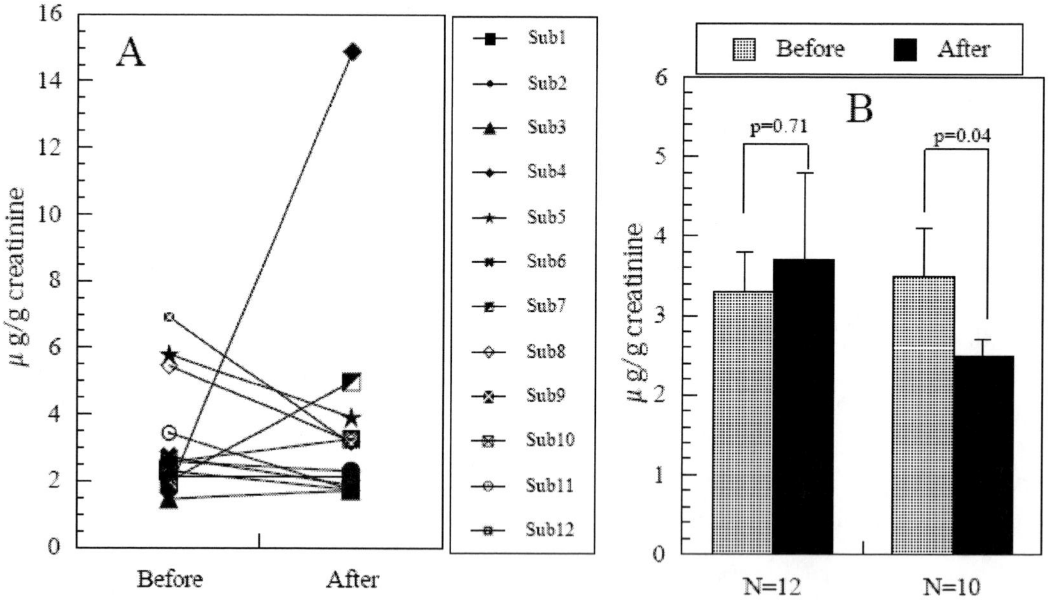

Cited from Li et al. J. Biol. Regul. Homeost Agents. 2010;24:157-65.

Figure 2. Effect of a day trip to a forest park on the urinary adrenaline concentration in male subjects. A: Urinary adrenaline shown in 12 subjects, B: data are presented as the mean+SE (n=12 for all subjects and n=10 for subjects who showed a decreased/unchanged urinary adrenaline). Significance was analyzed by the paired t-test.

EFFECT OF A DAY TRIP TO A FOREST PARK ON HUMAN URINARY ADRENALINE IN MALE SUBJECTS

Although a three-day/two-night trip to a forest park significantly reduced the concentration of urinary adrenaline, it is not clear whether a day trip to a forest park also reduced the concentration of urinary adrenaline. In this study [4], twelve healthy male subjects, aged 35-53 years (mean: 44.4±5.7), were selected from two large companies in Tokyo, Japan. Information gathered from a self-administered questionnaire, including age and lifestyle habits that asked about cigarette smoking, alcohol consumption, eating breakfast, sleeping hours, working hours, physical exercise, and nutritional balance, have been reported previously [1-3]. Written informed consent was obtained from all subjects after a full explanation of the study procedures. The Ethics Committee of the Nippon Medical School approved this study. The subjects experienced a day trip to a forest park in Saitama Prefecture located in the north of Tokyo, Japan on a Sunday at the end of July 2009. The subjects walked for two hours each in the morning and afternoon for a total of about 5 km in the forest park, and then returned to Tokyo. A questionnaire of POMS test was conducted before, during, and after the walk. The physical activity of the subjects was monitored with a pedometer [1-3]. The adrenaline concentration of urine and cortisol concentration in serum were determined before and after the trip.

As shown in Figure 2, the difference in the concentrations of adrenaline between before and after the trip was not significant (p=0.71, Figure 2b). However, when the subjects were divided into two groups: an increased group (2 subjects) and a decreased/unchanged group (10 subjects), it was found that there was a significant decrease in the decreased/unchanged group (10 subjects, p=0.04) (Figure 2b), indicating individual differences in the response to the trip (Figure 2a) [4].

These findings indicate that a day trip to a forest park also significantly decreased the concentrations of adrenaline in urine, confirming our previous findings [2-3], and suggesting that the subjects were under lower stress levels during the forest trip.

Moreover, a day trip to a forest park also significantly decreased the cortisol concentration in the serum [4]. In addition, forest bathing trip significantly decreased the scores for anxiety, depression, and anger in the POMS test [4], suggesting that the subjects were physiologically relaxed during the forest bathing trips [1, 3-4]. Phytoncides, such as α-pinene and β-pinene were detected in the forest air [4].

EFFECT OF FOREST ENVIRONMENTS ON URINARY ADRENALINE IN FEMALE SUBJECTS

Although it has been demonstrated that forest environments reduce urinary adrenaline in male subjects, it still remained to be resolved whether or not forest bathing trip (forest environment) also reduces urinary adrenaline in female subjects.

In this study [3], thirteen healthy nurses, aged 25-43 years, professional career 4-18 years, were selected with informed consent. The Ethics Committee of the Nippon Medical School approved this study. The subjects experienced a three-day/two-night trip to three different forest fields at Shinano town in Nagano prefecture located in northwestern Japan in early September, 2007. On the first day, the subjects walked for two hours in the afternoon in a forest field, and then stayed at a nearby hotel within the forest. On the second day, the subjects walked for 2 hours each in the morning and afternoon in two different forest fields, and then stayed at the same hotel. On day 3, the subjects finished the trip and returned to Tokyo after drawing blood and completing a questionnaire. Each walking course was 2.5 km, which closely resembled the normal physical activity for the subjects on an average working day. Daily physical activity of the subjects was monitored with a pedometer [2-3], and the level of background walking steps of the subjects on normal working days were monitored for a week, and the averaged walking distance of all the subjects was about 5.0 km/day. Then, we set the walking distances in the trip based on the result. Adrenaline and noradrenaline concentrations in urine were determined by HPLC before and after the trip. The schedule of the forest bathing trip and urine sampling was similar to that described previously [1-2]. The same control measurements were made before the trip on a normal working day. The concentrations of phytoncides in the forests were also measured.

The forest bathing trip significantly reduced urinary adrenaline and noradrenaline in female subjects [3] (Figure 3), which confirmed the previous findings in male subjects [2]. In addition, forest bathing trips significantly increased the score for vigour and decreased the scores for anxiety, depression, anger, fatigue, and confusion in the POMS test [3], suggesting that the subjects were physiologically relaxed during the forest bathing trips [1-4]. Phytoncides, such as α-pinene and β-pinene were detected in the forest air [3].

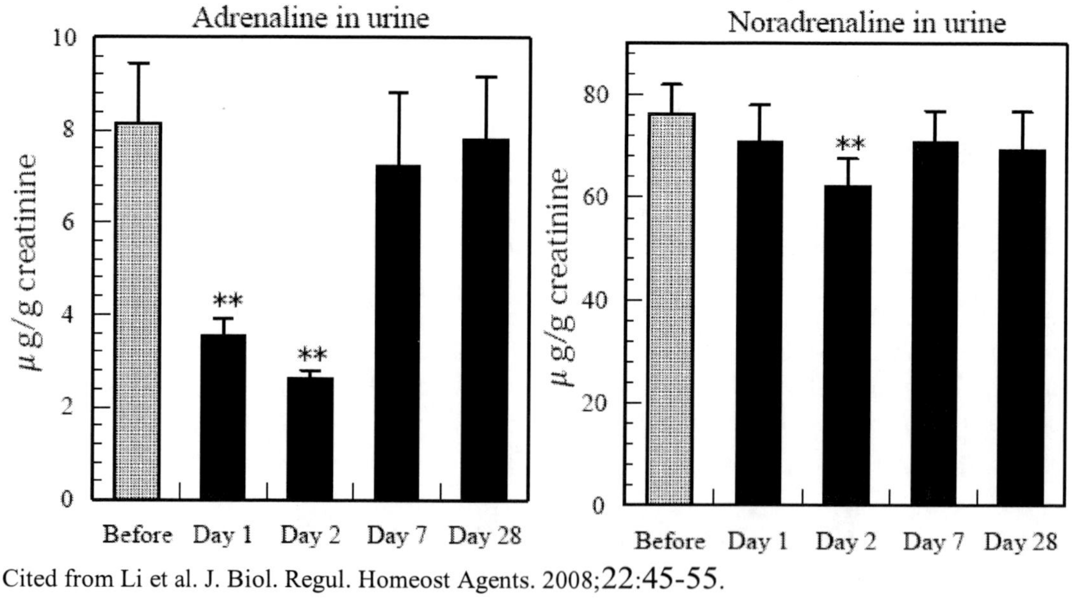

Cited from Li et al. J. Biol. Regul. Homeost Agents. 2008;22:45-55.

Figure 3. Effect of the forest bathing trip on the urinary adrenaline and noradrenaline concentrations in female subjects. Data are presented as the mean+SE (n=13). Two-way ANOVA with no-repeated measures indicated that the forest bathing trip significantly affected the adrenaline concentration in urine ($p<0.01$). *: $p<0.05$, **: $p<0.01$, significantly different from before the trip by paired t-test.

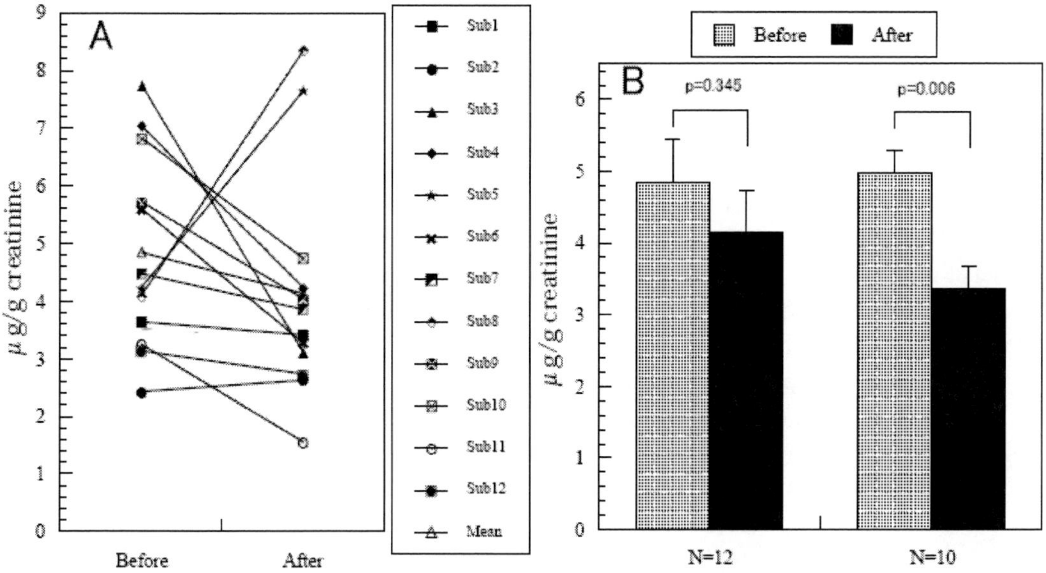

Cited from Li et al. Int. J. Immunopathol. Pharmacol. 2009;22:951-9.

Figure 4. Effect of phytoncide exposure on urinary adrenaline concentration. A: urinary adrenaline concentration shown in 12 subjects, B: Data are presented as the mean+SE (n=12 for all subjects and n=10 for subjects who showed a decreased urinary adrenaline). Statistical significances were analyzed by paired t-test.

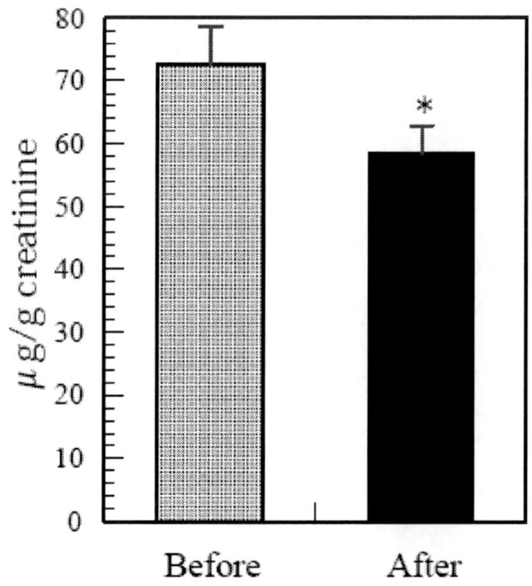

Cited from Li et al. Int. J. Immunopathol. Pharmacol. 2009;22:951-9.

Figure 5. Effect of phytoncide exposure on urinary noradrenaline concentration. Data are presented as the mean+SE (n=12). Statistical significances were analyzed by paired t-test.

MECHANISM OF FOREST ENVIRONMENT-INDUCED DECREASE IN URINARY ADRENALINE

Why did the forest environment reduce the urinary adrenaline? What kind of factors in the forest environment reduced the urinary adrenaline? We speculate that aromatic volatile substances (phytoncides) derived from trees such as α-pinene and limonene play an important role [8]. We detected several phytoncides, such as isoprene, α-pinene, β-pinene, and *d*-limonene, in the forest parks during the trips [1-4]. To investigate the effect of phytoncides on urinary adrenaline, twelve healthy male subjects, aged 37-60 years (51.8± 7.3), were selected from a medical school in Tokyo, Japan in the present study [21].

The sociodemographic information on the subjects, including age and lifestyle habits, was obtained by means of a self-administered questionnaire and has been reported previously [1-5, 18]. Written informed consent was obtained from all subjects after a full explanation of the study procedures. The Ethics Committee of the Nippon Medical School approved this study. The subjects stayed in an urban hotel in Tokyo, Japan for three consecutive nights. On the first day, urine was sampled at 0700 at their homes, blood was sampled at 0830 at a hospital, and several questionnaires including the POMS test, were completed as a control before the stay. The subjects then worked as usual. At 1900, all subjects stayed at the same hotel in the same type of room. During the 3 nights, aromatic volatile substances (phytoncides) were produced by vaporizing *Chamaecyparis obtusa* (hinoki in Japanese) stem oil with a humidifier in the hotel room for 3 consecutive nights. All subjects went to bed at 2300. During the experiment, the subjects got up at 0700, took a urine sample, completed the questionnaires and ate breakfast, and then went to their workplace and worked as usual. Daily physical activity of the subjects was monitored with a pedometer [1-3].

Adrenaline and noradrenaline concentrations in urine were determined. Although phytoncide exposure decreased the concentrations of adrenaline, this decrease was not significant ($p=0.345$). However, when the subjects were divided into two groups: the increased group (2 subjects) and the decreased group (10 subjects), it was found that there was a significant decrease in the decreased group (10 subjects, $p=0.006$) (Figure 4b), indicating individual differences on response to phytoncide exposure. On the other hand, phytoncide exposure significantly decreased the concentrations of noradrenaline in urine (Figure 5). In addition, phytoncide exposure significantly decreased the score for fatigue in the POMS test [21].

Komori et al. [22] also reported that citrus fragrance exposure affected the human endocrine system as analyzed by the measurement of urinary cortisol and dopamine levels, which supports our findings.

CONCLUSIONS

Taking together, forest environments may reduce stress and have relaxing effect on humans by reducing adrenaline and/or noradrenaline levels, and phytoncides from trees at least partially contribute to this effect.

ACKNOWLEDGMENTS

The authors are grateful to Ms. Kobayashi, Drs. Inagaki, Hirata, Katsumata, Wakayama and Li YJ, Mses. Hirata, Shimizu and Suzuki at the Department of Hygiene and Public Health, Nippon Medical School for their technical assistance in the experiments.

REFERENCES

[1] Li Q, Morimoto K, Nakadai A, Inagaki H, Katsumata M, Shimizu T, Hirata Y, Hirata K, Suzuki H, Miyazaki Y, Kagawa T, Koyama Y, Ohira T, Takayama N, Krensky AM, Kawada T. Forest bathing enhances human natural killer activity and expression of anti-cancer proteins. *Int. J. Immunopathol. Pharmacol.* 2007;20(S 2):3-8.

[2] Li Q, Morimoto K, Kobayashi M, Inagaki H, Katsumata M, Hirata Y, Hirata K, Suzuki H, Li YJ, Wakayama Y, Kawada T, Park BJ, Ohira T, Matsui N, Kagawa T, Miyazaki Y, Krensky AM. Visiting a forest, but not a city, increases human natural killer activity and expression of anti-cancer proteins. *Int. J. Immunopathol. Pharmacol.* 2008;21,117-128.

[3] Li Q, Morimoto K, Kobayashi M, Inagaki H, Katsumata M, Hirata Y, Hirata K, Shimizu T, Li YJ, Wakayama Y, Kawada T, Ohira T, Takayama N, Kagawa T, Miyazaki Y. A forest bathing trip increases human natural killer activity and expression of anti-cancer proteins in female subjects. *J. Biol. Regul. Homeost Agents*. 2008;22:45-55.

[4] Li Q, Kobayashi M, Inagaki H, Hirata Y, Hirata K, Li YJ, Shimizu T, Suzuki H, Wakayama Y, Katsumata M, Kawada T, Ohira T, Matsui N, Kagawa T. A day trip to a forest park increases human natural killer activity and the expression of anti-cancer proteins in male subjects. *J. Biol. Regul. Homeost. Agents.* 2010;24:157-65.

[5] Li Q. Effect of forest bathing trips on human immune function. *Environ. Health Prev. Med.* 2010;15:9-17.

[6] Li Q, Kawada T. Effect of forest environments on human natural killer (NK) activity. *Int. J. Immunopathol. Pharmacol.* (in press)

[7] Yamaguchi M, Deguchi M, Miyazaki Y. The effects of exercise in forest and urban environments on sympathetic nervous activity of normal young adults. *J. Int. Med. Res.* 2006;34:152-9.

[8] Li Q, Nakadai A, Matsushima H, Miyazaki Y, Krensky AM, Kawada T, Morimoto K. Phytoncides (wood essential oils) induce human natural killer cell activity. *Immunopharmacol Immunotoxicol*. 2006;28:319-33.

[9] Forestry Agency of Japan. http://www.rinya.maff.go.jp/toukei/genkyou/shinrin-jinkou.htm. 2002

[10] Morita E, Fukuda S, Nagano J, Hamajima N, Yamamoto H, Iwai Y, Nakashima T, Ohira H, Shirakawa T. Psychological effects of forest environments on healthy adults: Shinrin-yoku (forest-air bathing, walking) as a possible method of stress reduction. Public Health 2007;121:54-63.

[11] Park BJ, Tsunetsugu Y, Kasetani T, Kagawa T, Miyazaki Y. The physiological effects of Shinrin-yoku (taking in the forest atmosphere or forest bathing): evidence from field experiments in 24 forests across Japan. *Environ. Health Prev. Med.* 2010;15:18-26.

[12] Tsunetsugu Y, Park BJ, Miyazaki Y. Trends in research related to "Shinrin-yoku" (taking in the forest atmosphere or forest bathing) in Japan. *Environ. Health Prev. Med.* 2010;15:27-37.

[13] Li Q, Kobayashi M, Kawada T. Relationships between percentage of forest coverage and standardized mortality ratios (SMR) of cancers in all prefectures in Japan. *The Open Public Health Journal* 2008; 1, 1-7.

[14] Frankenhaeuser M. Experimental approach to the study of catecholamines and emotion. In: Levi L, editor. Emotions, Their Parameters and Measurement. New York: Raven Press; 1975. pp.209.

[15] Moleman P, Tulen JH, Blankestijn PJ, Man in 't Veld AJ, Boomsma F. Urinary excretion of catecholamines and their metabolites in relation to circulating catecholamines. Six-hour infusion of epinephrine and norepinephrine in healthy volunteers. *Arch Gen. Psychiatry.* 1992;49:568-72.

[16] Brown DE, James GD, Mills PS. Occupational differences in job strain and physiological stress: female nurses and school teachers in Hawaii. *Psychosom. Med.* 2006; 68:524-30.

[17] van der Beek AJ, Meijman TF, Frings-Dresen MH, Kuiper JI, Kuiper S. Lorry drivers' work stress evaluated by catecholamines excreted in urine. *Occup. Environ. Med.* 1995;52:464-9.

[18] Li Q, Morimoto K, Nakadai A, Qu T, Matsushima H, Katsumata M, Shimizu T, Inagaki H, Hirata Y, Hirata K, Kawada T, Lu Y, Nakayama K, Krensky AM. Healthy lifestyles are associated with higher levels of perforin, granulysin and granzymes A/B-expressing cells in peripheral blood lymphocytes. *Prev. Med.* 2007;44:117-23.

[19] Kawada T, Xin P, Kuroiwa M, Sasazwa Y, Suzuki S, Tamura Y. Habituation of sleep to road traffic noise as determined by polysomnography and accelerometer. *J. Sound Vib.* 2001;242:169-178.

[20] Mori H, Nishijo K, Kawamura H, Abo T. Unique immunomodulation by electro-acupuncture in humans possibly via stimulation of the autonomic nervous system. *Neurosci. Lett.* 2002;320: 21-4.

[21] Li Q, Kobayashi M, Wakayama Y, Inagaki H, Katsumata M, Hirata Y, Hirata K, Shimizu T, Kawada T, Ohira T, Park BJ, Kagawa T, Miyazaki Y. Effect of phytoncide from trees on human natural killer function. *Int. J. Immunopathol. Pharmacol.* 2009; 22:951-9.

[22] Komori T, Fujiwara R, Tanida M, Nomura J, Yokoyama MM. Effects of citrus fragrance on immune function and depressive states. *Neuroimmunomodulation.* 1995;2:174-80.

Reviewed by

Peiyu Wang, MD, PhD, Professor and Chair, Department of Social Medicine and Health Education, School of Public Health, Peking University, China.

In: Biochemistry Research Updates
Editor: Simon J. Baginski

ISBN 978-1-61209-700-8

Chapter 10

MICROARRAY-BASED ASSAY FOR SCREENING KINASE INHIBITORS BY BIOTINYLATED GOLD NANOPARTICLE PROBES

Tao Li and Zhenxin Wang*

State Key Laboratory of Electroanalytical Chemistry, Changchun Institute of Applied Chemistry, Chinese Academy of Sciences, Changchun, China
Graduate School of the Chinese Academy of Sciences, Beijing, China

ABSTRACT

Here, a microarray-based spectroscopic assay with three readout principles, fluorescence, surface resonance Raman (SERS) and resonance light scattering (RLS), for studying kinase functionality and screening kinase inhibitors has been reported. In this assay, the phosphorylation and inhibition events are marked by biotinylated anti-phosphoserinen antibodies, and gold nanoparticles are attached to the antibodies by standard streptavidin-biotin chemistry, followed by silver deposition for RLS and SERS signal enhancement. The streptavidin conjugated fluorescein is used as fluorescent probe. The utility of this assay to high-throughput screening was demonstrated with the interactions of the α-catalytic subunit of cyclic adenosine 5′-monophosphate (cAMP) dependent protein kinase (PKA) with a commercial inhibitor library, a collection of 80 potential kinase inhibitors, and satisfactory results were obtained. In addition, quantitative determination of binding strength and the inhibiting type (type I) of these inhibitors are also demonstrated by the adenosine 5′-triphosphate (ATP) competing assays.

Keywords: Microarray; Spectroscopic assay; Kinase; Inhibitors; Gold nanoparticles.

* Corresponding author. E-mail: wangzx@ciac.jl.cn; Fax:(+86) 431-85262243.

1. INTRODUCTION

The protein kinases (enzymes involved in protein phosphorylation reactions) are a diverse enzyme family that regulate many fundamental biochemical processes such as cell division, hormone responsiveness and gene expression, by catalyzing the transfer of the γ-phosphate of ATP to serine, threonine, or tyrosine residues.[1-8] The identification of kinases, their substrates and in particular, potential inhibitors have profound utility in drug discovery and biomedical research because mis-regulation of kinase activity has been implicated in a number of diseases such as cancers, diabetes, and inflammation.[4-8] Importantly, kinases are one of the major targets for drug discovery and several kinase inhibitors have been approved for clinical use. Because most potent kinase inhibitors bind in the highly conserved ATP-binding cleft, inhibitors must be tested experimentally against many kinases to determine their selectivity.

The radio-active γ-32P-ATP labeled technique is the standard method of measuring kinase activity and inhibition. However, this method has a number of drawbacks, such as cost of the radio-active γ-32P-ATP compound, time-consuming of assay, and threats to human health (handling of hazardous radioactive reagent). [8,9] Therefore, there is increasing interest on development of non-radioactive methods to monitor kinase activity and inhibition, such as immunoassays with fluorophore-labeled anti-phosphopeptide antibodies and nanoparticle-based biosensors. [10,11] Of the various available methods to study the functionality and inhibition of kinase, peptide microarrays have emerged as a powerful and versatile platform as it is compatible with many of the traditional solution-based kinase bioassays. [5,12,13] The salient features of the microarray make it very convenient to undertake thousands of kinase assays simultaneously.

Recently, we have also developed a kind of peptide microarray for studying kinase functionality and inhibition based on the three kinds of spectroscopic techniques, including fluorescence, surface resonance Raman (SERS) and resonance light scattering (RLS). [12,14,15] To investigate the feasibility of this microarray-based spectroscopic assay, this approach has been applied to screen both serine kinase (i.e., PKA) inhibitors from a commercial kinase inhibitor library which contains 80 potent kinase inhibitors. Quantitative determination of the inhibiting efficiencies (IC50 values) of these inhibitors is described and the outcomes of our assay are consistent with the IC50 values reported in literatures.

2. EXPERIMENTAL

2.1. Reagents

α-catalytic subunit cyclic adenosine 5′-monophosphate (cAMP) dependent protein kinase (PKA) was purchased from New England Biolabs (Beverly, MA, USA). Leukocyte-specific protein tyrosine kinase (LCK) and streptavidin, fluorescein conjugated (streptavidin-fluorescein) were purchased from Invitrogen Co. (USA). Inhibitor library was purchased from EMD Chemicals, Inc. (USA). Biotinylated anti-phosphoserinen antibody (anti-phosphoserine-biotin), adenosine 5′-triphosphate (ATP) and silver enhancer were purchased from Sigma-Aldrich Co. (USA). Peptides: CALNN, CALNNGK(biotin)G and kemptide

(CGGALRRASLG) were purchased from Scilight Biotechnology Ltd. (Beijing, China). Epoxy treated glass microscope slides was obtained from CapitalBio Ltd. (Beijing, China). Other chemicals were analytical grade and used as received. Milli-Q water (18.2 MΩ.cm) was used in all experiments.

2.2. Preparation of Gold Nanoparticles and Peptide-Stabilized Nanoparticles

The citrate stabilized 13 nm gold nanoparticles were synthesized by traditional Turkevich-Frens method. [16,17] Peptide-stabilized gold nanoparticles were prepared by our previously peptide capping procedure. [18,19] Generally, an aqueous solution of peptide mixture (CALNN: CALNNGK(biotin)G=9:1) was added to the solution of 2.7 nM 13 nm gold nanoparticles to give a final concentration of total peptide of 1.5 mM. Excess peptides were removed by repeated centrifugation at 13000 rpm, (~16100 g, 3 times) using an Eppendorf centrifuge (Eppendorf, Germany). The gold nanoparticles were re-suspended in PBS and stored at 4°C.

2.3. Peptide Microarray Fabrication and Phosphorylation

Peptide microarrays were manufactured by the standard procedure using a SmartArrayer 48 system (Capitalbio Ltd., China). 100 μM kemptide was spotted on epoxy functionalized slides in 0.3 M PBS (pH 8.5, 0.2 M NaCl) with 20 μg/mL BSA, following overnight incubation under vacuum at 25 °C. The slides were washed with 30 mL phosphate buffer (pH 7.5, 50 mM) containing 1% (w/v) BSA for 10 min, and then immersed in blocking buffer (pH 7.5, 50 mM PB, 0.15 M NaCl containing 3% w/v BSA) for 1 h to inactivate any free epoxy groups. After incubation, the slides were subjected to series washing steps: (1) 30 mL washing buffer (pH 7.5, 20 mM Tris, 0.15 M NaCl, 10 mM EDTA, 1 mM EGTA with 0.1% Triton X-100) for 10 min (3 times), (2) 30 mL kinase buffer (PKA buffer (pH 7.5, 50 mM Tris, 10 mM $MgCl_2$, 1 mM DTT) for 10 min, and (3) 30 mL kinase buffer containing 100 μM ATP for 10 min, respectively. Then, the slides were dried by centrifugation (200 g for 30 s) and divided into 12 independent 5 × 5 subarrays by a PTFE masker. [15] Subsequently, 10 μL kinase buffer with addition of 50 μM ATP, 100 units of PKA and desired amount of compound from the library was applied to each subarray, respectively. Following 1 h incubation period, the PTFE masker was removed and the slides were rinsed with 30 mL washing buffer for 5 min (3 times), 30 mL washing buffer without Triton X-100 for 5 min (3 times) and 30 mL water for 3 min (3 times), respectively. The slides were then dried by centrifugation (200 g for 30 s) and subjected to label procedure.

2.4. Recognization of the PKA -Catalyzed Phosphorylation Events

After phosphorylated reaction, the arrays were incubated with anti-phosphoserine-biotin (20 μg/mL) in 200 μL probe buffer, (pH 7.5, 50 mM PB, 0.15 M NaCl, supplemented with 0.1% Tween-20 (v/v) and 1% BSA (w/v)) for 1 h at 37 °C, respectively. Following a series of washing steps were applied to the arrays: (1) 30 mL PBS buffer with 1% Tween-20 for 5 min (3 times), (2) 30 mL PBS buffer for 5 min (3 times), and (3) 30 mL Milli-Q water for 3 min (18.2 MΩ, 3 times).

2.5. Attachment of Gold Nanoparticles

After recognized the PKA-catalyzed phosphorylation events by antibodies, the arrays were incubated with streptavidin-fluorescein (1 μM) in 200 μL probe buffer for 1 h at 37 °C. The arrays were then subjected to a series of washing and drying steps as previously described in recognization of the phosphorylation section. At this step, the microarrays were able to be read by conventional fluorescence microarray scanner (Luxscan-10K/A, Capitalbio Ltd., China). Then, the arrays were treated with 200 μL peptide-stabilized gold nanoparticles (5.0 nM) in probe buffer for 1 h at 37°C, washed and dried as described before.

2.6. Inhibition and IC_{50} Assays

To measure the IC_{50} values of inhibitors, different concentrations of the inhibitors (10^{-6} to 10^{2} μM, respectively) in the kinase solution were applied to individual subarray, following incubated, washed, spun-dried, recognized by antibodies, and probed by streptavidin-fluorescein and gold nanoparticles as described previously, respectively.

2.7. ATP Competition Assay

The ATP competition test was performed using the inhibition solution (kinase solution with 10 μM inhibitor) on different subarrays in the presence of desired concentrations of ATP (5 to 5×10^{6} μM, respectively). The arrays were then incubated, washed, spin-dried, recognized by antibodies, and probed by streptavidin-fluorescein and gold nanoparticles as before described, respectively.

2.8. Silver Enhancement and Detection

After labeled with gold nanoparticles, silver enhancer (1:1 mixed (1 mL total volume) solutions A ($AgNO_3$) and B (hydroquinone)) was applied to each microarray for 10 min and quickly rinsed with 50 mL Milli-Q water, then dried with centrifugation. After silver deposition, the slides were subjected to detect with a Renishaw 2000 Raman spectrophotometer (Gloucestershire, UK) equipped by an Ar^{+} ion laser and a CCD detector or ArrayIt SpotWare Colorimetric Microarray Scanner (TeleChem. International Inc. USA). All SERS spectra were collected with an excitation line at 514 nm with 10 s accumulation time, and 10 mW incident power. According to manufacturer's preset parameters, all RLS images were collected with broad spectrum white light source. The mean value and standard deviation of the signal were determined for the 50 spot replicates (two subarrays in different slides) per sample, respectively. The background originating from the slide was recorded and subtracted from each image/spectrum prior to evaluation.

3. Results and Discussion

3.1. Phosphorylation Assay

The kemptides was spotted and immobilised on epoxy treated glass microscope slides by a standard robotic procedure. The concept for the determination of kinase functionality and inhibition is shown in Figure 1.

In the presence of co-substrate, ATP, the kemptide is enabling to be phosphorylated by kinase PKA. Following the biotinylated antibody, anti-phosphoserine-biotin, is used to recognize the phosphorylation event. In a subsequent step, streptavidin-fluorescein was used as a bridge for attaching 13 nm peptide-stabilized gold nanoparticles to the immobilized peptides on the microarray spots. Here, the fluorescein plays as a flourescenc/Raman-active probe. In this step, the microarray can be detected by conventional fluorescence microarray scanner. For obtain high intensity SERS/RLS signal, a silver enhancement step was employed since the electromagnetic-field enhancement/ resonance light scattering of the small size (<30 nm) gold nanoparticles is relatively poor. [20]

For obtaining a good signal-to-noise ratio, 10 min exposure time is employed. [12] As shown in Figure 2, strong signals have been obtained at the spots which contain either kemptide, while no detectable reaction occurs on the control spots which contain control peptides.

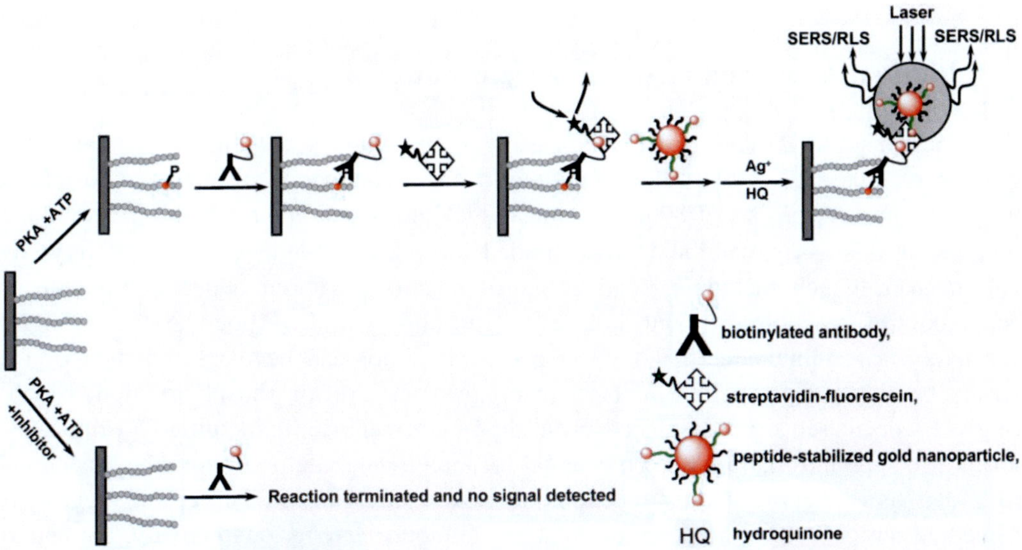

Figure 1. Schematic representation of the microarray-based assay showing from left to right, phosphorylation of the spotted peptide, probed by biotinylated antibody, then labeled by streptavidin-fluorescein and reading by detection of fluorescent singal, followed by the attachment of peptide stabilised gold nanoparticles, the silver enhancement step and reading by detection of SERS/RLS.

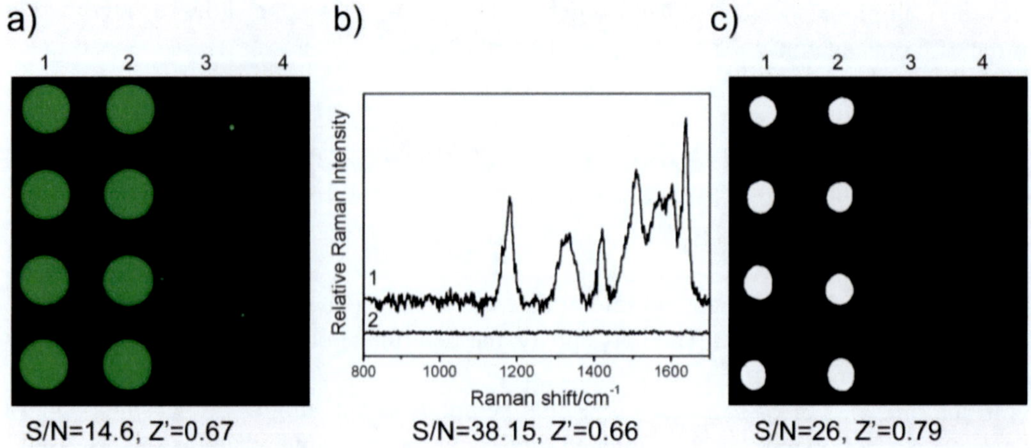

Figure 2. Fluorescence image (a), SERS spectra of fluorescein (b) and RLS image (c) acquired from spots with kemptide (column 1 and 2 in (a) and (c), and curve 1 in (b)) or contole peptide (column 3 and 4 in (a) and (c), and curve 2 in (b)) after phosphorylation by PKA, respectively.

This demonstrates the high selectivity of this detection method. Because the SERS intensity at 1638 cm^{-1} (xanthene ring stretch, and CC stretch of fluorescein molecule) is the strongest in the SERS spectrum, in the following assays, the SERS intensity at 1638 cm^{-1} is chosen for quantitative analysis. [21] To evaluate the performance of the assay, 100 identical enzymatic reactions were analyzed, and the signal-to-noise ratio (S/N) and quality factor (Z') were calculated. At given experimental condition, reasonable assay performance (S/N>14 and Z'>0.6) has been obtained. This result indicates that our assay is suitable for high throughput screening.

3.2. Inhibition Assay and IC_{50} Value Determination

To demonstrate the utility of this assay for inhibitor screening, we examined the inhibition of kinase PKA with the compounds from a commercial protein kinase inhibitor library. The library consists of 80 compounds which are potent and reversible protein kinase inhibitors. In this assay, each slide was divided into 12 independent 5 × 5 subarrays by a PTFE masker, which includes negative-control reactions without added compounds and positive-control reactions without kinase to set high and low boundaries for signals, respectively. The inhibited efficiency of specific inhibitor can be measured by the signal intensity because more efficient inhibitor leads to lower level of phosphorylation, which in turn gives a decreased siganl intensity. From the first screening, we identified 3 primary hits, defined as compounds that reduced the signal for kinase-catalytic phosphorylated reaction by >50% relative to negative control reactions (as shown in Figure 3). This result suggests that our method not only have the potential to screen inhibitor activity qualitatively, but can also be developed to yield quantitative data on the efficiency of different inhibitors.

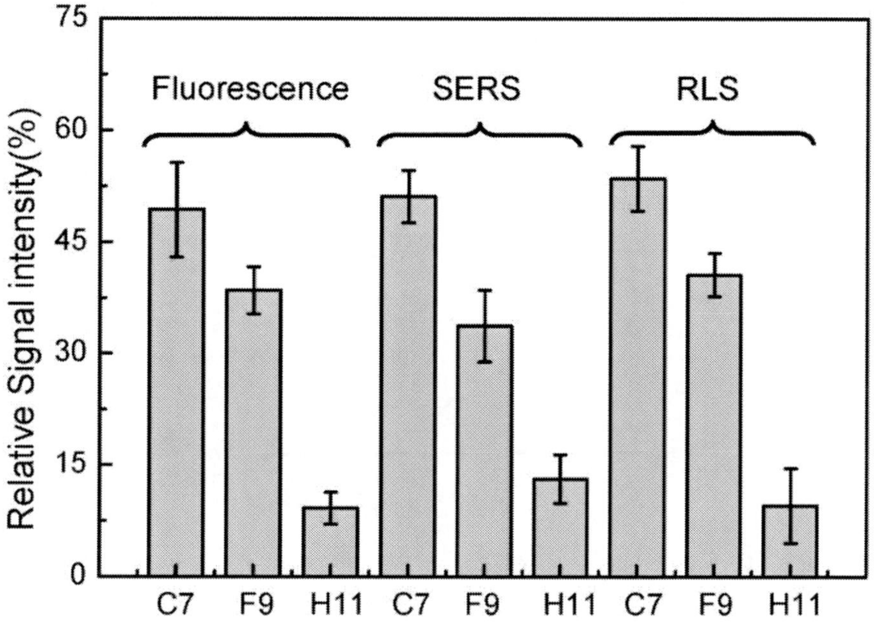

Figure 3. Data analysis of peptide microarrays with 10 μM inhibitors, respectively. The signals have been corrected for background noise (positive control) and normalized to the average intensity obtained in the absence of inhibitors (negative control). For SERS assay, the SERS intensity at 1638 cm^{-1} has been used.

Table 1. IC_{50} comparison and detection signal recovery

Inhibitors	IC_{50}, found (μM)	IC_{50}, reported (μM)	Recovery (%)	Reference
C7	32^{a} , 32^{b},44^{c}	44	>95	-
F9	500^{a}, 0.27^{b}, 430^{c}	0.43	>90	22
H11	0.016^{a}, 0.016^{b}, 0.012^{c}	7.0×10^{-3}	>75	23

The values are obtained by (a) fluorescence, (b) SERS and (c) RLS detection, respectively.

We determine the IC_{50} values for the primary kinase inhibitors to further demonstrate the utility of the microarray-based assay for a quantitative inhibition assay. The IC_{50} values of all inhibitors are summarized in Table 1. These IC_{50} values were comparable well with the literature reported values. [22,23] These results confirm that the microarray-based assay has potential to screen kinase inhibitors as well as determine the IC_{50} values. In particularly, the RLS-based approach circumvents the photobleaching of labels, thus providing significant improving on repeatability and reproducibility over fluorescence-based assay. Furthermore, the RLS-based assay can be detected by the scanner with white light source, which would greatly reduce the cost of instrumentation.

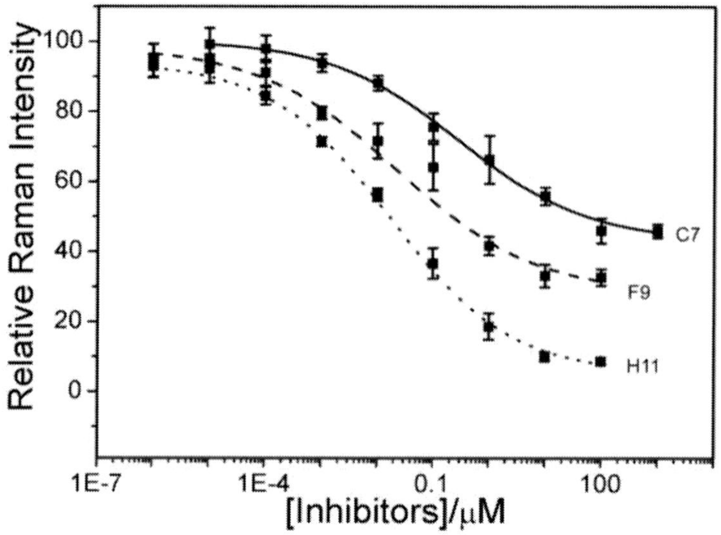

Figure 4. The effect of various concentrations of ATP in the PKA solution. The signals have been corrected for background noise (positive control) and normalized to the average intensity obtained in the absence of inhibitors (negative control). For SERS assay, the SERS intensity at 1638 cm^{-1} has been used. The concentration of H11 is 10 μM.

3.3. ATP Competition Assay

The effect of ATP on the inhibition efficiency of specific inhibitor was also tested because most of small molecule kinase inhibitors are ATP-competitors (type I inhibitor). [2,24] As shown in Figure 4, the detection signal is increased with increasing the concentration of ATP. This indicates that the inhibitors become less effective and the phosphorylation reaction can be recovered. This suggests that all of these inhibitors are bind with correspondent kinase in a type I binding mode, i.e., these inhibitors bind to the free enzyme and not to the enzyme-ATP complexes. In the presence of a typical concentration of ATP (5×10^5 μM) in the reaction mixture, the recoveries of these phosphorylation reactions are summarized in Table 1. Generally, the recovery rate is dependent on the inhibition efficiency of inhibitor, low inhibiting ability of inhibitor leads to high recovery rate. The experimental result indicates that the microarray-based assay has a great promise to study inhibition mechanism of inhibitors.

CONCLUSION

In conclusion, a microarray-based spectroscopic approach has been developed which can be used to perform quantitative assays for kinase inhibition. Compare to time-consuming traditional γ-^{32}P-ATP radiolabelled method, our assay does not pose any radioactive hazards and provides significant time-saving. The experimental results suggest that the assay is amenable to high throughput compound screening and accurate determination of IC_{50} values of inhibitors. This would open up possibilities for the future to develop microarray-based high throughput pharmacological tools in drug discovery.

ACKNOWLEDGMENT

The authors thank the NSFC (Grant No. 20675080) for financial support.

REFERENCES

[1] Beveridge, M., Park, Y.W., Hermes, J., Marenghi, A., Brophy, G. and Santos, A., *J. Biomol. Screening* 2000, 5, 205-212.

[2] Davies, S.P., Reddy, H., Caivano, M. and Cohen, P., *Biochem. J.* 2000, 351, 95-105.

[3] Hunter, T., *Cell* 2000,100, 113-127.

[4] Cohen, P., *Nat. Rev. Drug Discov*. 2002, 1, 309-315.

[5] Houseman, B.T., Huh, J.H., Kron S.J. and Mrksich, M., *Nat. Biotechnol*. 2002, 20, 270-274.

[6] Schutkowski, M., Reineke, U. and Reimer, U., *ChemBioChem*. 2005, 6, 513-521.

[7] Hill, Z.B., Perera, B.G.K. and Maly, D.J., *J. Am. Chem. Soc*. 2009, 131, 6686-6688.

[8] Jecklin, M.C., Touboul, D., Jain, R., Toole, E.N., Tallarico, J., Drueckes, P., Ramage, P. and Zenobi, R., *Anal. Chem*. 2009, 81, 408-419.

[9] Hilhorst, R., Houkes, L., Berg, A. and Ruijtenbeek, R., *Anal. Biochem*. 2009, 387, 150-161.

[10] Wang, Z.X., Levy, R., Fernig, D.G. and Brust, M., *J. Am. Chem. Soc*. 2006, 128, 2214-2215.

[11] Kerman, K. and Kraatz, H.B., *Chem. Commun*. 2008, 47, 5019-5021.

[12] Sun, L. L., Liu, D. J. and Wang, Z. X., *Anal. Chem*. 2007, *79*, 773–777.

[13] Uttamchandani, M., Wang, J.and Yao, S.Q., *Mol. BioSyst*. 2006, 2, 58-68.

[14] Li, T., Liu, D.J.and Wang, Z.X., *Biosens.Bioelectron*. 2009, 24, 3335-3339.

[15] Li, T., Liu, D.J. and Wang, Z.X., *Anal. Chem*. 2010, 82, 3067–3072.

[16] Frens, G., *Nat. Phy. Sci.* 1973, 241, 20-22.

[17] Turkevich, J., Stevenson, P.C., Hillier, J., 1951. *Discuss. Faraday Soc*. 11, 55-75.

[18] Levy, R., Nguyen, T.K., Thanh, R., Doty, C., Hussain, I., Nichols, R.J.,Schiffrin, D.J., Brust, M. and Fernig, D.G., *J. Am. Chem. Soc*. 2004, 126, 10076-10084.

[19] Wang, Z.X., Levy, R., Fernig, D.G. and Brust, M., *Bioconjugate Chem.* 2005,16, 497-500.

[20] Rosi, N.L. and Mirkin, C.A., *Chem. Rev*. 2005, 105, 1547-1562.

[21] Campbell, M., Lecomte, S. and Smith, W.E., *J. Raman Spectrosc*. 1999, 30, 37-44.

[22] Fabbro, D., Ruetz, S. and Bobis, S., *Anticancer Drug Des.* 2000, 15, 17-28.

[23] Couldwell, W.T., Hinton, D.R. and He, S., *FEBS Lett.* 1994, 345, 43-46.

[24] Young, P. R., McLaughlin, M. M., Kumar, S.; Kassis, S., Doyle, M. L., McNulty, D., Gallagher, T. F., Fisher, S., McDonnell, P. C., Carr, S. A., Huddleston, M. J.; Seibel, G., Porter, T. G., Livi, G. P., Adams, J. L. and Lee, J. C. *J. Biol. Chem.* 1997, *272*, 12116-12121.

Index

A

B

C

D

E

F

G

J

K

L

M

N

O

P

Q

R

T

U

V

W

X

Y

Z